In this series original contributions of physical research in all fields are contained, such as

— research monographs
— collections of papers to a single topic
— reports on congresses of exceptional interest for physical research.

This series is aimed at promoting quick information and communication between physicists of the various special branches.

In diese Reihe werden Originalbeiträge zu allen Gebieten der physikalischen Forschung aufgenommen wie

— Forschungsmonographien
— Sammlungen von Arbeiten zu einem speziellen Thema
— Berichte von Tagungen, die für die physikalische Forschung besonders aktuell sind.

Die Reihe soll die schnelle Information und gute Kommunikation zwischen den Physikern der verschiedenen Fachgebiete fördern.

Manuscripts
in English comprising at least 100 pages and not more than 500 pages can be admitted to this series. With respect to a quick publication the manuscripts are reproduced photomechanically. Authors who are interested in this series please turn directly to the Akademie-Verlag Berlin. Here you will get more detailed information about the form of the manuscripts and the modalities of publication.

Manuskripte
in englischer Sprache, die mindestens 100 Seiten und nicht mehr als 500 Seiten umfassen, können in diese Reihe aufgenommen werden. Im Interesse einer schnellen Publikation werden die Manuskripte auf fotomechanischem Wege reproduziert. Autoren, die an der Veröffentlichung entsprechender Arbeiten in dieser Reihe interessiert sind, wenden sich bitte direkt an den Akademie-Verlag Berlin. Sie erhalten dort genauere Informationen über die Gestaltung der Manuskripte und die Modalitäten der Veröffentlichung.

High Energy Ion Beam Analysis of Solids

Physical Research

Volume 6

High Energy Ion Beam Analysis of Solids

edited by
Gerhard Götz/Konrad Gärtner

High Energy Ion Beam Analysis of Solids

edited by
Gerhard Götz/Konrad Gärtner

With 178 Figures and 33 Tables

THEORIA CUM PRAXI

Akademie-Verlag Berlin
1988

Authors:

Dr. rer. nat. Achim Dittmar
Dr. sc. nat. Konrad Gärtner
Dr. rer. nat. Ernst Glaser
Prof. Dr. sc. nat. Gerhard Götz
Dr. rer. nat. Brigitte Weber
Dr. rer. nat. Werner Wesch
Dipl.-Phys. André Witzmann

Friedrich-Schiller-Universität Jena
Sektion Physik
Max-Wien-Platz 1
DDR-6900 Jena

Reproduction of the original authors' manuscript

ISBN 3-05-500497-3

Erschienen im Akademie-Verlag Berlin, DDR-1086 Berlin,
Leipziger Str. 3—4

Lizenznummer: 202 · 100/434/88
Printed in the German Democratic Republic
Gesamtherstellung: VEB Druckerei „Thomas Müntzer", 5820 Bad Langensalza
Einband: Karl Salzbrunn
LSV 1185
Bestellnummer: 763 829 8 (9117)
07200

Preface

The main effects on which ion beam analysis of solids is based were already discovered at the beginning of the twentieth century by E. Rutherford (1911, ion scattering by atoms) and J. Stark (1912, channeling). For a long time, high energy ion scattering was used only in the field of nuclear physics. Since about 1960 Rutherford backscattering spectrometry with high energy ion beams has also been successfully applied to the analysis of solid. In the early 1960s the channeling effect for ions moving in a single crystal was rediscovered. This offered the possibility of investigating the lattice structure and lattice defects of solids. Based on both the results of intensive research work in the field of atomic collision processes in solids and the improved experimental conditions (powerful accelerators with high energy stability, high resolution analyzers etc.), high energy ion beam analysis has become a very powerful quantitative method for the investigation of composition and lattice defects of solids.

The basic physical concepts and the experimental techniques for Rutherford backscattering spectrometry are reviewed in the books of J.W. Mayer and E. Rimini [1] and of W.K. Chu, J.W. Mayer, and M.-A. Nicolet [2]. A survey of the aspects of channeling, its theoretical description and application is given in Ref.[3] edited by D.V. Morgan. A lot of details concerning special problems and typical applications are published in the proceedings of the International Conferences on Ion Beam Analysis in Solids.

The idea of the authors in writing this book was to give the state-of-the-art of Rutherford backscattering spectrometry combined with channeling techniques in a single volume including the theoretical fundamentals, general methods, experimental equipment and examples of application. Chapter 1 contains the theoretical fundamentals where emphasis is put on the description of dechanneling and its application to defect analysis. In chapter 2 the experimental equipment and in chapter 3 the principal methods (Rutherford backscattering spectrometry, elastic recoil detection, nuclear reaction analysis) are represented. Applications to the analysis of layer composition and of the structure of crystals (including defects) are given in chapter 4 and 5, respectively.

This book is designed for students and scientists who want to become familiar with the method of ion beam analysis of solid. It is also intented to be a guide for practical work in this field and for the evaluation of experimental data.

The editors thank all of the authors contributing to this book. We are also grateful to F. Schwabe for his assistance in preparing chapter 2 and to U. Bornkessel, D. Stock, and E. Wenske for the technical completion of the manuscript. For the critical revision of the English we are indebted to E. Wagner and B. Thomas.

JENA, 1987

Gerhard Götz
Konrad Gärtner

Contents

Chapter 1

THEORETICAL FUNDAMENTALS

K. Gärtner

The typical situation for ion beam analysis of solids is shown in Fig.1.1. A beam of light ions (usually H^+ or He^+) with energy E_0 in the range of about 100 keV to some MeV is directed onto a solid target.

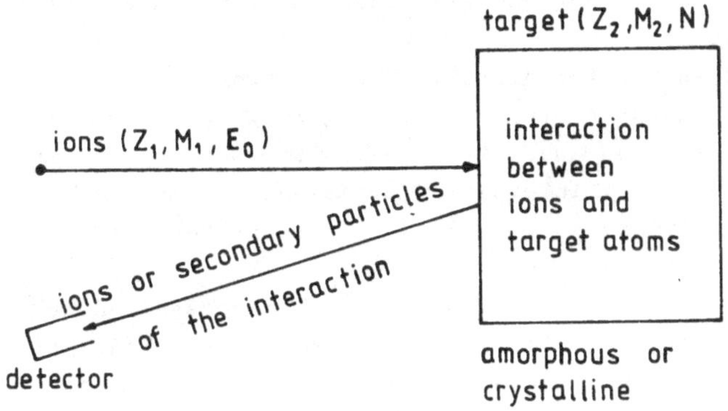

Fig.1.1. General arrangement for ion beam analysis of solids

The atomic number and the mass of the ions and of the target atoms are denoted by Z_1, M_1 and Z_2, M_2, respectively. Because their energy E_0 is very large compared with the binding energy of the target atoms ($\approx$ 10 eV) the interaction of the ions with the target can be treated as successive two-body collisions between the ion and the next nearest target atom. This is justified because the range of the interaction potential is smaller than or in the order of the interatomic distances in the solid target.

Due to the single collisions the ions change their direction, energy and charge state. For our considerations the charge state of the ions (changed by electron loss and capture) is of minor importance and is assumed to be well defined and known (e.g. also as an average value).

In amorphous solids the atoms are randomly distributed with N being the number of atoms per unit volume. Therefore the successive collisions can be treated as statistically independent two-body collisions. For monocrystalline solids the atoms are arranged on lattice sites. In this case the successive collisions are considerably correlated for ions moving approximately parallel to a major crystallographic direction (channeling). Thus, the ion motion in a crystalline target depends on the lattice structure and on the angle of incidence of the ion beam with respect to a major crystallographic direction.

1.1. INTERACTION PROCESSES

In the energy range considered, three different types of interaction processes take place: nuclear and electronic scattering and nuclear reactions. The elastic scattering of ions by the target atoms without changing the states of the two collision partners is called nuclear scattering for short. This process is described by an interatomic interaction potential the dominating part of which is given by the Coulomb interaction between the two nuclei. The electrons of the two collision partners influence the nuclear scattering only by screening the interaction between the two nuclei. Electronic scattering and nuclear reactions are inelastic interaction processes changing the electronic states of the colliding atoms and the states of their nuclei, respectively.

In general, all the different types of interaction processes can take place during a single collision and they are more or less coupled. For example the electronic scattering process depends on the ion trajectory determined by nuclear scattering. Nevertheless in most cases of application (especially for ion beam analysis) the different types of interaction processes mentioned above can be treated as statistically independent processes. Therefore a separate theoretical treatment of the different interaction processes is possible.

Neglecting inelastic processes, the scattering kinematics and dynamics, the interaction potential and the differential cross section for nuclear scattering are given in Sect.1.1.1. The differential cross section for electronic scattering is briefly discussed in the succeeding section. Since the cross sections for nuclear reactions are very small compared with those for electronic and nuclear scattering in the energy range considered, from the statistical point of view the nuclear reactions are not important for the ion motion in the solid target. Like Rutherford backscattering and X-ray production they are of interest for the detection of ion collision processes with very small impact parameters and are therefore mentioned in Sect.3.3.

1.1.1. Nuclear scattering

1.1.1.1. Scattering kinematics

Let us consider the elastic scattering between an ion (atom) of velocity $\vec{v}$ in the laboratory system and a target atom initially at rest. First, we give some general results obtained only from the conservation of the total momentum and the total energy of the two collision partners and being independent of the special type of the interaction. As usual, the center-of-mass system is used in the following because many features are simpler in this frame. In our case the velocity of the center-of-mass is given by

$$\vec{v}_{cm} = \vec{v}\, M_1/(M_1+M_2). \tag{1.1}$$

The initial velocities of the ion and the target atom in the center-of-mass system are

$$\vec{v}_{1,cm} = \vec{v} - \vec{v}_{cm} = \vec{v}\, M_2/(M_1+M_2) \tag{1.2}$$

and

$$\vec{v}_{2,cm} = -\vec{v}_{cm} = -\vec{v}\, M_1/(M_1+M_2), \tag{1.3}$$

respectively. As a consequence of the conservation of momentum and energy in the center-of-mass system (non-relativistic case) the velocities of the ion in the center-of-mass system before and after the collision have the same absolute value. This is also true for the target atom. The direction of the velocities of the two particles after the collision must be antiparallel because the total momentum is zero (as before the collision). In this way the scattering process is simply described in the center-of-mass system by a joint twisting of the velocity vectors $\vec{v}_{1,cm}$ and $\vec{v}_{2,cm}$ by an angle ϑ without changing their lengths as shown in Fig.1.2. Within the scattering kinematics the scattering angle ϑ in the center-of-mass system is a free parameter.

As can be seen in Fig.1.2, the final scattering angles in the laboratory system can be expressed in terms of the scattering angle ϑ in the center-of-mass system. From the triangle 1) we get the scattering angle of the ion in the laboratory system

$$\tan\Theta = \frac{v_{1,cm} \sin\vartheta}{v_{cm} + v_{1,cm} \cos\vartheta} = \frac{M_2 \sin\vartheta}{M_1 + M_2 \cos\vartheta} \tag{1.4}$$

where Eqs.(1.1 – 1.3) are used.

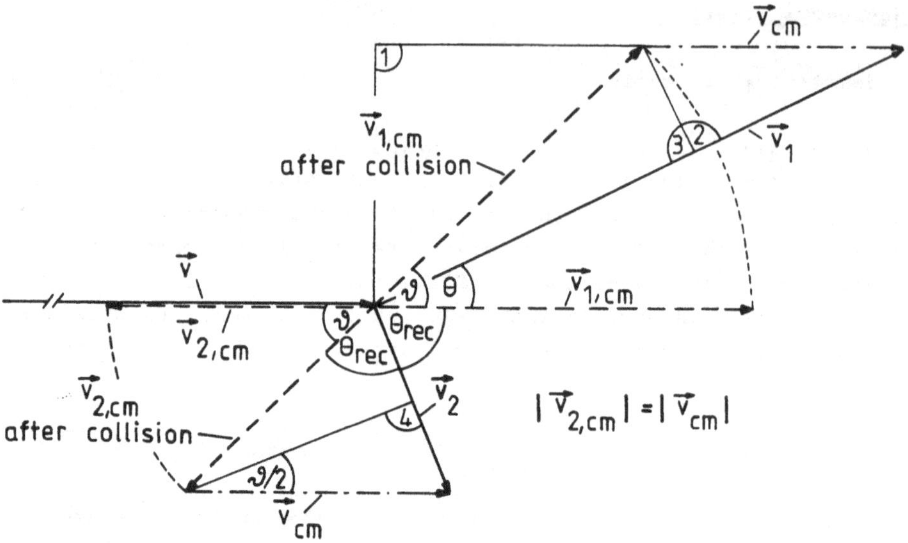

Fig.1.2. Scattering kinematics for the collision between an ion with velocity $\vec{v}$ and a target atom at rest

The scattering angle θ_{rec} of the target atom (recoil) in the laboratory system is immediately provided by

$$\theta_{rec} = (\pi - \vartheta)/2 \quad . \tag{1.5}$$

For application in the following sections, the ion velocity v_1 after the collision and the velocity v_2 of the target atom after the collision as a function of θ and ϑ, respectively, are of interest. Using the triangles 2) and 3) in Fig.1.2 we find

$$v_1 = v_{cm} \cos\theta + (v_{1,cm}^2 - v_{cm}^2 \sin^2\theta)^{1/2}$$

$$= v \, \frac{\cos\theta + [(M_2/M_1)^2 - \sin^2\theta]^{1/2}}{1 + M_2/M_1} \quad . \tag{1.6}$$

According to the triangle 4) the velocity v_2 is given by

$$v_2 = 2\, v_{cm} \sin(\vartheta/2) = v\, [2\, M_1/(M_1+M_2)] \sin(\vartheta/2). \tag{1.7}$$

With exception of Eq.(1.5), which is independent of the masses, the relations given above depend on the mass ratio which ranges from $M_2/M_1 \approx 1$ to $M_2/M_1 \gg 1$ in the case of ion beam analysis with light ions.

For the two limits we obtain

$$\left.\begin{aligned} \Theta &= \vartheta/2 \\ v_1 &= v\cos\Theta \\ v_2 &= v\sin\vartheta/2 \end{aligned}\right\} \quad \text{for } M_2/M_1 = 1 \tag{1.8}$$

and

$$\left.\begin{aligned} \Theta &\leq \vartheta \\ v_1 &\leq v \\ v_2 &\ll v \end{aligned}\right\} \quad \text{for } M_2/M_1 \gg 1 \quad . \tag{1.9}$$

1.1.1.2. Kinematic factors

After an elastic scattering event an ion with an energy E before the scattering process will have a kinetic energy $\frac{1}{2}M_1v_1^2$ and the target atom initially at rest moves with an energy $\frac{1}{2}M_2v_2^2$.

In the case of Rutherford backscattering spectrometry the laboratory scattering angle Θ of the ion is fixed by the position of the detector. Therefore the ion energy after a single collision with the scattering angle Θ is of interest. Using Eq.(1.6) we find

$$\frac{1}{2} M_1 v_1^2 = KE \quad , \tag{1.10}$$

with the kinematic factor for the ion

$$K = (v_1/v)^2 = \left\{\frac{\cos\Theta + [(M_2/M_1)^2 - \sin^2\Theta]^{1/2}}{1 + M_2/M_1}\right\}^2 \quad . \tag{1.11}$$

For fixed Θ, the kinematic factor depends only on the mass ratio M_2/M_1. The main behaviour of this dependence can already be seen for $\Theta = \pi$. In this case the kinematic factor

$$K = [(M_2-M_1)/(M_2+M_1)]^2 \tag{1.12}$$

is zero for $M_1 = M_2$ and increases with increasing or decreasing mass ratio. The kinematic factors for H and He ions as functions of the scattering angle Θ and the mass M_2 of the target atoms are represented in Appendices 2 and 3.

Especially for the detection of light elements in the target an ion beam analysis using the foreward scattering of the recoils is very useful. In this case the scattering angle Θ_{rec} of the target atom (recoil) is fixed by the detector. The corresponding energy $\frac{1}{2} M_2v_2^2$ of the target atom after collision is given by Eqs.(1.5) and (1.7)

$$\frac{1}{2} M_2 v_2^2 = kE \quad , \tag{1.13}$$

with the corresponding kinematic factor for the recoils

$$k = \frac{4M_1 M_2}{(M_1+M_2)^2} \cos^2\theta_{rec} \quad , \tag{1.14}$$

which has a maximum value at $M_1 = M_2$.

1.1.1.3. Scattering dynamics

Within the validity of classical mechanics, the scattering angle ϑ in the center-of-mass system is determined by the impact parameter s as introduced in Fig.1.3. Assuming the interaction potential V(r) to depend

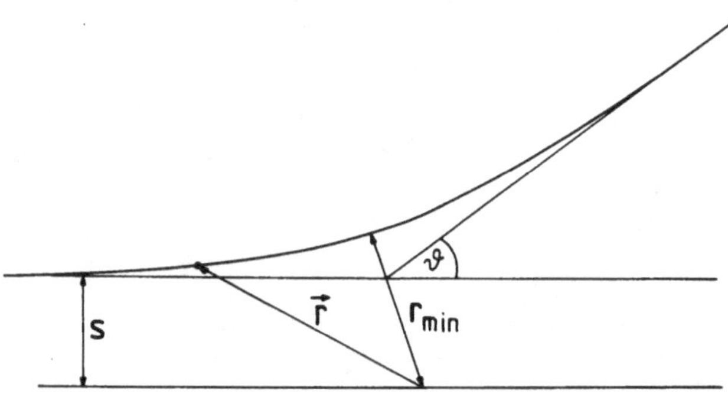

Fig.1.3. Scattering geometry

only on the absolute value r of the distance between the two colliding atoms the classical equations of motion can be solved and provide

$$\vartheta(s) = \pi - 2s \int_{r_{min}}^{\infty} dr/[r^2 f(r)] \quad , \tag{1.15}$$

with

$$f(r) = [1 - V(r)/E_{cm} - s^2/r^2]^{1/2} \quad , \tag{1.16}$$

E_{cm} being the energy of the two collision partners in the center-of-mass system

$$E_{cm} = E\, M_2/(M_1+M_2) \quad , \tag{1.17}$$

and the minimum distance r_{min} is determined by

$$f(r_{min}) = 0 \quad . \tag{1.18}$$

In the special case in which the interaction is given by the Coulomb potential

$$V(r) = \frac{Z_1 Z_2 e^2}{4\pi \varepsilon_o r} \quad , \tag{1.19}$$

(ε_o is the permittivity of vacuum and e is the elementary charge) we find the well known expression

$$\tan \frac{\vartheta(s)}{2} = \frac{b}{2s} \quad . \tag{1.20}$$

The collision diameter

$$b = \frac{Z_1 Z_2 \; e^2}{4\pi \varepsilon_o E_{cm}} \tag{1.21}$$

is the minimum distance of approach r_{min} for the impact parameter equal to zero. On the other hand b represents a critical impact parameter because $\vartheta > \pi/2$ for $s < b/2$ and $\vartheta < \pi/2$ for $s > b/2$. Therefore all ions incident within a circle of the diameter b will be backscattered. In the energy range considered, the collision diameter b is very small compared with the atomic radii (e.g. $b = 4.6*10^{-5}$ nm for scattering of 1 MeV He ions on Si). As expected $\vartheta(s)$ starts with $\vartheta(0) = \pi$ and decreases with increasing impact parameter s.

In general, $\vartheta(s)$ must be calculated by numerical integration. For small scattering angles Θ in the laboratory system (which is of interest for small angle multiple scattering processes and channeling) the small angle or momentum approximation is very useful. Starting from the undisturbed motion of the ion $z = vt$ (see Fig.1.4) the change of the

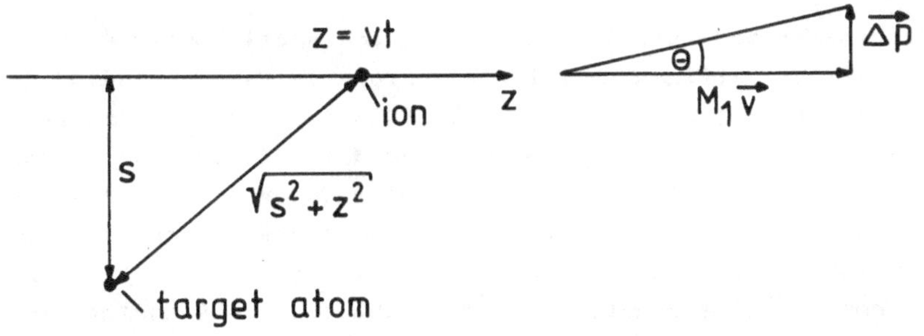

Fig.1.4. Scattering geometry for the momentum approximation

transverse momentum Δp caused by the component of the force transverse to the z-axis is calculated by the integral

$$\Delta p = - \int_{-\infty}^{\infty} dt \, \frac{\partial}{\partial s} V(\sqrt{s^2 + z^2}) = - \int_{-\infty}^{\infty} \frac{dz}{v} \frac{\partial}{\partial s} V(\sqrt{s^2 + z^2}) \quad . \tag{1.22}$$

According to Fig.1.4 the scattering angle in the laboratory system is given by

$$\theta(s) = \frac{p}{M_1 v} = - \frac{1}{E} \int_0^{\infty} dz \, \frac{\partial}{\partial s} V(\sqrt{s^2 + z^2}) \qquad \text{for } \theta \ll 1 \; . \tag{1.23}$$

Using Eq.(1.4) for small ϑ we find the corresponding scattering angle in the center-of-mass system

$$\vartheta(s) = \frac{M_1+M_2}{M_2} \theta = - \frac{1}{E_{cm}} \int_0^{\infty} dz \, \frac{\partial}{\partial s} V(\sqrt{s^2 + z^2}) \quad \text{for } \vartheta \ll 1. \tag{1.24}$$

If V(r) is given by the Coulomb potential (Eq.(1.19)) Eq.(1.24) provides

$$\vartheta(s) = b/s \quad \text{for } \vartheta \ll 1 \quad . \tag{1.25}$$

Comparing with the exact expression (Eq.(1.20)) for the Coulomb potential we find that the momentum approximation is valid to the same extent as $\tan(\vartheta/2)$ can be approximated by $\vartheta/2$ (error of 1% for $\vartheta = 0.35 = 20^o$).

1.1.1.4. Interatomic interaction potential

The Coulomb interaction between the nuclei of the moving ion (atom) and the target atom is screened by their electrons. This is expressed by the screening function φ introduced by

$$V(r) = \frac{Z_1 Z_2 e^2}{4\pi \varepsilon_o r} \varphi(r, Z_1, Z_2, E). \tag{1.26}$$

For distances r between the nuclei of the two atoms small compared with the atomic radii the influence of the electrons is small and φ is expected to approach unity, i.e. $\varphi(0, Z_1, Z_2, E) = 1$. With increasing r the screening function decreases and tends to zero for distances larger than the atomic radii. In general, the screening function also depends on the kinetic energy E of the ion, e.g. introduced by the energy dependent deformation of the electronic shells and by the energy dependent enhancement of the kinetic energy of the electrons during the collision. As has been shown in Ref.[4] these effects can be neglected for the application of the interatomic interaction potential to problems in ion beam analysis of solids. Therefore in the following the energy dependence of φ is neglected and the interaction potential is determined in the approximation of superimposed free-atom electron densities. In this way the interaction potential is given by the difference of the quantum mechanical energies of the quasi-molecule with superimposed free-atom electron densities $E_{12}(r,Z_1,Z_2)$ and the two separated free atoms ($r \rightarrow \infty$) $E_1(Z_1)$ and $E_2(Z_2)$

$$V(r) = E_{12}(r,Z_1,Z_2) - E_1(Z_1) - E_2(Z_2). \tag{1.27}$$

In order to study the general behaviour of the interatomic interaction potential or of the corresponding screening function (Eq.(1.26)) the results obtained using the simple statistical model of atoms for the calculation of E_1, E_2 and E_{12} are discussed first.

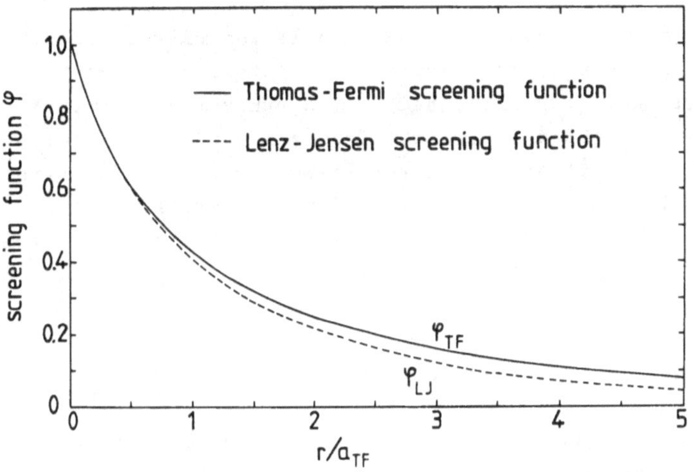

Fig.1.5. Universal screening functions derived by Thomas and Fermi, and Lenz and Jensen.

The screening function as derived by Thomas and Fermi [5] for one atom and extended by Lindhard [6] to the interaction of two atomic systems

$$\varphi(r,Z_1,Z_2) = \varphi_{TF}(r/a_{TF}(Z_1,Z_2)) \tag{1.28}$$

is only a function of the reduced distance r/a_{TF} with the Thomas-Fermi screening length

$$a_{TF}(Z_1,Z_2) = 0.8853\ a_o(Z_1^{2/3} + Z_2^{2/3})^{-1/2} , \tag{1.29}$$

where a_o = 0.05292 nm is the atomic unit of length (Bohr radius). The dependence of φ_{TF} on the reduced distance r/a_{TF} as has been calculated numerically (Ref.[7]) is depicted in Fig.1.5. For small and large values of r/a_{TF} the Thomas-Fermi screening function is approximately given by $\varphi_{TF} \approx 1 - 1.588\,(r/a_{TF})$ and $\varphi_{TF} \approx 144\,(a_{TF}/r)^3$, respectively. Another universal interaction potential was derived by Lenz and Jensen [8]. They found an analytical expression for the screening function

$$\varphi_{LJ}(r/a_{TF}) = (1 + x + 0.3344\,x^2 + 0.0485\,x^3 + 0.002647\,x^4)\,e^{-x} \quad , \quad (1.30)$$

with $x = (9.67\,r/a_{TF})^{1/2}$. As can be seen in Fig.1.5 the Lenz-Jensen screening function agrees with the Thomas-Fermi screening function for small values of r/a_{TF} and it is smaller than φ_{TF} for large r/a_{TF}. The exponential decrease of φ_{LJ} for large r is in accordance with the exponential decrease of the atomic wave functions for large r and therefore it is more realistic than the corresponding r^{-3}-behaviour of φ_{TF}.

Using more refined atomic models for the calculation of the quantum mechanic energies E_1, E_2 and E_{12}, the interatomic interaction potential (Eq.(1.27)) can no longer be expressed by a universal screening function. It must be calculated separately for each projectile target combination Z_1 and Z_2. The values of the resulting screening functions $\varphi(r, Z_1, Z_2)$ are generally between those of the two universal screening functions $\varphi_{LJ}(r/a_{TF})$ and $\varphi_{TF}(r/a_{TF})$. An example is briefly discussed in the following section in comparison with experimental data.

1.1.1.5. Differential scattering cross section

In terms of the impact parameter s the differential cross section for elastic scattering is given by

$$d\sigma = 2\pi s\,ds \ . \quad (1.31)$$

The differential cross section per unit solid angle $d\Omega_{cm} = 2\pi \sin\vartheta\, d\vartheta$ in the center-of-mass system

$$\left(\frac{d\sigma}{d\Omega}\right)_{cm} = \frac{s(\vartheta)}{\sin\vartheta}\left|\frac{ds(\vartheta)}{d\vartheta}\right| \quad (1.32)$$

can be calculated from the relation between the scattering angle ϑ and the impact parameter s as given by Eq.(1.15). For the calculation of the number of ions scattered in a solid angle $d\Omega$ in the laboratory system the differential cross section per unit solid angle $d\Omega = 2\pi \sin\theta\, d\theta$ is introduced. Using the dependence of the laboratory scattering angle θ on the center-of-mass scattering angle ϑ as given in Eq.(1.4) we find

$$\frac{d\sigma}{d\Omega} = \left(\frac{d\sigma}{d\Omega}\right)_{cm} \frac{(1 - \cos^2\vartheta)^{1/2}[(M_1/M_2) + \cos\vartheta]^2}{\sin\theta \cos^2\theta \, [1 + (M_1/M_2)\cos\vartheta]} , \tag{1.33}$$

with

$$\cos\vartheta = \cos\theta \, [\, 1 - (M_1/M_2)^2 \sin^2\theta \,]^{1/2} - (M_1/M_2) \sin^2\theta \quad . \tag{1.34}$$

Note that for $M_1 > M_2$ the laboratory scattering angle θ is restricted by $\sin\theta \le M_2/M_1$.

In some cases the investigation of foreward scattered target atoms (recoils) is used for analysing thin layers. For this purpose the differential cross section per unit solid angle $d\Omega_{rec} = 2\pi \sin\theta_{rec} d\theta_{rec}$ of the recoils scattered to an angle θ_{rec} in the laboratory system is presented. Eqs.(1.31) and (1.32) provide

$$\left(\frac{d\sigma}{d\Omega}\right)_{rec} = \begin{cases} \left(\frac{d\sigma}{d\Omega}\right)_{cm} 4\cos\theta_{rec} & \text{for} \quad \theta_{rec} \le \pi/2 \\ 0 & \text{for} \quad \theta_{rec} > \pi/2 \end{cases} , \tag{1.35}$$

with the relation $\vartheta = \pi - 2\theta_{rec}$ directly given by Eq.(1.5).

In the special case in which the interaction is given by the Coulomb potential (Eq.(1.19)) equations (1.20) and (1.32) immediately provide the differential scattering cross section per unit solid angle in the center-of-mass system

$$\left(\frac{d\sigma}{d\Omega}\right)^R_{cm} = \left(\frac{b}{4}\right)^2 \frac{1}{\sin^4\vartheta/2} = \left[\frac{Z_1 Z_2 e^2 (M_1+M_2)}{16\pi\varepsilon_o M_2 E}\right]^2 \frac{1}{\sin^4\vartheta/2} , \tag{1.36}$$

which is well known as the Rutherford scattering cross section. As can be seen this cross section is proportional to E^{-2} and small angle scattering is heavily preferred.

Taking into account the screening of the Coulomb interaction due to the electrons of the ion and the target atom (Eq.(1.26)) the cross section changes mainly for small scattering angles ϑ (large impact parameters), especially the pole for $\vartheta = 0$ as found in the Rutherford scattering cross section disappears. In general, the scattering cross section $(d\sigma/d\Omega)_{cm}$ must be calculated numerically for screened Coulomb potentials. In some cases approximative methods can be used to reduce the expense. Two of them are discussed in the following.

(i) Correction to the Rutherford scattering cross section

For large energies and large scattering angles the minimum distance of approach r_{min} becomes very small. In this case the main contribution to the integral (Eq.(1.15)) is provided from the region of small values of r. Therefore a linear approximation of the screening function φ by

$$\varphi_{app}(r,Z_1,Z_2) = \begin{cases} 1 - r/r_{max}(Z_1,Z_2) & \text{for} \quad r \le r_{max} \\ 0 & \text{for} \quad r > r_{max} \end{cases} \tag{1.37}$$

is possible, with r_{max} given by the first derivative of φ at $r = 0$

$$r_{max}(Z_1,Z_2) = -\left[\frac{d\varphi(r,Z_1,Z_2)}{dr}\bigg|_{r=0}\right]^{-1} . \tag{1.38}$$

Using the screening function φ_{app}, the scattering angle ϑ (Eq.(1.15)) can be calculated analytically (Refs.[9,10]) which provides the scattering cross section per unit solid angle in the center-of-mass system

$$\left(\frac{d\sigma}{d\Omega}\right)_{cm} = \left(\frac{d\sigma}{d\Omega}\right)^R_{cm} \left[1 - \frac{b}{r_{max}(Z_1,Z_2)}\right] \quad \text{for } b \ll r_{max} , \tag{1.39}$$

with b being the collision diameter as given in Eq.(1.21). Note that this expression holds only for large angle scattering as mentioned above. For the determination of the parameter r_{max} (Eq.(1.38)) the Thomas-Fermi screening function (Eq.(1.28)) can be used because there are no remarkable uncertainties for small r. In this case we find for light ions

$$\frac{b}{r_{max}(Z_1,Z_2)} = 0.049\ Z_1 Z_2^{4/3}\ \frac{M_1+M_2}{M_2} \left(\frac{E}{\text{keV}}\right)^{-1} , \tag{1.40}$$

which means that for 1 MeV He ions the correction exceeds 1% for $Z_2 > 35$ and reaches about 4% for $Z_2 = 90$.

(ii) Wide angle extrapolation of small angle scattering cross section

For small scattering angles ($\vartheta \le 20^\circ$) $\vartheta(s)$ can be calculated according to the momentum approximation (Eq.(1.24)). Using the universal Thomas-Fermi potential we obtain

$$\vartheta(s) = \frac{b}{s} F(s/a_{TF}(Z_1,Z_2)), \tag{1.41}$$

with the deviation F from the Coulomb scattering expression (Eq.(1.25)) being also a universal function given by

$$F(x) = \int_0^{\pi/2} d\alpha \ [\cos\alpha \ \varphi_{TF}(x/\cos\alpha) - x \ \varphi'_{TF}(x/\cos\alpha)] , \tag{1.42}$$

where φ'_{TF} is the first derivative of φ_{TF}. The wide angle extrapolation of Eq.(1.41) as suggested by Lindhard et al. [11] reads

$$2\sin\frac{\vartheta(s)}{2} = \frac{b}{\sqrt{s^2 + s_o^2}} F\left(\frac{\sqrt{s^2 + s_o^2}}{a_{TF}}\right) \quad . \tag{1.43}$$

Note that the addition of a constant s_o^2 to s^2 does not change the differential scattering cross section $d\sigma = \pi\, d(s^2)$ but gives the opportunity of an approximative fitting of Eq.(1.43) for large angles ($\vartheta > 20^o$). The value of s_o for each energy E is determined by adjusting the wide angle extrapolation at the largest scattering angle ϑ to the correct value according to $\vartheta(0) = \pi$. Using the relation $\vartheta(s)$ given by Eq.(1.43) the differential cross section can be written in the following form

$$d\sigma = \pi\, d(s^2) = \pi\, a_{TF}^2\, d(t^{1/2})\, t^{-1}\, f(t^{1/2}), \tag{1.44}$$

with the Lindhard scattering funtion $f(t^{1/2})$ being a universal function of the parameter

$$t^{1/2} = \varepsilon_r \sin(\vartheta/2), \tag{1.45}$$

and ε_r the reduced energy

$$\varepsilon_r = \frac{a_{TF}}{b} = \frac{a_{TF} 4\pi\varepsilon_o}{Z_1 Z_2 e^2} \frac{M_2}{M_1+M_2} E \quad . \tag{1.46}$$

The scattering function $f^{TF}(t^{1/2})$ calculated from the universal function F(x) as given in Eq.(1.42) (based on the Thomas-Fermi screening function) is depicted in Fig.1.6 in comparison with $f^R(t^{1/2}) = (2t^{1/2})^{-1}$ corresponding to the Rutherford scattering cross section (Eq.(1.36)). For $t^{1/2} \geq 10$ there is no remarkable difference, while for low scattering angles and energies ($t^{1/2} < 1$) the cross section calculated using the Thomas- Fermi potential is much smaller than the Rutherford scattering cross section. For comparison, the scattering function $f^{LJ}(t^{1/2})$ calculated using the universal Lenz-Jensen screening function is added in Fig.1.6.

In order to provide details of the electron structure of the colliding atoms in the differential cross section, a more refined interaction potential than the universal one must be used. For the interaction potential (Eq.(1.27)) calculated within a refined atomic model of Climenti and Roetti [12] the corresponding scattering function $f(t^{1/2})$ shows pronounced oscillations which are correlated with the electron shell structure of the colliding atoms. As an example the resulting function $f(t^{1/2})$ for He ions scattered on Xe atoms divided by

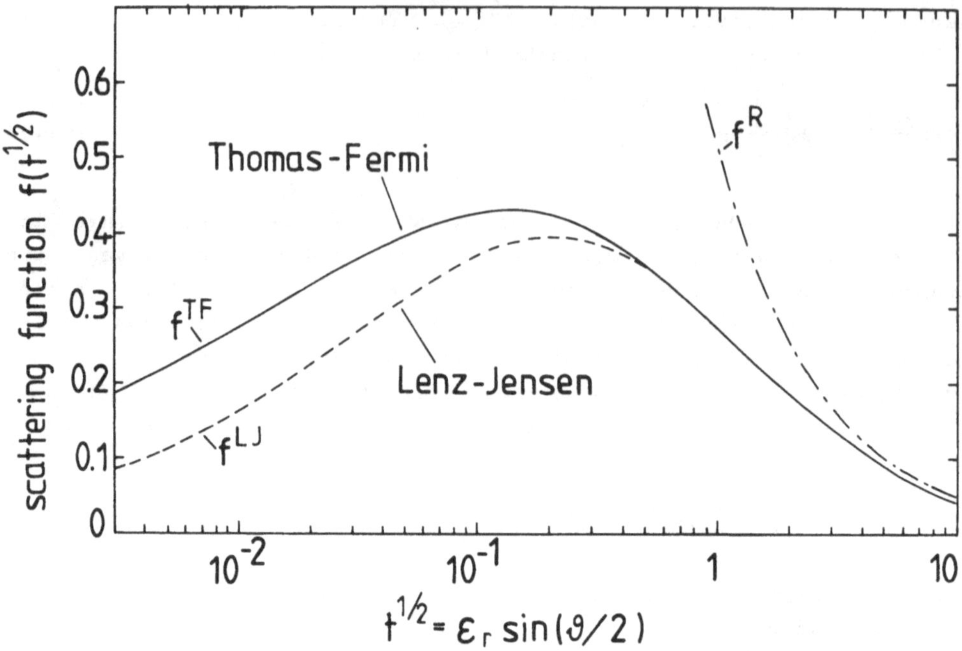

Fig.1.6. Universal elastic scattering cross sections calculated using the Thomas-Fermi screening function (Ref.[11]) and the Lenz-Jensen screening function (Ref.[13]), respectively

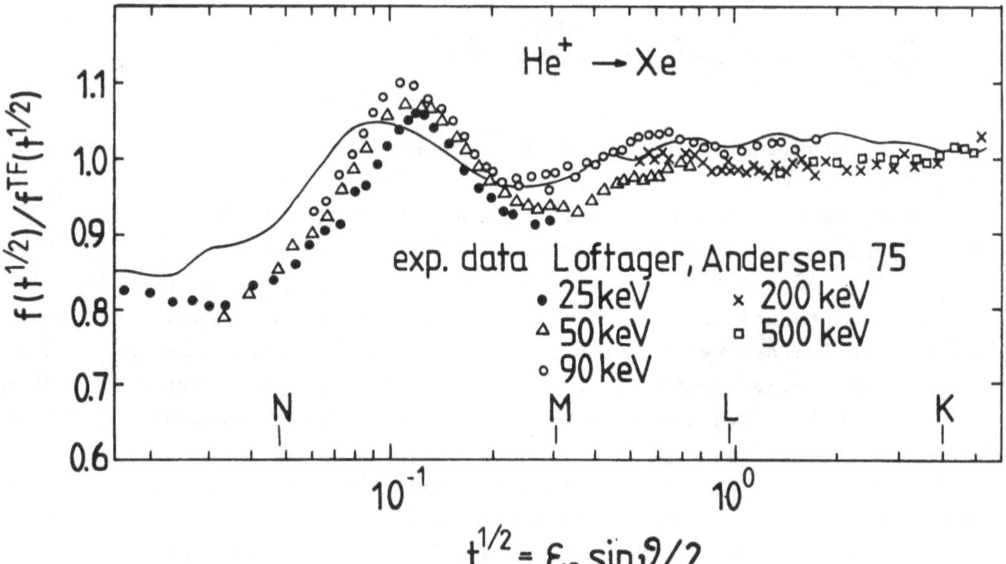

Fig.1.7. Scattering function for He^+ scattering on Xe related to f^{TF}. The theoretical results are obtained using the interaction potential (Ref.[4]) based on the atomic model of Clementi and Roetti. Experimental data of Loftager and Andersen [13] are used for comparison.

the scattering function $f^{TF}(t^{1/2})$ obtained for the Thomas-Fermi potential is depicted in Fig.1.7. As can be seen, the measured structure effects are well described by the calculated scattering function.

1.1.1.6. Nuclear energy loss

The energy loss of a moving ion during the elastic collision (nuclear energy loss) T is given by the kinetic energy of the target atom after the collision (see Eq.(1.7))

$$T = \frac{1}{2} M_2 v_2^2 = T_{max} \sin^2(\vartheta/2), \tag{1.47}$$

with the maximum transferred energy

$$T_{max} = \frac{4M_1M_2}{(M_1+M_2)^2} E . \tag{1.48}$$

For $\vartheta(s)$ determined by Eq.(1.15) the differential cross section $d\sigma_n(T)$ for transferring an energy T during a nuclear collision is provided from T(s) (Eq.(1.47)) by

$$d\sigma_n(T) = \pi \left| \frac{d(s^2(T))}{dT} \right| dT . \tag{1.49}$$

In the case of Coulomb interaction $\vartheta(s)$ is given by Eq.(1.20) which provides

$$T(s) = T_{max} \frac{(b/2)^2}{(b/2)^2 + s^2} \tag{1.50}$$

and

$$d\sigma_n(T) = \begin{cases} B_n T^{-2} dT & \text{for } T \le T_{max} \\ 0 & \text{for } T > T_{max} \end{cases} , \tag{1.51}$$

with

$$B_n = \pi \left(\frac{b}{2}\right)^2 T_{max} = \pi \frac{M_1}{M_2} \left(\frac{Z_1 Z_2 e^2}{4\pi\varepsilon_0}\right)^2 \frac{1}{E} . \tag{1.52}$$

Starting from a large value $T(0) = T_{max}$ for $s = 0$, which can be a considerable amount of the ion energy E depending on the mass ratio, the transferred energy T(s) decreases rapidly with increasing impact parameter s. Note that the characteristic length b/2 is very small compared with the mean distances of the target atoms for the conditions of ion beam analysis (e.g. $b/2 = 2.3*10^{-5}$ nm for 1 MeV He ions scattered on silicon atoms).

The determination of the nuclear stopping power in an amorphous material using the differential cross section $d\sigma_n(T)$ for an energy transfer T in a single collision is described in Sect.1.2.1.1.

1.1.2. Electronic scattering

Due to the interaction between the moving ion (atom) and the target atom the electronic states of the two atomic systems can be changed (electronic scattering). The corresponding cross sections must be calculated within a quantum mechanical treatment of scattering. This is a complicated many-body problem and there is no general solution in contrast to nuclear scattering. With respect to ion beam analysis of solids we are interested in the differential cross section $d\sigma_e(T)$ for an energy loss T due to electronic scattering and in the cross section for X-ray production, which are briefly discussed in the following sections.

1.1.2.1. Electronic energy loss

The electronic energy loss is determined by all possible electronic excitations of the target atom and the ion. Because of the large number of contributions from different excitations the special features of their cross sections are heavily suppressed. Therefore simplified differential cross sections for excitation will provide useful values of the electronic energy loss.

For a simple understanding of electronic energy loss, particularly in comparison with nuclear energy loss, let us consider the collision of a totally ionized projectile with a free electron initially at rest. This approximates the situation for large ion velocities, compared with the orbital velocities of the electrons of the target atom. Since for Coulomb interaction there is no difference between the quantum mechanical and the classical cross section we can immediately use the corresponding results obtained for nuclear scattering, except for M_2 and Z_2 being replaced by the electronic mass $M_2 = m$ and $Z_2 = 1$, respectively. Taking into account $m << M_1$ we obtain the electronic energy loss from Eqs.(1.50), (1.48), and (1.21)

$$T(s) = T_{max,e} \frac{(b_e/2)^2}{(b_e/2)^2 + s^2} , \qquad (1.53)$$

with the maximum transferred energy

$$T_{max,e} = 2\, mv^2 , \qquad (1.54)$$

the collision diameter

$$b_e = \frac{Z_1 e^2}{4\pi\varepsilon_o E} \frac{M_1}{m} , \qquad (1.55)$$

and s being the impact parameter of the ion with respect to the electron considered. The differential cross section for an electronic energy loss T is provided from Eqs.(1.51) and (1.52) by the same procedure as used for the energy loss

$$d\sigma_e(T) = \begin{cases} B_e T^{-2} dT & \text{for} \quad T \le T_{max,e} \\ 0 & \text{for} \quad T > T_{max,e} \end{cases} , \qquad (1.56)$$

with

$$B_e = \pi \frac{M_1}{m} \left(\frac{Z_1 e^2}{4\pi\varepsilon_o} \right)^2 \frac{1}{E} . \qquad (1.57)$$

Comparing with the corresponding expressions for nuclear scattering we get

$$\frac{T_{max,e}}{T_{max}} = \frac{m}{M_2} \left(\frac{M_1+M_2}{M_1} \right)^2 \ll 1 ,$$

$$\frac{b_e}{b} = \frac{1}{Z_2} \frac{M_1 M_2}{m(M_1+M_2)} \gg 1 , \qquad (1.58)$$

$$\frac{B_e}{B_n} = \frac{1}{Z_2^2} \frac{M_2}{m} \gg 1 .$$

Therefore the nuclear energy loss for s = 0 is much larger than the electronic one but it decreases more rapidly with increasing impact parameter s then the electronic energy loss does. This behaviour is demonstrated in Fig.1.8a with the example of 1 MeV He ions scattered on Si atoms. Fig.1.8b shows the corresponding differential cross sections for electronic and nuclear energy loss. Because of $B_e \gg B_n$ the cross section for an energy loss T is much larger for electronic scattering than for nuclear scattering but $d\sigma_e$ is much more restricted in T ($T_{max,e} << T_{max}$).

The impact parameter dependence of the electronic and nuclear energy loss as shown in Fig.1.8a justifies the assumption of statistical independence of electronic and nuclear scattering. For very small impact parameters ($s < (M_2/m)^{1/2} M_1 b/(2M_2)$) the energy loss is determined by nuclear scattering and for larger values of s the electronic contribution to the energy loss heavily dominates. Therefore there is hardly a correlation between the two processes and the total differential cross section can be expressed by a sum

$$d\sigma = d\sigma_n + d\sigma_e \ . \tag{1.59}$$

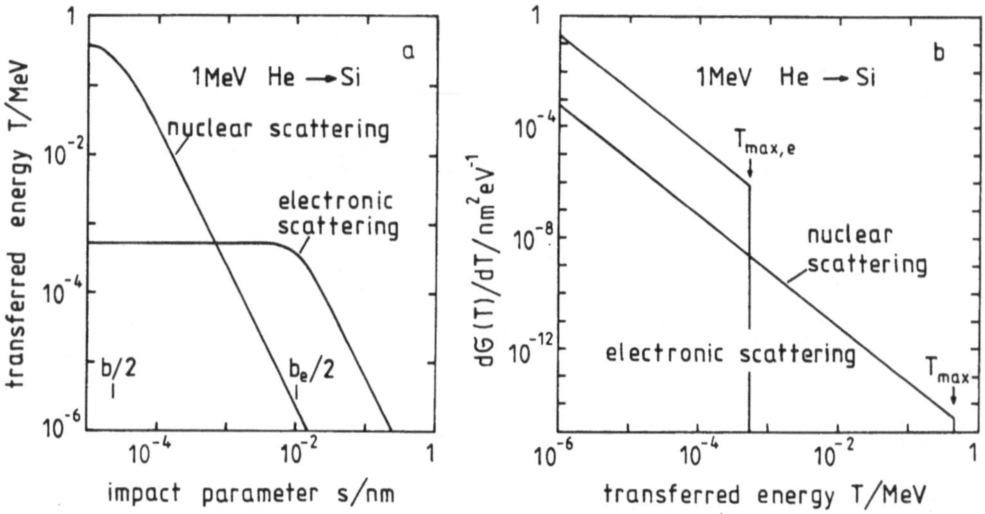

Fig.1.8. The impact parameter dependence of the transferred energy (a) and the differential cross section as a function of the transferred energy (b) for nuclear and electronic scattering in the case of 1 MeV He ions scattered on Si atoms

With the differential cross section $d\sigma_e$ for an electronic energy loss T as given in Eq.(1.56) only a qualitative discussion of the electronic scattering processes is possible. In the energy region used for ion beam analysis of solids the binding of the electrons and their orbital velocity must be taken into account. Usually, the so-called electronic stopping cross section $\int d\sigma_e(T)\, T$ is calculated directly, rather than the differential cross section $d\sigma_e(T)$. Therefore, with respect to a more detailed description of electronic energy loss we refer to Sects.1.2.1.2 and 1.2.1.3.

1.1.2.2. X-ray cross section

In contrast to energy loss, only special excitations of the colliding atoms are responsible for the generation of X-rays. With respect to the application of particle induced X-ray emission (PIXE) to the analysis of solids we consider only light ions, especially protons. In the energy region considered (some 100 keV and above) light ions moving in solid targets can be assumed to be totally ionized. In this case X-ray emission is induced by inner shell vacancy production due to the Coulomb interaction between the point charged ion and the inner shell electron of a target atom. In principle, the transitions of the electron considered to all unoccupied bound states and continuum states (with the other electrons remaining in their initial states) contribute to the vacancy production, but the contributions due to excitations to unoccupied bound states are negligible as was shown by Merzbacher and Lewis [14]. Assuming the moving ion to be described by a plane wave the differential cross section can be determined in a first Born approximation in terms of the momentum $\hbar q$ transferred for ionization into a given final continuum state and provides (Ref.[15])

$$d\sigma_{fi}(q) = 16\pi \left(\frac{Z_1 e^2}{4\pi\varepsilon_o \hbar v}\right)^2 \frac{dq}{q^3} \left| \int d^3\vec{r}\, e^{i\vec{q}\vec{r}}\, \psi_f^*(\vec{r})\, \psi_i(\vec{r}) \right|^2 , \qquad (1.60)$$

where $\psi_i(\vec{r})$ and $\psi_f(\vec{r})$ are the wave functions of the initial state (atomic wave function) and the final state (plane wave), respectively. To obtain the cross section for inner shell vacancy production Eq.(1.60) must be integrated over all momentum transfers compatible with the production of a final state ψ_f and summed up over all final states compatible with energy and momentum conservation.

Using the wave functions $\psi_i(\vec{r})$ according to the atomic model of Slater [16] (hydrogenic wave function) the cross section $d\sigma_{v,i}$ for the production of vacancies in an inner shell i as provided from Eq.(1.60) can be expressed by

$$\sigma_{v,i} = \left(\frac{Z_1}{E_{B,i}}\right)^2 f_i\left(\frac{m\,E}{M_1 E_{B,i}}, \frac{M_1}{m}\right) , \qquad (1.61)$$

with f_i being a universal function of $mE/(M_1 E_{B,i})$ for a given shell i and $E_{B,i}$ being the binding energy of the electron in the shell i. Since not all the vacancies produced result in X-ray emission the corresponding X-ray cross section $\sigma_{X,i}$ is smaller than $\sigma_{v,i}$. The fraction of vacancies in a given shell i which lead to X-ray emission is usually called fluorescence yield $\omega_i(Z_2)$ depending on the atomic number Z_2. In this way the X-ray cross section reads

$$\sigma_{X,i} = \omega_i \sigma_{v,i} = \left(\frac{Z_1}{E_{B,i}}\right)^2 \omega_i \ f_i\left(\frac{m\ E}{M_1 E_{B,i}}, \frac{M_1}{m}\right) \quad . \tag{1.62}$$

As has been shown by calculations the fluorescence yield is small ($\omega_i(Z_2) << 1$) for small Z_2 and approaches unity for large Z_2. The other fraction $1 - \omega_i$ of the vacancies is responsible for the production of Auger electrons.

For example, the reduced K-shell X-ray cross section $\omega_K f_K$ for proton impact is depicted in Fig.1.9. The theoretical results are in good agreement with the experimental data for $1 \le Z_2 \le 73$, which proves the scaling law (Eq.(1.62)) to be a good approximation. For further information about X-ray cross sections see the review of Garcia et al. [15].

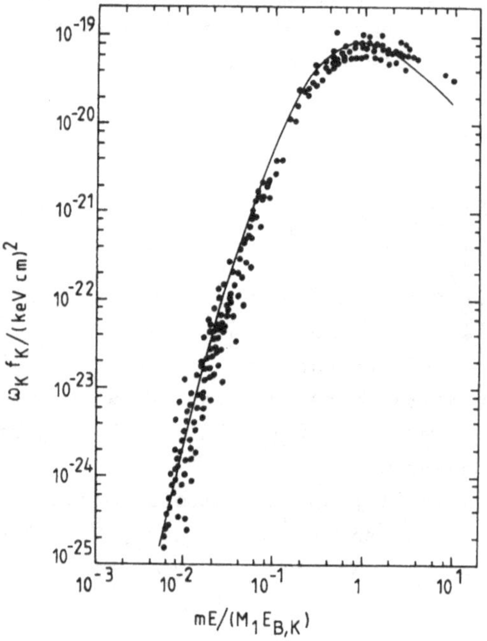

Fig.1.9. Universal K-shell X-ray cross section $\omega_K f_K = \sigma_{X,K}(E_{B,K}/Z_1)^2$ as a function of the reduced ion energy $mE/(M_1 E_{B,K})$ (Ref.[15]). The result of the binary encounter approximation (solid line) is compared with experimental data for $1 \le Z_2 \le 73$.

1.2. AMORPHOUS AND POLYCRYSTALLINE SOLIDS

In this section random materials (with randomly distributed atoms) are considered. Besides amorphous solids, polycrystalline solids without texture and with small crystallite sizes can also be treated as random materials. The typical experimental situation is illustrated in Fig.1.10. A beam of ions with an energy E_0 penetrates a thin amorphous layer of thickness Δz. Due to two-body collisions with the atoms of this layer an ion loses a certain amount of energy ΔE and is deflected by an angle φ. This is a statistical process and provides no fixed values ΔE and φ but a probability distribution $W(\Delta E,\varphi,\Delta z)\, d(\Delta E)\, 2\pi \sin\varphi d\varphi$. In general, the measured energy loss distribution depends on the angular interval $\varphi \dots \varphi + \delta\varphi$ fixed by the detector and the angular distribution depends on the energy loss interval registered by the detector. As will be seen later for high energy ion beam analysis the width of the angular distribution is smaller than or in the order of $\delta\varphi$ given by the detector. Therefore, in most cases only the integral energy loss distribution $P(\Delta E,\Delta z)\, d(\Delta E)$ (integral of W over φ) is of interest.

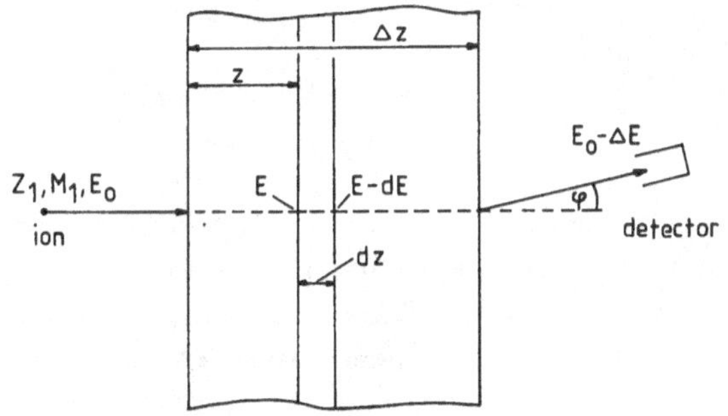

Fig.1.10. Energy and scattering angle of an ion after passing a thin amorphous layer of thickness Δz

Before discussing the full distribution its first and second moments, the mean energy loss $\langle\Delta E\rangle$ and the energy straggling $\Omega^2 = \langle(\Delta E - \langle\Delta E\rangle)^2\rangle$ are considered in Sects.1.2.1 and 1.2.2. With respect to the depth resolution in ion beam analysis of solids the angular distribution $P(\varphi,\Delta z)\, 2\pi \sin\varphi d\varphi$ (integral of W over ΔE) is of interest. It is discussed in Sect.1.2.5.

1.2.1. Mean energy loss

In a random material, the two-body collisions of the ions with the target atoms are statistically independent. In this case the differential cross section for an energy loss T of the ion due to the collision with a target atom is equal for each collision with the same ion energy E and it is denoted by $d\sigma(T,E)$. Because of the energy dependence of $d\sigma$, the mean energy loss dE of the ions in a layer of differential small thickness dz is considered first (see Fig.1.10). The probability $dw(T,E)$ for an energy transfer T due to the collision of an ion of energy E with a target atom in the layer of thickness dz is given by

$$dw(T,E) = N\, dz\, d\sigma(T,E) \quad , \tag{1.63}$$

where N is the number of target atoms per unit volume. Taking into account the correct sign (dE < 0) the mean energy loss reads (Ref.[17])

$$dE = - \int dw\, T = - N\, dz \int d\sigma\, T \quad . \tag{1.64}$$

The mean energy loss per unit path length or stopping power

$$\frac{dE}{dz} = - N\, \varepsilon(E) \, , \tag{1.65}$$

with

$$\varepsilon(E) = \int d\sigma(T,E)\, T \quad ,$$

is a product of the stopping cross section $\varepsilon(E)$ (which is determined by the structure of the target atoms) and the atomic density N (which is the only information about the structure of the target). This simple structure of Eq.(1.65) reflects the statistical independence of the two-body collisions.

If the stopping cross section $\varepsilon(E)$ is known, the mean energy loss $\langle\Delta E\rangle$ for a layer of thickness Δz can be derived implicitly by solving the differential equation (1.65)

$$z = \int_{E_o - \langle\Delta E\rangle}^{E_o} dE/[N\varepsilon(E)] \quad . \tag{1.66}$$

As discussed in Sect.1.1 there are nuclear (elastic) and electronic (inelastic) collisions of the ions with the target atoms. Assuming the nuclear and electronic collisions to be statistically independent (Eq.(1.59)) the probability dw for an energy transfer T is given by a sum of the two contributions

$$dw(T,E) = dw_n + dw_e = N\, dz\, [d\sigma_n(T,E) + d\sigma_e(T,E)] \quad , \tag{1.67}$$

which provides

$$\varepsilon(E) = \varepsilon_n(E) + \varepsilon_e(E). \tag{1.68}$$

In the following, the nuclear and electronic contributions to the stopping cross section are considered separately.

1.2.1.1. Nuclear stopping

If the interatomic interaction potential is known, the nuclear stopping cross section

$$\varepsilon_n(E) = \int d\sigma_n(T,E)\, T \tag{1.69}$$

can be calculated from the differential cross section $d\sigma_n$ given by Eq.(1.49).

In the case of high energies (reduced energy $\varepsilon_r >> 1$ as discussed in Sect.1.1.1.5) the interatomic interaction potential can be approximated by the Coulomb potential of the two nuclei. The corresponding differential cross section (Eq.(1.51)) provides

$$\varepsilon_n(E) = B_n \int dT/T = B_n \ln(T_{max}/T_{min}) \tag{1.70}$$

with $T_{max} = 4M_1M_2E/(M_1+M_2)^2$. Because of the long range of the Coulomb potential, ε_n becomes infinite for $T_{min} = 0$. Therefore the screening of the Coulomb potential due to the electrons must be taken into account. The most simple approach is given by cutting the Coulomb potential at a distance $r = a_{TF}$. This provides a minimum transferred energy $T_{min} = T(s = a_{TF}) = T_{max}/(2\varepsilon_r)^2$ according to Eq.(1.50) (note that $a_{TF} >> b/2$) and the nuclear stopping cross section reads

$$\varepsilon_n(E) = \varepsilon_{n,o} \frac{1}{\varepsilon_r} \ln(2\varepsilon_r) \qquad \text{for } \varepsilon_r >> 1 , \tag{1.71}$$

with the factor

$$\varepsilon_{n,o} = \frac{2\pi Z_1 Z_2 e^2}{4\pi\varepsilon_o} a_{TF} \frac{M_1}{M_1+M_2} \tag{1.72}$$

and the reduced energy as introduced in Eq.(1.46)

$$\varepsilon_r = \frac{E}{E_\varepsilon} , \quad E_\varepsilon = \frac{Z_1 Z_2 e^2}{4\pi\varepsilon_o a_{TF}} \frac{M_1+M_2}{M_2} . \tag{1.73}$$

For H and He ions and different target materials the values of $\varepsilon_{n,o}$ and E_ε are given in Appendix 4. As can be seen for $\varepsilon_r >> 1$ ε_n decreases with increasing energy.

The determination of the nuclear stopping cross section for smaller energies ($\varepsilon_r \leq 1$) requires a more realistic interatomic interaction potential taking into account the screening of the Coulomb potential due to the electrons. Using the universal differential cross section (Eq.(1.44)) as found, for example, for the Thomas-Fermi screening function (see Fig.1.6) the nuclear stopping cross section (Eq.(1.69)) can be expressed by

$$\varepsilon_n(E) = \varepsilon_{n,o}\, g_n(\varepsilon_r) \quad , \tag{1.74}$$

with $g_n(\varepsilon_r)$ being a universal function. The dependence of the nuclear stopping function $g_n(\varepsilon_r)$ on the reduced energy ε_r as calculated by Lindhard et al. [11] is depicted in Fig.1.11 in comparison with the results of the Coulomb approximation $g_n^C = \ln(2\varepsilon_r)/\varepsilon_r$. As can be seen, the nuclear stopping cross section increases rapidly with increasing energy, reaches a maximum value at about $\varepsilon_r = 0.4$ and decreases with further increasing energy.

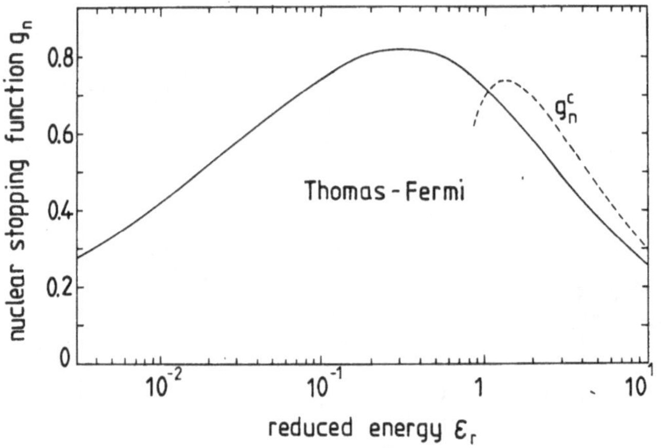

Fig.1.11. Universal nuclear stopping cross section (stopping function) calculated using the Thomas-Fermi screening function (Ref.[11])

Using a more precise screening function than the Thomas-Fermi one, the function g_n depends on E, Z_1 and Z_2 separately and no longer on one generalized parameter ε_r. There are remarkable deviations of the resulting $g_n(E,Z_1,Z_2)$ from $g_n(\varepsilon_r(E,Z_1,Z_2))$ calculated from the Thomas-Fermi potential for $\varepsilon_r < 1$ but they are not of interest with respect to ion beam analysis because of the high energies used ($\varepsilon_r > 1$). In most cases even the Coulomb approximation can be used which is valid for $\varepsilon_r > 10$.

1.2.1.2. High energy electronic stopping

For a rough approximation of the ion energy loss caused by inelastic scattering processes (electronic scattering) let us start with a simple classical consideration as performed by Bohr [17]. If the ion energy is sufficiently high, the screening of the Coulomb interaction between the nucleus of the projectile and a target electron (due to the other electrons) can be neglected and the target electron can be assumed to be free and to be initially at rest. In this case the contribution of the i-th electron of a target atom to the electronic stopping cross section calculated within classical mechanics is given by

$$\varepsilon_e^i(E) = \int d\sigma_e T = B_e \ln(T_{max,e}/T_{min}) \tag{1.75}$$

where $d\sigma_e$ is the cross section (Eq.(1.56)) for the collision of the completely ionized projectile with a free electron at rest as given in Sect.1.1.2.1 and $T_{max,e} = 2mv^2$ is the corresponding maximum transferred energy. The binding of the electron is approximately taken into account afterwards by introducing a minimum transferred energy $T_{min} = I_i$ with I_i being the ionization energy of th i-th electron. Now the classical electronic stopping cross section for one atom reads

$$\varepsilon_{e,cl}(E) = \sum_{i=1}^{Z_2} \varepsilon_e^i(E) = \frac{2\pi Z_1^2 Z_2}{mv^2} \left(\frac{e^2}{4\pi\varepsilon_o}\right)^2 \ln\left(\frac{2mv^2}{I}\right), \tag{1.76}$$

with

$$\ln I = \frac{1}{Z_2} \sum_{i=1}^{Z_2} \ln(I_i) \quad . \tag{1.77}$$

A quantum mechanical treatment as necessary was performed first by Bethe [18] and Bloch [19] and it provides

$$\varepsilon_e(E) = \frac{4\pi Z_1^2 Z_2}{mv^2} \left(\frac{e^2}{4\pi\varepsilon_o}\right)^2 \ln\left(\frac{2mv^2}{I}\right), \tag{1.78}$$

with a mean excitation energy I determined by the electronic structure of the target atom

$$\ln I = \sum_n f_{nn_o} \ln(E_n - E_{n_o}). \tag{1.79}$$

Here, the dipole oscillator strength f_{nn_o} is the weight factor and is given by

$$f_{nn_o} = [2m/(3\hbar^2 Z_2)]\,(E_n - E_{n_o})\left|\langle n|\sum_{i=1}^{Z_2} \vec{r}_i|n_o\rangle\right|^2 , \tag{1.80}$$

where $|n_o\rangle$ and $|n\rangle$ are the wave vectors of the ground state and the excited state of the target atom, respectively, and E_n and E_{n_o} are the corresponding energies. Within the statistical model of the atoms the mean excitation energy depends monotonously on Z_2

$$I = I_o Z_2, \tag{1.81}$$

with $I_o \approx 10$ eV. The Bethe-Bloch formula (Eq.(1.78)) holds for large energies. The restriction is mainly given by applying the first Born approximation ($E > 25\ M_1 Z_1^{4/3}$ keV/amu required) and by the assumption of a completely ionized projectile. As can be seen, the classical approach (Eq.(1.76)) provides already the main features of the energy dependence of ε_e for large energies but differs from the correct value by a factor of 2. It was shown by a number of detailed investigations (e.g. Ref. [20]) that the energy loss can be divided into two contributions related to free collisions and resonant collisions. For large energies the two contributions are approximately equal (equipartition rule). Therefore the classical estimation of the stopping cross section, which does not include resonant collisions, provides only half of the value given by Bethe and Bloch.

For comparison with the nuclear stopping cross section (Eq.(1.71)) the Bethe-Bloch formula is rewritten

$$\varepsilon_e(E) = \varepsilon_{e,o}\frac{1}{\eta_r}\ln(5.44\eta_r) , \tag{1.82}$$

with

$$\varepsilon_{e,o} = 4\pi Z_1^2 a_o \frac{e^2}{4\pi\varepsilon_o} = 9.575\ Z_1^2 * 10^{-15}\ \text{eVcm}^2 \tag{1.83}$$

and the dimensionless energy parameter

$$\eta_r = \frac{v^2}{v_o^2\, Z_2} = \frac{E}{E_\eta} , \tag{1.84}$$

($v_o = e^2/(4\pi\varepsilon_o\hbar)$) determined by a characteristic energy E_η for electronic scattering

$$E_\eta = \frac{1}{2} M_1 v_o^2 Z_2 = 24.98\ M_1 Z_2\ \text{keV/amu} . \tag{1.85}$$

For H and He ions and different target materials the energy E_η is tabulated in Appendix 4. As can be seen, the electronic stopping cross section ε_e for $\eta_r = E/E_\eta > 1$ and the nuclear stopping cross section ε_n for $\varepsilon_r = E/E_\varepsilon > 1$ (Eqs.(1.71) and (1.82)) show a similar energy dependence, only the scaling is different.

$$E_\eta = \frac{M_1 M_2}{2Z_1 m(M_1+M_2)} \frac{a_{TF}}{a_o} E_\varepsilon >> E_\varepsilon \tag{1.86}$$

$$\varepsilon_{e,o} = \frac{2Z_1(M_1+M_2)}{Z_2 M_1} \frac{a_o}{a_{TF}} \varepsilon_{n,o} > \varepsilon_{n,o} \tag{1.87}$$

For $\eta_r < 1$ the Bethe-Bloch formula is no longer applicable because the first Born approximation does not hold and the projectile is not completely ionized. With decreasing energy more and more states $|n\rangle$ of the target atoms (with large E_n) cannot be excited because the energy transfer is too small. Therefore the energy loss due to excitation of the target atoms is expected to decrease rapidly with decreasing energy ($\varepsilon_e = 0$ for $\eta_r = 0.18$ in the Bethe-Bloch formula can be understood in this way) and other processes become more and more important for the energy loss at low energies.

1.2.1.3. Low energy electronic stopping

Firsov [21] investigated the electronic stopping in the limit of $\eta_r << 1$. In this case the excitation of the target atoms is of minor importance and the projectile is only weakly ionized. Because of the low velocity of the projectile, the duration time of the collision is assumed to be large enough for the fusion of the projectile atom and the target atom in a quasi-molecule. During the collision time the electrons of the projectile atom and of the target atom are able to exchange. The transition of projectile electrons with the mean momentum mv to the target atom and the capture of target electrons by the moving projectile atom (connected with an acceleration of these electrons) results in an energy transfer from the projectile to the target atom. For a simple description of these processes Firsov introduced a plane, which perpendicularly bisects the line between the projectile and target atom (see Fig.1.12). This plane divides the quasi-molecule into two regions, let us say the P-region (projectile) and the T-region (target atom). The loss and capture of electrons is then expressed by the flux of electrons through the Firsov plane. In this way the energy loss for a collision with the impact parameter s reads

$$T(s) = mv \int_{-\infty}^{\infty} dz\ \Phi_e(\sqrt{z^2+s^2}) \quad , \tag{1.88}$$

where $\Phi_e(r)$ is the flux of the electrons moving from the P-region to the T-region across the Firsov plane which depends on the distance r of the two atoms of the quasi-molecule.

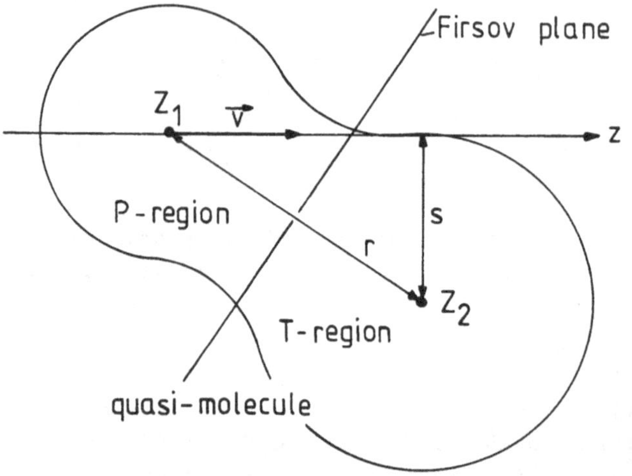

Fig.1.12. Quasi-molecule formed during the collision of the moving ion and a target atom. It is divided into two regions by the Firsov plane.

Using the atomic model of Thomas and Fermi for the calculation of the electron flux Φ_e Firsov [21] got an approximation for the transferred energy

$$T(s) = \frac{0.35(Z_1+Z_2)^{5/3}\ \hbar/a_o}{[1+0.16(Z_1+Z_2)^{1/3}s/a_o]^5}\ v \quad , \tag{1.89}$$

which provides (with $d\sigma_e = 2\pi s ds$)

$$\varepsilon_e(E) = \int_0^{\infty} 2\pi s\ ds\ T(s) = 1.03\ (Z_1+Z_2)\left(\frac{E}{M_1}\ \frac{amu}{keV}\right)^{1/2} * 10^{-15}\ eV cm^2\ . \tag{1.90}$$

Lindhard and Scharff used a different approach for the determination of the electronic energy loss for low energies. They started from the motion of a projectile in a uniform electron gas with projectile velocities v small compared to the Fermi velocity v_F of the target electrons. In the following a simplified description of this approach as given by Bonderup [20] is briefly discussed. Because of the large mass of the projectile compared with the electron mass, the center-of-mass system coincides with the system fixed to the projectile. In this system the velocity distribution of the target electrons is given by a sphere

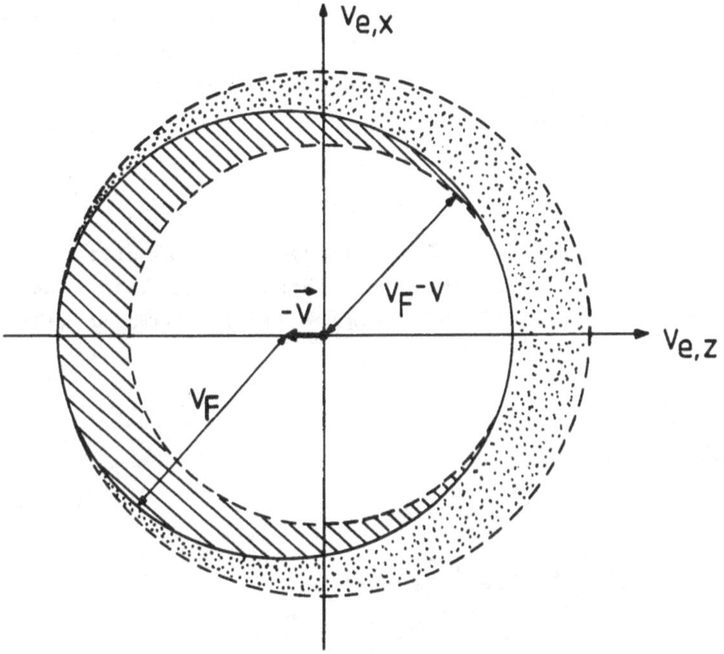

Fig.1.13. Distribution of the velocity $\vec{v}_e$ $(v_{e,x}, v_{e,y}, v_{e,z})$ of the target electrons in the system fixed to the moving ion (center-of-mass system)

with radius v_F shifted by $-\vec{v}$ as shown in Fig.1.13. As known from Sect.1.1.1.1 the elastic scattering does not change the absolute value of the velocities in the center-of-mass system. Therefore the Pauli principle allows only scattering of the electrons of the hatched part to states in the dotted part of the velocity space. Within these considerations the energy dependence of the energy loss can easily be found. For $v \ll v_F$ the relative part of electrons which take part in scattering $[v_F^3-(v_F-v)^3]/v_F^3$ is proportional to v/v_F. The velocity of these electrons is approximately given by v_F. Now, the stopping power can be written in the following form

$$\frac{dE}{dz} = \frac{d}{dz}\left(\frac{1}{2} M_1 \vec{v}^2\right) = \vec{v}\,\frac{d(M_1\vec{v})}{dz} = \frac{dp_\parallel}{dt} \sim \Delta p_\parallel \Phi_e \sigma(v_F) \quad . \tag{1.91}$$

The momentum transfer parallel to the projectile velocity $\vec{v}$ per unit time $dp_\parallel/dt$ is given by the product of the corresponding momentum transfer $\Delta p_\parallel$ of one electron, the flux Φ_e of electrons which take part in the scattering, and the cross section $\sigma(v_F)$ for the scattering of these electrons with velocity v_F. As can be seen in Fig.1.13 the momentum transfer $\Delta p_\parallel$ (hatched part → dotted part) is in the order of mv_F. The flux Φ_e is given by the full flux $n_e v_F$ (n_e is the number of electrons per unit volume) multiplied by the relative part of electrons which take part in the scattering (proportional to v/v_F), i.e. $\Phi_e \sim n_e v$. Inserting these expressions in Eq.(1.91) we find the stopping power in a uniform electron gas of density n_e

$$\frac{dE}{dz} \sim m v_F n_e \sigma(v_F) v \quad \text{for } v << v_F \tag{1.92}$$

to be proportional to the ion velocity. In a more detailed investigation (which never has been published) Lindhard and Scharff applied these considerations to the calculation of the energy loss in atomic targets. The result is given in Ref.[22].

$$\begin{aligned} \varepsilon_e(E) &= 8\pi a_o \frac{e^2}{4\pi\varepsilon_o} \frac{Z_1^{7/6} Z_2}{(Z_1^{2/3} + Z_2^{2/3})^{3/2}} \frac{v}{v_o} \\ &= 3.83 \frac{Z_1^{7/6} Z_2}{(Z_1^{2/3} + Z_2^{2/3})^{3/2}} \left(\frac{E}{M_1} \frac{\text{amu}}{\text{keV}}\right)^{1/2} * 10^{-15} \text{ eVcm}^2 \end{aligned} \tag{1.93}$$

In good agreement with the experimental data the theories of Firsov and Lindhard provide an electronic stopping cross section for low energies ($E << 25\ M_1 Z_1^{4/3}$ keV/amu) proportional to the square root of the energy. Because Firsov and Lindhard used the atomic model of Thomas and Fermi their results showed a monotonous dependence of ε_e (Eqs.(1.90, 1.93)) on the atomic numbers Z_1 and Z_2. This is not verified by the experimental data which show pronounced oscillations in the Z_1- and Z_2-dependence of ε_e (e.g. Refs.[23,24]). On average the Lindhard formula (Eq.(1.93)) shows a better agreement with the measured electronic stopping cross section than the Firsov formula (Eq.(1.90)) does. However, the Firsov model is a good starting point for further improvements of the theory. Using a quantum mechanical expression for the electron flux in Eq.(1.88) (Ref.[25]) within a refined atomic model (e.g. the Hartree-Fock-Slater model) the numerically calculated electronic stopping cross sections show close agreement with the experimental data also with respect to the Z_1- and Z_2-oscillations (e.g. Refs.[26,27]).

1.2.1.4. General formulae for electronic stopping

As follows from previous sections the general energy dependence of the electronic stopping cross section is given by $\varepsilon_e \sim E^{1/2}$ for low projectile energies $E << E_m = 25\ M_1 Z_1^{4/3}$ keV/amu, by $\varepsilon_e \sim E^{-1}(\ln E + \text{const})$ for high energies $E >> E_m$ and between these energy regions ε_e must reach

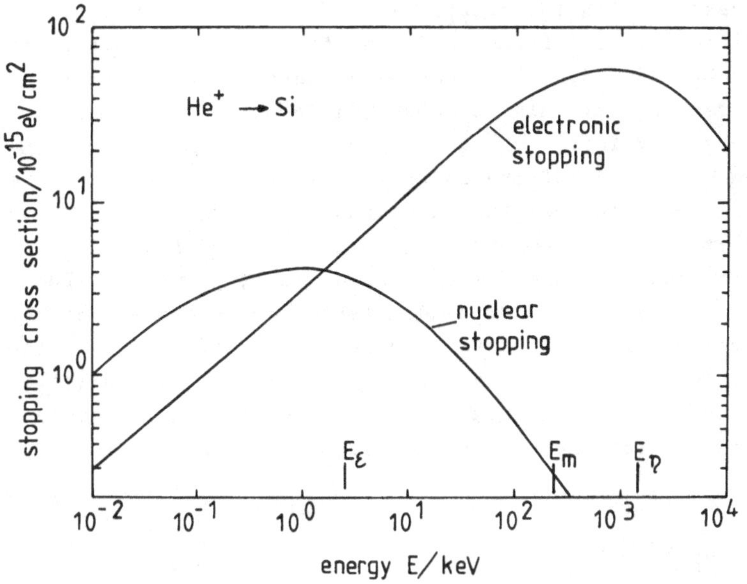

Fig.1.14. Nuclear and electronic stopping cross section for He ions in silicon

a maximum value. This is visualized in Fig.1.14 for the example of He ions incident on a Si target. As can be seen, the electronic stopping cross section has a maximum value in the vicinity of $E \approx E_m$. For comparison, the nuclear stopping cross section ε_n is included in Fig.1.14. As has already been mentioned in Sect.1.2.1.2 there is a similar energy dependence as for ε_e but with different scaling (the scaling energies E_ε and E_η for nuclear and electronic scattering, respectively, are marked in Fig.1.14). The maximum value of ε_n is considerably smaller than the maximum value of ε_e and the maximum of ε_n is positioned at energies small compared to E_m. Therefore with respect to high energy ion beam analysis the nuclear contribution to the energy loss can be neglected.

For application in ion beam analysis it is very useful to have general formulae for the electronic stopping cross section $\varepsilon_e(E,Z_1,Z_2)$ as a function of the projectile energy and the atomic numbers Z_1 and Z_2. As shown in Sects.1.2.1.2 and 1.2.1.3 there are general formulae in the limits of low energies (Eqs.(1.90), (1.93)) and high energies (Eq.(1.78)). These formulae are derived by using the atomic model of Thomas and Fermi and therefore they provide smooth and monotonous Z_1- and Z_2-dependences which do not express the experimentally observed Z_1- and Z_2-oscillations. A lot of work has been done in order to improve the theoretical description by using a more refined atomic model. The electronic stopping cross sections for low and high energies calculated in this way provide the structural effects in the Z_1- and Z_2-dependence and are in good agreement with the experimental data but they no longer can be presented as general formulae (they have to be calculated separately for each Z_1 and Z_2). In the energy range $E \approx E_m$ there is no general formula for ε_e at all because different mechanisms are responsible for the energy loss.

Up to now the usual way of providing a general formula for the electronic stopping power for all (nonrelativistic) energies has been given by modification and semiphenomenological extension of the theoretical descriptions in the low and high energy limits. These procedures introduce more or less free adjustable parameters which are fixed by fitting the experimental data. Brice [28] modified the Firsov theory by introducing a quantum mechanical expression for the electron flux and extended the formalism to higher energies ($E > E_m$). The result is an analytical formula for the electronic stopping power

$$\varepsilon_e(E) = \frac{4\hbar^2}{5m} (Z_1+Z_2)\ g_1(\varepsilon)\ g_2(v/v_o) \quad , \tag{1.94}$$

with

$$g_1(\varepsilon) = \varepsilon^{1/2}\ \frac{30\varepsilon^3 + 83\varepsilon^2 + 74\varepsilon + 21}{3(1+\varepsilon)^3} + (10\varepsilon + 1)\arctan(\varepsilon^{1/2}) \quad , \tag{1.95}$$

$$g_2(v/v_o) = [1 + (a\,\frac{v}{v_o})^n]^{-1} \quad , \tag{1.96}$$

and

$$\varepsilon = \frac{v^2}{4v_o^2 Z^2} = \frac{E}{99.92\ M_1 Z^2}\ \frac{\text{amu}}{\text{keV}} \quad . \tag{1.97}$$

The parameters Z (effective nuclear charge), a, and n are fixed by fitting the measured values of $\varepsilon_e(E)$ for each projectile target combination separately. Unfortunately, the values of Z, a, and n are only available for hydrogen and helium stopping in some gaseous targets and therefore this formula is hardly applicable to ion beam analysis of solids at present.

Another treatment starting from the high energy limit was given in Ref.[9]. In this case the simple classical description of electronic stopping as given in Sect.1.2.1.2 is extended in a phenomenological manner to low energies by taking into account the orbital velocity of the electrons, the screening of the Coulomb interaction between the projectile and a target electron and the fact that with decreasing energy more and more electrons cannot contribute to ionization. The formula and all the parameters needed are given in Ref.[9] and will not be presented here. As an example the result for He^+ stopping in a Si target is depicted in Fig.1.15 in comparison with experimental data.

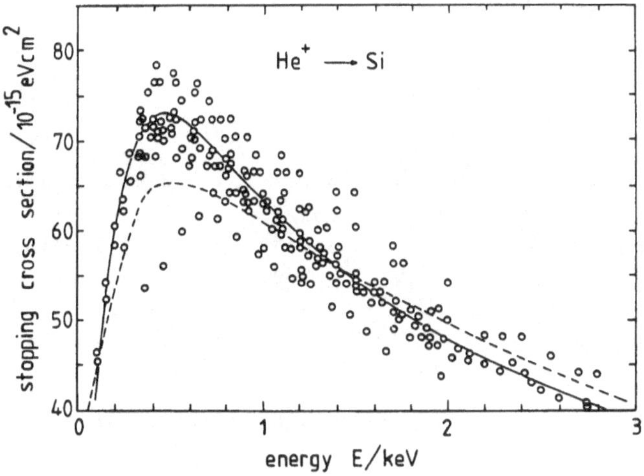

Fig.1.15.Cross section for He^+ stopping in Si. The solid line represents the fit data of Ziegler given by Eqs.(1.98-1.100) and the dashed line is determined from the general formula of Ref.[9]. The experimental data (Ref.[31]) are represented by the open circles.

The simplest method for providing an electronic stopping cross section formula valid for all nonrelativistic energies is given by a combination of the formulae for low and high energy stopping as e.g. proposed by Biersack and Fink [29]

$$\varepsilon_e(E)^{-1} = \varepsilon_{low}(E)^{-1} + \varepsilon_{high}(E)^{-1} \quad . \tag{1.98}$$

Ziegler [30,31] used this formula and determined the low and high energy stopping cross sections from all experimental stopping data available. Guided by the results of the theoretical descriptions he found

$$\varepsilon_{low}(E) = A_1 \ (E/keV)^{A_2} * 10^{-15} \ eVcm^2 \tag{1.99}$$

and

$$\varepsilon_{high}(E) = \frac{A_3}{E/MeV} \ln(1 + \frac{A_4}{E/MeV} + A_5 E/MeV) * 10^{-15} \ eVcm^2 \tag{1.100}$$

to provide a good fit to the experimental data for He stopping with the parameters $A_1 \ldots A_5$ given for all elemental targets in Ref.[31]. In the case of other projectiles similar expressions are achieved.

The Ziegler stopping data are the most reliable ones and usually used for ion beam analysis of solids. The corresponding values for H^+ and He^+ stopping for all elements are tabulated in Appendices 5 and 6. For He^+ stopping in silicon the data are also depicted in Fig.1.15.

1.2.2. Energy straggling

In analogy to the mean energy loss the straggling contribution $d\Omega^2 = \langle(dE - \langle dE\rangle)^2\rangle$ must be calculated first for a very thin layer (thickness dz, see Fig.1.10). According to Bohr [17] this contribution is given by

$$d\Omega^2 = \int dw\, T^2 = N\, dz \int d\sigma\, T^2 = N\, dz\, W(E) \quad , \qquad (1.101)$$

with the probability dw (Eq.(1.63)) and the straggling cross section

$$W(E) = \int d\sigma(T,E)\, T^2 \quad , \qquad (1.102)$$

which is introduced analogously to the stopping cross section $\varepsilon(E)$. The assumption of statistical independence of the nuclear and electronic processes as expressed in Eq.(1.67) provides

$$W(E) = W_n(E) + W_e(E), \qquad (1.103)$$

with

$$W_n(E) = \int d\sigma_n\, T^2 \quad \text{and} \quad W_e(E) = \int d\sigma_e\, T^2. \qquad (1.104)$$

For sufficiently high energies ($E \gg E_\varepsilon$ for nuclear scattering, $E \gg E_\eta$ for electronic scattering) the Coulomb approximation of the differential cross section $d\sigma_n$ (Eq.(1.51)) and $d\sigma_e$ (Eq.(1.56)) can be used (Bohr approximation). With T_{max} given by Eqs.(1.48) and (1.54) for nuclear and electronic scattering, respectively, we find (index B refers to Bohr approximation)

$$W_n^B = B_n \int dT = B_n T_{max} = 4\pi Z_1^2 Z_2^2 \left(\frac{e^2}{4\pi\varepsilon_o}\right)^2 \left(\frac{M_1}{M_1+M_2}\right)^2$$

$$= 260.5\ Z_1^2 Z_2^2 \left(\frac{M_1}{M_1+M_2}\right)^2 * 10^{-15}\ eV^2 cm^2 \qquad (1.105)$$

and

$$W_e^B = Z_2 B_e \int dT = Z_2 B_e T_{max} = 4\pi Z_1^2 Z_2 \left(\frac{e^2}{4\pi\varepsilon_o}\right)^2$$

$$= 260.5\ Z_1^2 Z_2 * 10^{-15}\ eV^2 cm^2 \quad . \tag{1.106}$$

First, we note that the energy straggling is not so strongly influenced by the screening of the Coulomb interaction as stated in the case of the mean energy loss because for energy straggling the contributions from large transferred energies (i.e. small impact parameters) are of more importance. Therefore it is not necessary to introduce a minimum transferred energy T_{min} as it has to be done for the determination of the mean energy loss (otherwise it becomes infinite). Second, we find the two energy straggling contributions to be independent of the energy in the energy ranges considered. Therefore the energy straggling for a layer with finite thickness Δz can immediately be derived by integration of Eq.(1.101)

$$\Omega^2 = N\Delta z\ (W_n^B + W_e^B). \tag{1.107}$$

The ratio

$$W_e^B / W_n^B = [(M_1+M_2)/M_1]^2 / Z_2 \tag{1.108}$$

indicates that the nuclear contribution to the energy straggling can be of the same order of magnitude as the electronic one in contrast to the mean energy loss, where the nuclear contribution is negligible for $E >> E_\varepsilon$. This is understood by the fact that large transferred energies T are more important for straggling, and large T are more probable for nuclear scattering.

For small energies $E << E_\eta$ the Bohr approximation of W_e is no longer valid. As can be shown, the deviation of the straggling cross section W_e from the Bohr value can be expressed approximately by a factor w_e which depends only on the energy parameter $\eta_r = E/E_\eta$.

$$W_e(E) = W_e^B\ w_e(\eta_r). \tag{1.109}$$

Using the dielectricity formalism Lindhard and Scharff [32] derived the approximative expression

$$w_e(\eta_r) = \begin{cases} 0.68\ \eta_r^{1/2} - 0.008\ \eta_r^{3/2} & \text{for } \eta_r < 3 \\ 1 & \text{for } \eta_r > 3 \end{cases} \quad . \tag{1.110}$$

More detailed calculations of $w_e(\eta_r)$ were performed by Chu and Mayer [33] based on the theoretical description of straggling by Bonderup and Hvelplund [34]. The results are tabulated in Appendix 7.

For still smaller energies $E \lesssim E_\varepsilon$ also the nuclear straggling contribution W_n deviates from the corresponding Bohr value because the screening of the Coulomb interaction becomes more and more important. Using the universal Thomas-Fermi screening function and the wide angle approximation of the differential cross section (Eq.(1.44)) Lindhard determined the nuclear straggling cross section. The results can be written as

$$W_n(E) = W_n^B \, w_n(\varepsilon_r), \tag{1.111}$$

where the deviation w_n from the Coulomb expression is only a function of the reduced energy $\varepsilon_r = E/E_\varepsilon$ which is presented in Fig.1.16.

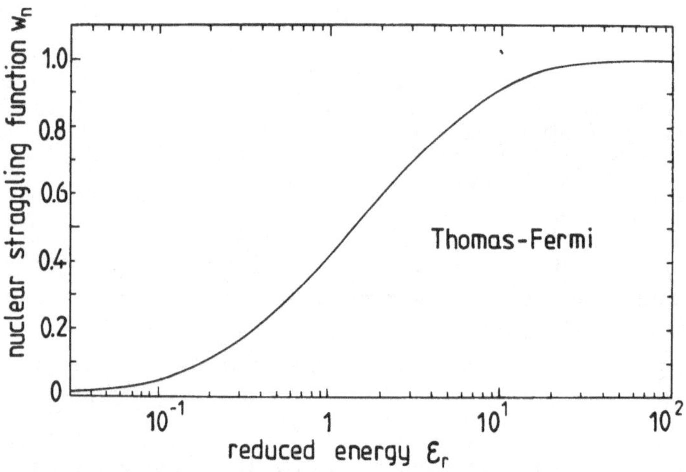

Fig.1.16. Universal nuclear straggling cross section (straggling function) calculated using the Thomas-Fermi screening function (Ref.[11])

Because of the weak energy dependence of $W_e(E)$ and $W_n(E)$ (except of $\varepsilon_r \ll 1$) the energy straggling for layers with not too large thickness Δz is given by

$$\Omega^2 = N \, \Delta z \, (W_n(\bar{E}) + W_e(\bar{E})) \tag{1.112}$$

with a mean energy $\bar{E} = E_o - \frac{1}{2} \langle \Delta E \rangle$, otherwise Eq.(1.101) must be integrated.

1.2.3. Mean energy loss and energy straggling in compound targets

The number of atoms per unit volume of the component i of the target is denoted by $N^{(i)}$. Assuming statistically independent collisions of the ion with the target atoms, the probability dw(T,E) for an energy transfer T as defined in Eq.(1.63) is given by a sum

$$dw(T,E) = \sum_i N^{(i)} \, dz \, d\sigma^{(i)}(T,E) \; , \tag{1.113}$$

where $d\sigma^{(i)}$ is the cross section for ion scattering on an atom of component i. In this way the stopping power (Eq.(1.65)) and energy straggling (Eq.(1.107)) applied to compound targets read

$$\frac{dE}{dz} = -\sum_i N^{(i)} \; \varepsilon^{(i)}(E) \tag{1.114}$$

with

$$\varepsilon^{(i)}(E) = \int d\sigma^{(i)}(T,E) \; T \; ,$$

and

$$\Omega^2 = \Delta z \sum_i N^{(i)} W^{(i)}(E) \tag{1.115}$$

with

$$W^{(i)}(E) = \int d\sigma^{(i)}(T,E) \; T^2 \; .$$

Introducing the relative composition of the target

$$c^{(i)} = N^{(i)}/N \tag{1.116}$$

we find

$$\frac{dE}{dz} = - N \; \varepsilon(E), \tag{1.117}$$

with the stopping cross section (Bragg's rule)

$$\varepsilon(E) = \sum_i c^{(i)} \; \varepsilon^{(i)}(E), \tag{1.118}$$

and

$$\Omega^2 = \Delta z \; N \; W(E), \tag{1.119}$$

with the straggling cross section

$$W(E) = \sum_i c^{(i)} W^{(i)}(E). \tag{1.120}$$

The stopping cross section $\varepsilon^{(i)}$ and the straggling cross section $W^{(i)}$ depend on the electronic structure of the atoms of component i which is influenced by the chemical binding of the target atoms. Therefore the contributions $\varepsilon^{(i)}$ and $W^{(i)}$ from one atom are not independent of the kind of the surrounding atoms. This violates the statistical independence of the two-body collisions and causes deviations from Eqs.(1.118) and (1.120). However, in most cases these deviations are below 5%. Only for gaseous elements there arise errors up to about 30%.

1.2.4. Energy loss distribution

Assuming statistically independent two-body collisions, Landau [35] derived a general expression for the energy loss distribution which can be written as a Fourier transform

$$P(\Delta E,\Delta z) = \frac{1}{2\pi}\int d\alpha \; e^{i\alpha \Delta E - \Delta z N \sigma(\alpha,E)} \tag{1.121}$$

with

$$\sigma(\alpha,E) = \int d\sigma(T,E)\,(1 - e^{-i\alpha T}) \; . \tag{1.122}$$

This formula is valid for layer thicknesses small enough that the variation of $d\sigma(T,E)$ with the ion energy E is not important. A series expansion of the exponential function in Eq.(1.122) provides

$$\sigma(\alpha,E) = i\alpha\varepsilon(E) + \Delta\sigma(\alpha,E) \; , \tag{1.123}$$

with

$$\Delta\sigma(\alpha,E) = -\sum_{m=2}^{\infty} \frac{(-i\alpha)^m}{m!}\int d\sigma(T,E)\; T^m \; . \tag{1.124}$$

The separate consideration of the first term of the series is useful because the calculation of the stopping cross section $\varepsilon(E)$ needs more care than the determination of the higher order moments $\int d\sigma\; T^m$ $(m \geq 2)$ as discussed in Sects.1.2.1 and 1.2.2. This can easily be understood because the uncertainties in $d\sigma(T,E)$ for small transferred energies T have less influence on the higher order moments than on the first moment $\varepsilon = \int d\sigma\; T$.

There are nuclear and electronic contributions to the moments

$$\int d\sigma\; T^m = \int d\sigma_n\; T^m + \int d\sigma_e\; T^m \; . \tag{1.125}$$

As already has been discussed, the electronic contribution dominates the nuclear one in a wide energy region for m = 1 (stopping cross section) and is in the same order of magnitude as the nuclear contribution for m = 2 (straggling cross section). With increasing m the electronic contribution $\int d\sigma_e \, T^m$ becomes more and more negligible compared with the nuclear one because large transferred energies T become more and more important.

Inserting Eq.(1.123) into Eq.(1.121) we obtain

$$P(\Delta E,\Delta z) = \frac{1}{2\pi} \int_{-\infty}^{\infty} d\alpha \; e^{i\alpha(\Delta E - \langle\Delta E\rangle) - \Delta z \, N \Delta\sigma(\alpha,E)} \tag{1.126}$$

depending on $\Delta E - \langle\Delta E\rangle$. If only the first term of the series in $\Delta\sigma$ is taken into account

$$\Delta\sigma(\alpha,E) = \frac{1}{2}\alpha^2 \, (W_n(E) + W_e(E)) \; , \tag{1.127}$$

the integration over α can be performed analytically and provides the symmetric Gaussian distribution

$$P(\Delta E,\Delta z) = (2\pi \, \Omega^2)^{-1/2} \; e^{-[(\Delta E - \langle\Delta E\rangle)/\Omega]^2/2} \quad , \tag{1.128}$$

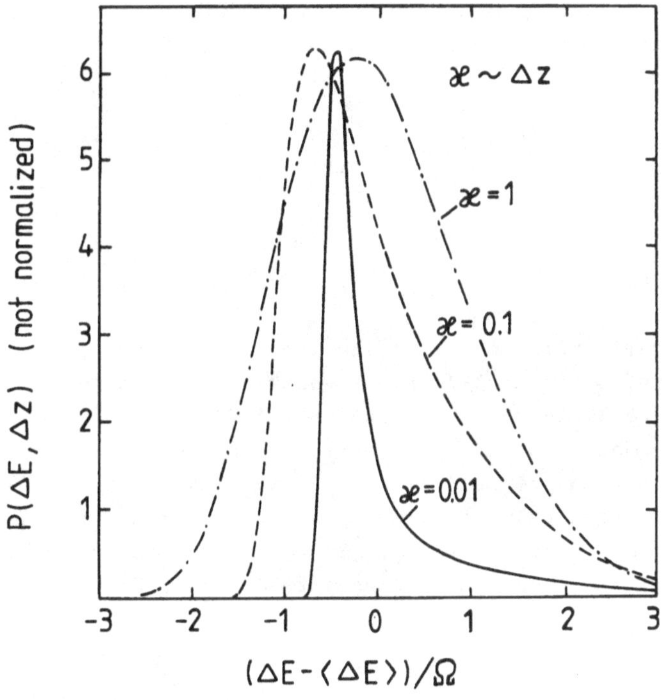

Fig.1.17. Electronic energy loss distribution for ions having penetrated an amorphous layer of thickness Δz

with the half width $2\,(2\ln 2)^{1/2}\Omega$ ($\Omega = [\Delta z\, N\, (W_n + W_e)]^{1/2}$). Taking into account higher order terms in $\Delta\sigma$ the distribution in ΔE becomes asymmetrical. For Coulomb interaction the summation in $\Delta\sigma$ (Eq.(1.124)) can be performed analytically (Ref.[36]). Applied to electronic scattering, the distribution $P(\Delta E, \Delta z)$ obtained by numerical integration of Eq.(1.126) (Ref.[36]) is depicted in Fig.1.17 as a function of $(\Delta E - \langle\Delta E\rangle)/\Omega$, where $\varkappa = \Delta z\, N\, W_e^B/(2mv^2)$ is a dimensionsless parameter proportional to the layer thickness Δz. There is a strong asymmetry for small Δz ($\varkappa << 1$) introduced by the higher order terms in $\Delta\sigma$. For $\varkappa \gtrsim 1$ the distribution $P(\Delta E, \Delta z)$ becomes more and more symmetrical and tends to the Gaussian one (Eq.(1.128)). In this case the number of collisions of an ion passing the layer of thickness Δz is large and $P(\Delta E, \Delta z)$ is more determined by statistics than by the concrete differential cross section. Therefore the shape of $P(\Delta E, \Delta z)$ for $\varkappa >> 1$ is independent of the special features of the single collision processes and it becomes symmetrical.

1.2.5. Angular distribution

Assuming statistically independent two-body collisions and neglecting energy loss, Bothe [37] derived a general expression for the angular distribution of an ion beam after traversing a thin layer of thickness Δz

$$P(\varphi, \Delta z) = \frac{1}{2\pi} \int_0^\infty d\alpha\, \alpha\, J_0(\alpha\varphi)\, e^{-\Delta z\, N\, \sigma(\alpha, E)} \quad , \tag{1.129}$$

with

$$\sigma(\alpha, E) = \int_0^\infty d\sigma(\theta, E)\, [1 - J_0(\alpha\theta)] \, , \tag{1.130}$$

where φ is the total deflection angle, J_0 is the zero-order Bessel function of the first kind, θ is the laboratory scattering angle for the single collision and $d\sigma(\theta, E)$ is the corresponding differential cross section. Because of the small mass of the electrons, the deflection due to electronic scattering can be neglected for the ion motion in amorphous targets. Therefore only the nuclear contribution to the differential cross section $d\sigma(\theta, E) = d\sigma_n(\theta, E)$ must be taken into account. Using the generalized cross section as introduced by Eq.(1.44)

$$d\sigma_n(\theta, E) = \pi\, a_{TF}^2\, d(t^{1/2})\, t^{-1}\, f(t^{1/2}) \quad , \tag{1.131}$$

with

$$t^{1/2} = \frac{Ea_{TF}}{2Z_1Z_2} \frac{4\pi\varepsilon_o}{e^2} \theta \qquad (1.132)$$

for small scattering angles θ the angular distribution reads

$$P(\varphi,\Delta z) = \frac{1}{2\pi} \left(\frac{Ea_{TF}}{2Z_1Z_2} \frac{4\pi\varepsilon_o}{e^2}\right)^2 f_1(\tilde{\varphi},\tau) \quad , \qquad (1.133)$$

with

$$f_1(\tilde{\varphi},\tau) = \int_0^\infty d\beta\ \beta\ J_o(\beta\tilde{\varphi})\ e^{-\tau\,\Delta(\beta)} \qquad (1.134)$$

and

$$\Delta(\beta) = \int_0^\infty d(t^{1/2})\ t^{-1}\ f(t^{1/2})\ [1 - J_o(\beta t^{1/2})] \quad , \qquad (1.135)$$

depending only on the reduced deflection angle

$$\tilde{\varphi} = \frac{Ea_{TF}}{2Z_1Z_2} \frac{4\pi\varepsilon_o}{e^2} \varphi = c_\varphi \varphi \qquad (1.136)$$

and on the reduced thickness

$$\tau = \pi a_{TF}^2\ N\Delta z = c_z\ N\Delta z \qquad (1.137)$$

(values of the factors c_φ and c_z for 1 MeV H ions and He ions incident on Si are given in Appendix 8). Numerical calculations of the function $f_1(\tilde{\varphi},\tau)$ were performed by Meyer [38] and Sigmund and Winterbon [39] using the scattering function $f^{TF}(t^{1/2})$ (Fig.1.6) obtained from the Thomas-Fermi interaction potential. The results are tabulated in Refs.[38,39]. For illustration the ratio $f_1(\tilde{\varphi},\tau)/f_1(0,\tau)$ is depicted in Fig.1.18 as a function of $\tilde{\varphi}$ given in units of the reduced half width $\tilde{\varphi}_{1/2}$. As can be seen the shape of the angular distribution remarkably depends on the thickness parameter τ for $\tau \leq 1$ and it is hardly influenced by the value of τ for $\tau >> 1$. In the limit of very large τ the angular distribution approaches a Gaussian one. This is understood in the same way as for the energy loss distribution. Large values of τ are equivalent to a large number of single collisions. In this case the angular distribution is determined by statistics rather than by special features of the single collision processes. The half width of the calculated angular distribution reads

$$\varphi_{1/2}(\Delta z) = \tilde{\varphi}_{1/2}(\tau)/c_\varphi \qquad (1.138)$$

with the factor c_φ and the reduced half width $\tilde{\varphi}_{1/2}(\tau)$ (Ref.[39]) given

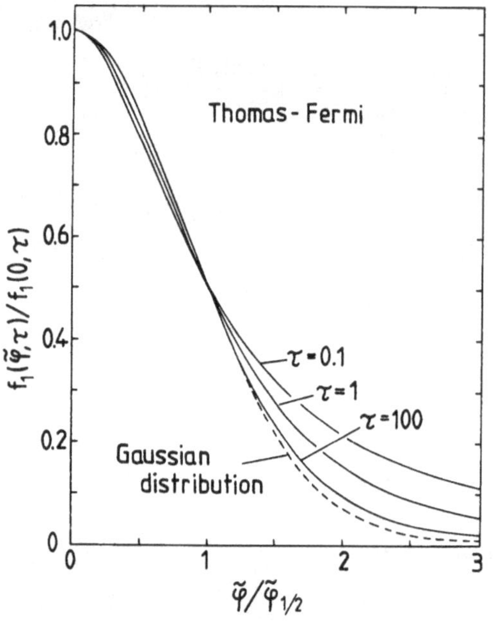

Fig.1.18. Universal angular distribution of ions having penetrated an amorphous layer of thickness Δz ($\tau \sim \Delta z$). The curves are obtained (Ref.[39]) using the Thomas-Fermi screening function and they are normalized to unity at $\tilde{\varphi} = 0$.

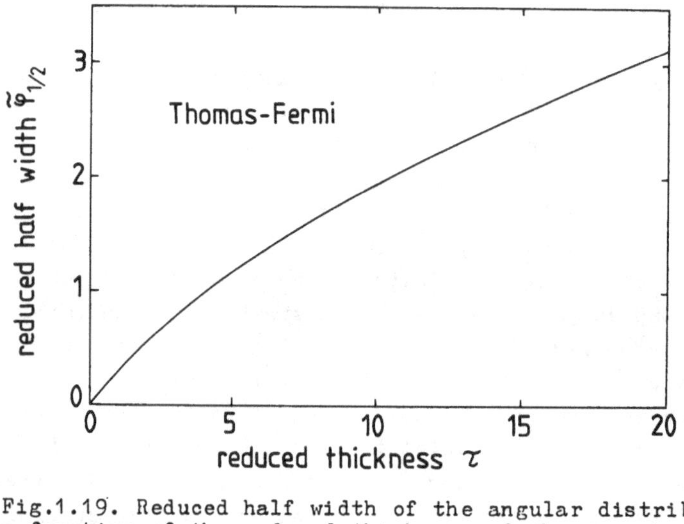

Fig.1.19. Reduced half width of the angular distribution (Fig.1.18) as a function of the reduced thickness of the amorphous layer

in Eq.(1.136) and in Fig.1.19. For ion beam analysis, H and He ions are of special interest. Therefore the corresponding factors $c_z = \tau/N\Delta z$ and $c_\varphi = \tilde{\varphi}_{1/2}/\varphi_{1/2}$ are listed in Appendix 8. If the Lenz-Jensen interaction potential is used (Ref.[39]) instead of the Thomas-Fermi one the resulting reduced half width changes remarkably for $\tau \ll 1$ (about a factor 2 at $\tau = 0.1$) and is hardly influenced for $\tau > 1$.

In general, the measured half widths agree with the calculated ones for $\tau > 1$ and are between the values obtained with the two universal potentials for $\tau < 1$. Since a universal differential cross section is used, the calculated half width smoothly depends on Z_1 and Z_2 and does not provide Z_1-oscillations as have been found e.g. by Andersen et al. [40] for very small values of τ.

For the application in ion beam analysis, the depth intervals of interest ($\Delta z > 20$ nm) mostly correspond to $\tau \geq 1$. Therefore the angular distribution calculated using the Thomas-Fermi interaction potential will be sufficient.

1.3. MONOCRYSTALLINE TARGETS

1.3.1. Channeling and blocking phenomena

In contrast to the amorphous target the atoms in a monocrystal are well ordered and successive two-body collisions are no longer statistically independent. Especially, there is a strong correlation of the collision processes if the angle ψ between the actual ion velocity and a low index direction of the crystal lattice is small. To gain a basic understanding of the effects caused by the lattice order we start with a simple model for an idealized crystal without electrons, thermal vibrations and defects.

(i) Channeling

First, we consider a beam of ions all of which are incident parallel to a low index direction of the crystal ($\psi_{in} = 0$) with different distances s_1 to the atomic strings (all atoms positioned on a line parallel to the low index direction form a string). An ion with the impact parameter s_1 is scattered on the first atom by an angle $\Theta(s_1)$ (see Fig.1.20a). The impact parameter s_2 for the following collision is given by

$$s_2 = s_1 + d\,\Theta(s_1) \qquad (1.139)$$

where d is the interatomic spacing for the corresponding low index direction. As shown in Fig.1.20b, impact parameters s_2 below a critical value s_c are excluded because Θ decreases with increasing s_1. The corresponding distribution $I(s_2)$ of ions with respect to s_2 is depicted in Fig.1.20c.

Second, let us consider an ion beam with inclined incidence. With increasing angle of incidence ψ_{in} the distribution $I(s_2)$ is shifted in s_2 by $d\psi_{in}$. While processes which need very small impact parameters like

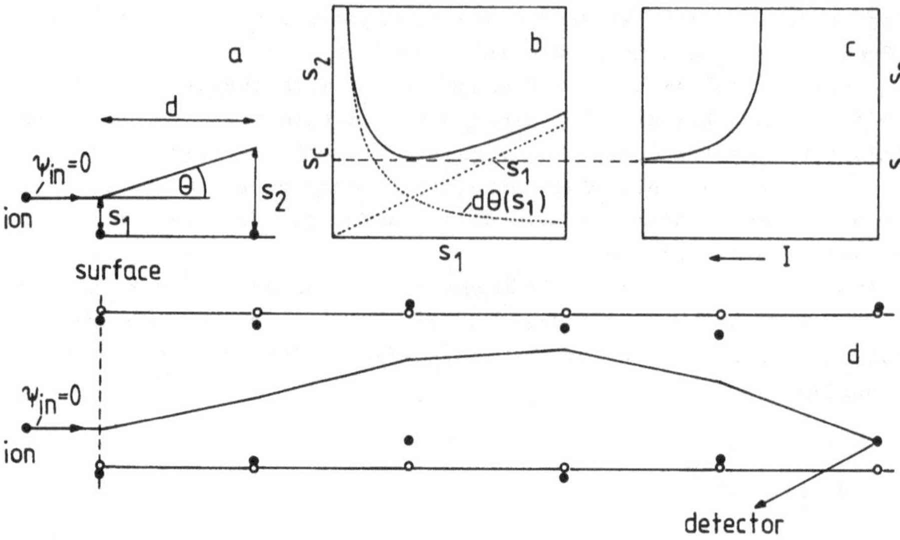

Fig.1.20. Simplified representation of the channeling effect for ions incident at an angle $\psi_{in} = 0$ with respect to a low index direction (string) of a crystal. (a) Scattering of the ion on a surface atom. (b) Relation between the impact parameters of the first and second collision. (c) Ion distribution with respect to the impact parameter for the second collision. (d) Successive collisions in the crystal including thermal vibrations

Rutherford backscattering are excluded for small values of ψ_{in}, they are possible for $\psi_{in} > \psi_c$, where the critical angle ψ_c is given by $\psi_c = s_c/d$. In order to estimate the value of ψ_c, the small angle approximation of $\theta(s)$ for the Coulomb potential (Eq.(1.23)) is used in Eq.(1.139). The resulting minimum value s_c provides

$$\psi_c = \left(\frac{4Z_1Z_2}{Ed}\frac{e^2}{4\pi\varepsilon_0}\right)^{1/2} , \qquad (1.140)$$

where E is the ion energy in the laboratory system. Typical values of ψ_c for the conditions of the ion beam analysis are in the order of one degree or below (e.g. $\psi_c = 1.1^\circ$ for 1 MeV He ions incident on <111> Si).

This simple consideration already provides the main feature of the Rutherford backscattering yield $Y(\psi_{in})$ which is proportional to the ion distribution $I(s_2 = d\psi_{in})$ (see Fig.1.21). There is a pronounced dip for small values of ψ followed by a shoulder at $\psi_{in} \geq \psi_c$. For $\psi_{in} \gg \psi_c$ the yield becomes constant and equal to the yield for amorphous targets (random position of the second atom, the Rutherford backscattering cross section is approximately constant for the small variation of ψ_{in}), i.e. the lattice order has no considerable influence on the ion motion in this case. Therefore, this part is called the random yield for short.

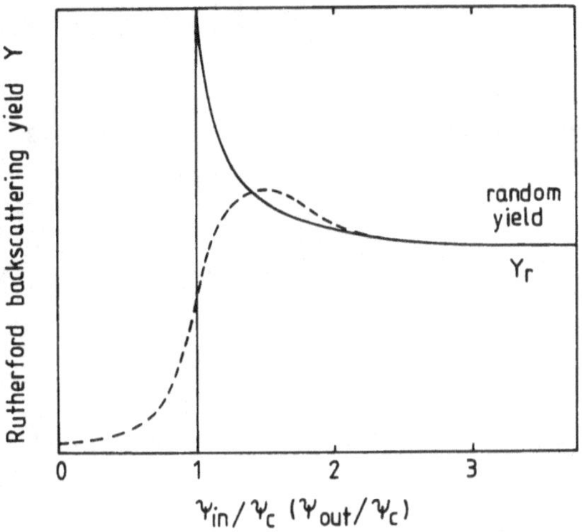

Fig.1.21. Rutherford backscattering yield as a function of the angle of incidence (or exit angle) corresponding to the ion distribution given in Fig.1.20c

The simplified angular dependence $Y(\psi_{in})$ discussed above is mainly modified by the influence of the thermal vibrations and of the collisions of the ions with the following atoms of the string (see Fig.1.20d). While the thermal vibrations can be taken into account easily by an average of $Y(\psi_{in})$ with a Gaussian distribution of the position of the second atom, the influence of the collisions of the ions with the following atoms of the string cannot be included in a simple way but it is more easily discussed within a suitable theoretical description as given in Sect.1.3.2. The result is a smoothed angular distribution as depicted in Fig.1.21 by the dashed line. There remains a pronounced dip with a small shoulder.

Because of the lattice order, small impact parameters for the ion collisions with the following atoms of a string are also suppressed for all the ions with the actual angle $\psi < \psi_c$. This indicates that there exist ion trajectories which are only determined by successive small angle deflections as shown in Fig.1.20d. After some depth, the angle ψ will be increased by collisions with thermally displaced atoms, electrons or defects which finally provides an opportunity for large angle scattering. In this way the Rutherford backscattering yield $Y(\psi_{in}, z)$ also depends on the depth z.

The ion motion governed by the atomic strings is called axial channeling and axial dechanneling denotes the disturbance of the idealized channeled motion caused by the scattering of the ions on electrons, thermally vibrating atoms or lattice defects. Analogously, there is also a governed ion motion along a low index plane of the crystal for small angles of incidence with respect to the plane, which is called planar channeling.

(ii) Blocking

Now let us consider ions starting approximately from lattice sites, e.g. provided by Rutherford backscattering with very small impact parameters (see Fig.1.22b).

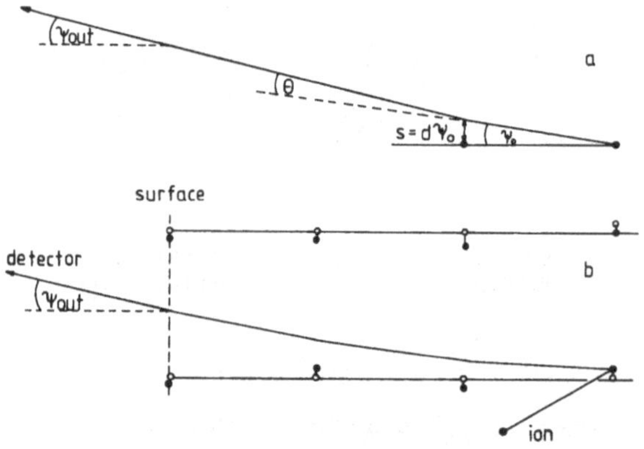

Fig.1.22. Simplified representation of the blocking effect.
(a) Scattering of an ion starting from a lattice site on the next lattice atom. (b) Successive collisions in the crystal including thermal vibrations

If an ion is emitted at an angle ψ_o with respect to a low index direction (Fig.1.22a) the corresponding angle ψ_{out} after the collision with the next atom of the string is given by

$$\psi_{out} = \psi_o + \theta(d\psi_o) , \tag{1.141}$$

wich can be transferred to Eq.(1.139) by substituting $s_2 \rightarrow d\psi_{out}$ and $s_1 \rightarrow d\psi_o$. This means that the distribution of the ions with respect to the angle is given by the same distribution $Y(\psi_{out}) \sim I(s_2 = d\psi_{out})$ shown in Fig.1.21, with the same critical angle ψ_c introduced by Eq.(1.140). Since angles $\psi_{out} < \psi_c$ of the emerging ions are excluded for all ψ_o, this effect is called blocking or shadow effect. The fact that the blocking angular distribution of backscattered ions is equal to the channeling backscattering yield as a function of the angle of incidence

$$Y_{blocking}(\psi_{out}) = Y_{channeling}(\psi_{in}) \quad \text{for} \quad \psi_{in} = \psi_{out} , \tag{1.142}$$

is well known as the reversibility rule. As will be shown later, the reversibility rule also holds when thermal vibrations and the collisions with the following string atoms are included, as far as dissipative processes can be neglected.

As with channeling, axial and planar blocking is possible. For planar channeling or blocking, the shoulder in the angular distribution is more pronounced than that in the axial case. This is easily understood within our simple consideration as a dimension effect. The shoulder must contain all the ions pushed away from the dip region, i.e. $\int d\psi_{in}(Y(\psi_{in}) - Y_r) = 0$ for the planar case and the corresponding integral over the solid angle $\int d\psi_{in}\, \psi_{in}\, (Y(\psi_{in}) - Y_r) = 0$ for the axial case. Because of the factor ψ_{in}, the shoulder must be lower in the latter case.

As already mentioned, processes with small impact parameters ($s < d\psi_c \approx 0.005$ nm) are suppressed for ions with $\psi < \psi_c$ which also influences the energy loss. Because the decrease of the transferred energy T(s) within $s = 0...d\psi_c$ is very strong for nuclear scattering and weak for electronic scattering ($b << d\psi_c << b_e$, see Eq.1.58) the nuclear energy loss is reduced much more under channeling conditions than the electronic one. With respect to high energy ion beam analysis the nuclear stopping is negligible in comparison to the electronic stopping. In this case the energy loss of channeled ions can be reduced to about half of the value found for ions moving in random materials.

1.3.2. Basic description of channeling

As was shown in the previous section, the channeled ion motion in a crystal can be qualitatively understood considering only two of the successive collisions with the target atoms. A quantitative inclusion of all the successive collisions is only possible by Monte Carlo computer simulation as used by some authors (Refs.[41,42]) for special investigations. However, these calculations are very time consuming and expensive and the results are always restricted to a special parameter set. A more convenient approach with respect to the analytical treatment and the application was given by Lindhard [43]. It is the basis for the quantitative description of all channeling processes outlined in the following sections.

1.3.2.1. Transverse energy conservation

Let us consider an ion moving in an idealized crystal (without electrons, thermal vibrations and defects) along a string of atoms (axial channeling). As estimated in the previous section, channeling effects (correlated collisions) arise only for small angles ψ between the vector of the ion velocity and the string of atoms. Therefore, with respect to the description of channeling, the momentum (or small angle) approximation can be used without any restriction. According to Eq.(1.22) the change of the transverse momentum $\vec{p}$ due to the collision of the ion with one atom of the string at an impact parameter r ($r = |\vec{r}|$, see Fig.1.23)

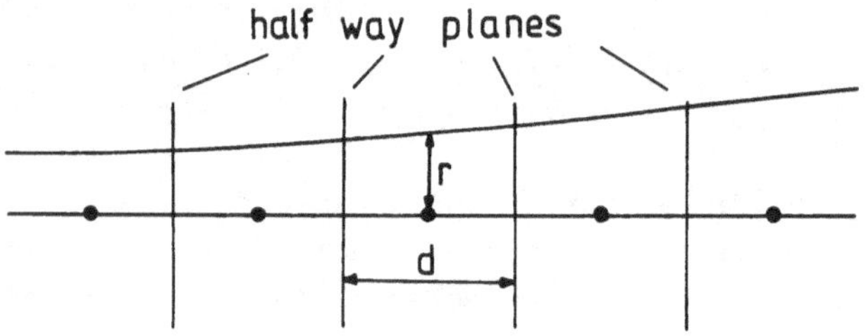

Fig.1.23. Successive collisions of the ion with the atoms of a string

is given by

$$\Delta\vec{p} = - \int_{-\infty}^{\infty} dz \frac{1}{v} \frac{\partial}{\partial\vec{r}} V(\sqrt{r^2 + z^2}) = - \frac{d}{v} \frac{\partial U(r)}{\partial\vec{r}} \quad , \tag{1.143}$$

where

$$U(r) = \frac{1}{d} \int_{-\infty}^{\infty} dz \; V(\sqrt{r^2 + z^2}) \tag{1.144}$$

is the potential of a string averaged over the depth z or string potential for short. Introducing the time of flight for one collision $\Delta t = d/v$ we get

$$\frac{\Delta\vec{p}}{\Delta t} = - \frac{\partial U(r)}{\partial\vec{r}} \quad . \tag{1.145}$$

If the ion trajectory is determined by a lot of single collisions (smoothly curved as shown in Fig.1.23) the change of the transverse momentum in a fixed time interval can be replaced by a continuous one

$$\frac{d\vec{p}}{dt} = - \frac{\partial U(r)}{\partial\vec{r}} \quad . \tag{1.146}$$

As a consequence the transverse energy

$$E_\perp = \frac{p^2}{2M_1} + U(r) = E\psi^2 + U(r) \tag{1.147}$$

is conserved ($p = M_1 v \sin\psi = M_1 v\psi$ because $\psi << 1$). A more detailed investigation by Lindhard [43] shows that the transverse energy conservation holds at the half-way planes between the atoms of the string for $E_\perp \leq E\psi_1^2$. The critical angle

$$\psi_1 = \left(\frac{2Z_1 Z_2}{Ed} \frac{e^2}{4\pi\varepsilon_o}\right)^{1/2} \qquad (1.148)$$

defining a critical transverse energy $E\psi_1^2$ differs only by a factor of $2^{1/2}$ from the critical angle ψ_c for the angular distribution introduced in the previous section. From Eq.(1.147) follows that the distance of an ion (with the transverse energy $E_\perp$) from the string cannot be smaller than a minimum distance r_{min} given by

$$E_\perp = U(r_{min}) \longrightarrow r_{min}(E_\perp). \qquad (1.149)$$

This means that collisions with small impact parameters are suppressed which is a consequence of the strong correlation between the collisions.

1.3.2.2. Statistics for ions with fixed transverse energy

1.3.2.2.1. Axial channeling

With respect to the ion motion, the crystal is considered as a sum of all the strings at positions $\vec{r}_n$ with the total potential

$$U_{tot}(\vec{r}) = \sum_n U(|\vec{r} - \vec{r}_n|) . \qquad (1.150)$$

The question arises if a special statistical treatment for the channeled ion motion in the crystal is possible starting from the interaction with one string and taking into account the conservation of the transverse energy. After entering the crystal (depth z = 0) the ions have a certain distribution with respect to the transverse phase space coordinates $\vec{r}$ and $\vec{p}$. If the transverse momentum $\vec{p}$ is not parallel to another low index direction (otherwise we get planar channeling) there is a trend towards the statistical equilibrium on the transverse energy shell due to the collisions of the ions with the strings. A measure for this process is the change of the average transverse momentum $\langle\vec{p}\rangle$ with depth z. Lindhard [43] found

$$\langle\vec{p}\rangle = \langle\vec{p}\rangle_o \, e^{-z/\lambda_\perp} , \qquad (1.151)$$

with $\lambda_\perp$ being of the order of 10^3 atomic layers or less. From this follows, that an initial average transverse momentum $\langle\vec{p}\rangle_o$ rapidly vanishes and after some 0.1 μm there remains no preferred direction of the transverse momentum. Therefore, except for such small depths, an equilibrium on the transverse energy shell can be assumed.

Because of this ergodic behaviour the classical statistics for microcanonical systems can be applied. Then, the phase density for ions with a given transverse energy $E_\perp$ in the 4-dimensional phase space (axial channeling) reads

$$\rho_{ph}(\vec{r},\vec{p}) = \rho_o\ \delta(E_\perp - p^2/(2M_1) - U_{tot}(\vec{r})) \quad , \qquad (1.152)$$

with ρ_o being a normalization constant and δ the well known δ-function. The statistical average of a physical quantity $f(\vec{r},\vec{p})$ is provided by

$$\overline{f} = \frac{\int_{A_o} d^2\vec{r} \int d^2\vec{p}\ \rho_{ph}(\vec{r},\vec{p})\ f(\vec{r},\vec{p})}{\int_{A_o} d^2\vec{r} \int d^2\vec{p}\ \rho_{ph}(\vec{r},\vec{p})} \quad . \qquad (1.153)$$

Taking into account the symmetry, the integration over the space is reduced to the unit cell with the area A_o (see Fig.1.24).

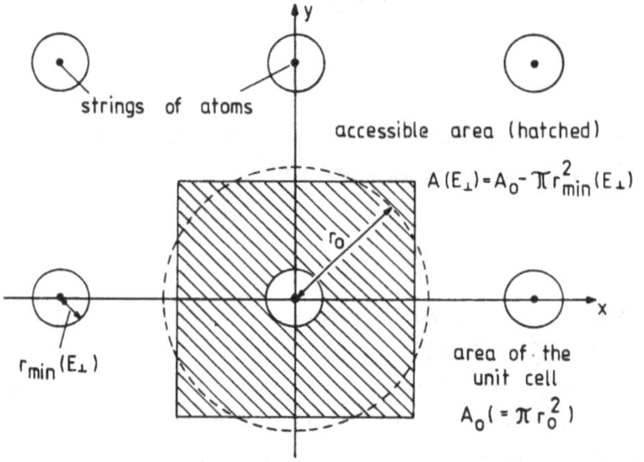

Fig.1.24. Transverse plane for axial channeling with the unit cell for one string of atoms. The hatched area is the accessible portion of the unit cell for ions with a given transverse energy $E_\perp$.

The integration over the momentum space can be performed using polar coordinates $\vec{p} = (p,\chi)$

$$\overline{f}(E_\perp) = \int_{A(E_\perp)} \frac{d^2\vec{r}}{A(E_\perp)} \int_0^{2\pi} \frac{d\chi}{2\pi}\ f(\vec{r}, p=\sqrt{2M_1[E_\perp - U_{tot}(\vec{r})]}, \chi) \quad , \qquad (1.154)$$

where $A(E_\perp) = \int_{A_o} d^2\vec{r}\ \Theta(E_\perp - U_{tot}(\vec{r}))$ (Θ is the step function) is the accessible area in the unit cell for ions with given transverse energy $E_\perp$ (hatched area in Fig.1.24). As can be seen in Eq.(1.154) the probability distribution in transverse space

$$F(\vec{r},E_\perp) = \Theta(E_\perp - U_{tot}(\vec{r}))/A(E_\perp) \qquad (1.155)$$

is constant within the accessible area and zero otherwise.

Because of the short range of the string potential, the total potential in the unit cell is mainly given by the potential of the string within the unit cell, i.e. $U_{tot}(\vec{r}) = U(r)$ for not-too-low transverse energies $E_\perp$. In the same approximation the unit cell can be replaced by a circle with radius r_o determined by $\pi r_o^2 = A_o$ and the potential is normalized by $U(r_o) = 0$. Using polar coordinates $\vec{r} = (r,\varphi)$ we find

$$\bar{f}(E_\perp) = \int_{r^2_{min}(E_\perp)}^{r_o^2} \frac{d(r^2)}{r_o^2 - r^2_{min}(E_\perp)} \int_0^{2\pi} \frac{d\varphi}{2\pi} \int_0^{2\pi} \frac{d\chi}{2\pi}\, f(r,\varphi,p=\sqrt{2M_1[E_\perp - U(r)]},\chi), \qquad (1.156)$$

which is a good approximation of Eq.(1.154) and widely used (in the following denoted as single string approximation). The deviations for very small $E_\perp$ are not important for the results in most cases of interest except for flux peaking as will be discussed later.

1.3.2.2.2. Planar channeling

For planar channeling an analogous treatment is possible. If the ion moves nearly parallel to a low index plane of a crystal (and not parallel to a low index direction) the trajectory is determined by many correlated small angle collisions with the atoms of the plane. Similar to the axial case (Eqs.(1.146, 1.147)) the ion motion governed by a plane can be described by the potential of all atoms of the plane averaged over the coordinates (x,z) parallel to the plane

$$Y(y) = N\, d_p \int_{-\infty}^{\infty}\int_{-\infty}^{\infty} dx\, dz\; V(\sqrt{x^2+y^2+z^2}\,)\ , \qquad (1.157)$$

with d_p being the distance between the planes and N the number of atoms per unit volume as usual. The plane potential is normalized by $Y(d_p/2) = 0$. Note that the y-dependence of the plane potential is somewhat weaker than the r-dependence of the string potential, in particular Y(0) is finite while U(0) and V(0) are infinite. The conservation law of the corresponding transverse energy reads

$$E_\perp = \frac{p_y^2}{2M_1} + Y(y) = E\,\psi^2 + Y(y) = \text{const.}\quad , \qquad (1.158)$$

where ψ is now the angle between the ion velocity vector and the plane. As a consequence there exists a minimum distance y_{min} for the ion motion (see Fig.1.25) given by

$$Y(y_{min}) = E_\perp \longrightarrow y_{min}(E_\perp) \quad . \tag{1.159}$$

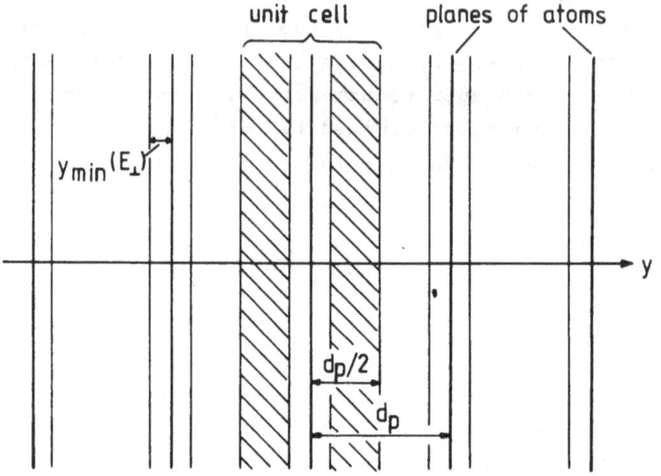

Fig.1.25. Transverse plane for planar channeling with the unit cell for one plane of atoms. The hatched area is the accessible portion of the unit cell for ions with a given transverse energy $E_\perp$.

While there is a two-dimensional statistical ion motion for axial channeling, in the planar channeling case only one dimension (y-coordinate) remains for the statistical motion. The phase density for ions with a given transverse energy $E_\perp$ in the 2-dimensional phase space

$$\rho_{ph}(y,p_y) = \rho_o \; \delta(E_\perp - p_y^2/(2M_1) - Y(y)) \quad , \tag{1.160}$$

provides for the statistical average of a physical quantity $f(y,p_y)$

$$\overline{f}(E_\perp) = \frac{\int_0^{d_p/2} dy \int_{-\infty}^{\infty} dp_y \; \rho_{ph}(y,p_y) \; f(y,p_y)}{\int_0^{d_p/2} dy \int_{-\infty}^{\infty} dp_y \; \rho_{ph}(y,p_y)}$$

(1.161)

$$= \int_{y_{min}(E_\perp)}^{d_p/2} dy \; \frac{K(E_\perp)}{[E_\perp - Y(y)]^{1/2}} \; f(y, p_y = \sqrt{2M_1[E_\perp - Y(y)]}) \quad ,$$

with

$$K(E_\perp) = \left\{ \int_{y_{min}(E_\perp)}^{d_p/2} dy \frac{1}{[E_\perp - Y(y)]^{1/2}} \right\}^{-1} . \tag{1.162}$$

As can be seen in Eq.(1.161) the probability distribution in transverse space

$$F(y,E_\perp) = \frac{K(E_\perp)}{[E_\perp - Y(y)]^{1/2}} \; \theta(E_\perp - Y(y)) \tag{1.163}$$

is no longer constant within the accessible area as for axial channeling but varies widely with y. For $y = y_{min}$ the probability becomes infinite because of the infinite duration time of the ion (the transverse velocity p_y/M_1 is zero for $y = y_{min}$). It must be emphasized that the striking difference between the probability distribution in transverse space for axial and planar channeling is purely a dimension effect.

Apart from Sect.1.3.2.5, in the following we will restrict the discussion to axial channeling. Except for the differences in the statistics the theoretical treatment for planar channeling is very similar to that for axial channeling.

1.3.2.3. Transverse energy distribution

In reality, instead of a fixed value of $E_\perp$ we get an initial distribution of transverse energies $g(E_\perp,0)$ at depth z = 0 (which refers to the beginning of the governed ion motion in the crystal). For the idealized crystal considered up to now, the distribution does not change with depth z. This is no longer true if the interactions of the ions with the electrons, thermal vibrating atoms and defects of the real crystal are taken into account. The transverse energy conservation is violated by these processes, which causes a change of the transverse energy distribution $g(E_\perp,z)$ with depth z and is called dechanneling. A good description of the dechanneling processes is very important for the application of RBS (combined with channeling) to defect analysis and will be treated in a separate section. We assume that these interaction processes do not destroy the equilibrium on the transverse energy shells or the equilibrium is quickly reattained, i.e. $g(E_\perp,z)$ is assumed to be the equilibrium transverse energy distribution for each depth z. Taking into account the $E_\perp$-distribution, the (measurable) mean value of a physical quantity f is given by an additional average

$$\langle f \rangle(z) = \int_0^\infty dE_\perp \; g(E_\perp,z) \, \overline{f}(E_\perp) \; , \tag{1.164}$$

with $\int_0^\infty dE_\perp \; g(E_\perp,z) = 1$ and $\bar{f}$ is defined in Eqs.(1.156) or (1.161). The main problem is the calculation of $g(E_\perp,z)$. First, the initial distribution is determined for channeling and blocking conditions.

1.3.2.3.1. Initial distribution for channeling

A homogeneous beam of ions is directed onto a crystal surface having a small angle of incidence ψ_{in} with respect to a low index direction. Entering the crystal the ions have the initial transverse energy $E_\perp = p^2/(2M_1) + U_{tot}(\vec{r})$ with a fixed value of the transverse momentum $p = M_1 v \, \psi_{in}$ and uniformly distributed $\vec{r}$. This provides the initial transverse energy distribution

$$g(E_\perp,0;\psi_{in}) = \int_{A_o} \frac{d^2\vec{r}}{A_o} \, \delta(E_\perp - E\psi_{in}^2 - U_{tot}(\vec{r})) = T_{in}(E_\perp, E\psi_{in}^2) \tag{1.165}$$

with the normalization $\int_0^\infty dE_\perp \; g(E_\perp,0;\psi_{in}) = 1$. Because of the symmetry (see Fig.1.24) the integration can be reduced to the unit cell with the area A_o. As can be seen in Eq.(1.165) the variation of the angle of incidence ψ_{in} provides only a shift of the transverse energy distribution by $E\,\psi_{in}^2$, i.e. only $T_{in}(E_\perp,0)$ must be calculated.

$$T_{in}(E_\perp, E\psi_{in}^2) = \begin{cases} T_{in}(E_\perp - E\psi_{in}^2, 0) & \text{for} \quad E_\perp \geq E\psi_{in}^2 \\ 0 & \text{for} \quad E_\perp < E\psi_{in}^2 \end{cases} \quad . \tag{1.166}$$

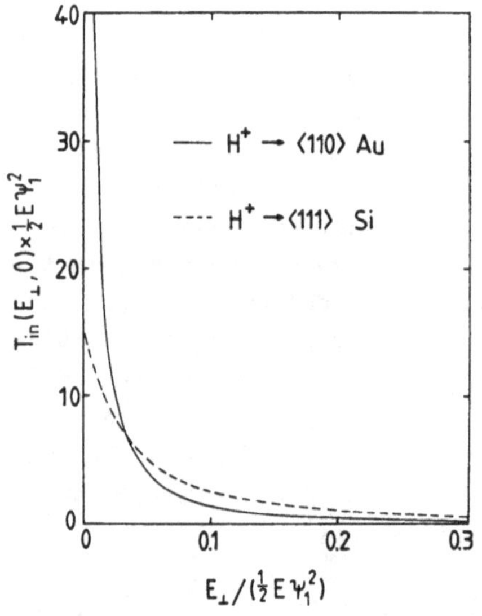

Fig.1.26. Initial transverse energy distribution for H ions incident on <110> Au and <111> Si

In the single string approximation ($U_{tot}(\vec{r}) = U(r)$ and a spherical unit cell with the radius r_o) we get

$$T_{in}(E_\perp,0) = \int_0^{r_o} dr\, \frac{2r}{r_o^2}\, \delta(E_\perp - U(r)) = \frac{2r_{min}(E_\perp)}{r_o^2} \left|\frac{dU}{dr}\right|^{-1}_{r=r_{min}(E_\perp)} , \qquad (1.167)$$

which is heavily peaked at $E_\perp = 0$ and decreases rapidly with increasing $E_\perp$ as shown in Fig.1.26. This is easily understood because only a small part of the ions enter the crystal near a string and get a large potential energy while all the others get only a small potential energy.

1.3.2.3.2. Initial distribution for blocking

In this case the ions start their governed motion from lattice sites (e.g. after Rutherford backscattering or generated by nuclear reactions) with transverse energy $E_\perp = p^2/(2M_1) + U_{tot}(\vec{r})$, where neither $p = M_1 v\psi$ nor $\vec{r}$ are fixed. For the small interval of the initial angle ψ of interest an isotropic distribution can be assumed (even in the case of Rutherford backscattering). The initial distance $\vec{r}$ is distributed via the thermal vibration of the emitting lattice atom. For simplicity, a Gaussian type distribution

$$dP(r) = d(r^2/\rho_\perp^2)\; e^{-(r^2/\rho_\perp^2)} \qquad (1.168)$$

is used, with $\rho_\perp^2$ being the average of r^2 e.g. given by the Debye theory

$$\rho_\perp^2 = \int_0^\infty dP(r)\, r^2 = \frac{6\hbar^2}{M_2 k T_D}\left[\frac{1}{4} + \left(\frac{T}{T_D}\right)^2 \int_0^{T_D/T} d\alpha\, \alpha\, (e^\alpha - 1)^{-1}\right]. \qquad (1.169)$$

The Debye temperatures T_D and mean square thermal vibration amplitudes $\rho_\perp^2(T)$ as functions of the temperature T are depicted in Fig.1.27 for some materials (k is the Boltzman constant). Because of $\rho_\perp \ll r_o$ only very small values of r are important and the single string approximation can be used from the beginning. In this way the initial distribution $g(E_\perp,0)$ is proportional to

$$\Pi_{out}(E_\perp) = \int_0^{r_o} dP(r) \int_0^\infty d(E\psi^2)\; \delta(E_\perp - E\psi^2 - U(r))$$

$$= \int_{r_{min}(E_\perp)}^{r_o} dP(r) = e^{-r_{min}^2(E_\perp)/\rho_\perp^2} , \qquad (1.170)$$

where the second term after the integration is neglected because of

$\rho_\perp \ll r_o$. As shown in Fig.1.28 for 1.4 MeV H ions incident on <111> Si, the function $\Pi_{out}(E_\perp)$ is small for small $E_\perp$, increases rapidly at $E_\perp \approx U(\rho_\perp)$ and tends to unity for large $E_\perp$.

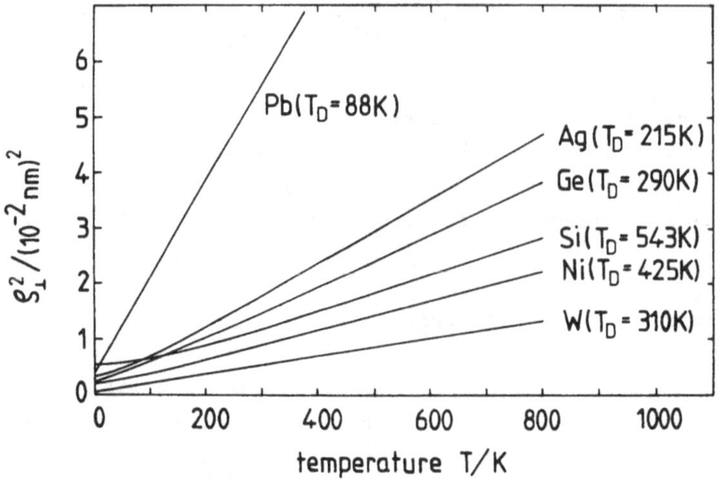

Fig.1.27. Mean square thermal vibration amplitude $\rho_\perp^2$ as a function of the temperature T. The Debye temperature T_D is added in brackets.

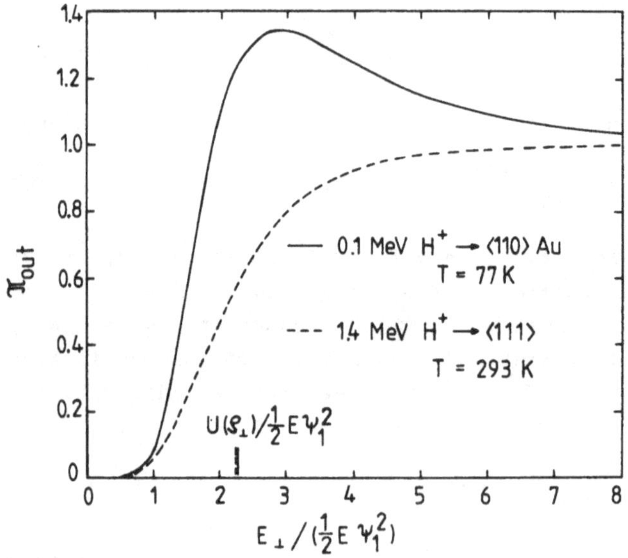

Fig.1.28. Initial transverse energy distribution $\Pi_{out}(E_\perp)$ for blocking (Ref.[44])

Note that $\Pi_{out}(E_\perp)$ is not a normalized function. For randomly distributed atoms an isotropic beam of ions remains isotropic, which can be described by $U(r) = 0$. In this case we get $\Pi_{out}(E_\perp) = 1$. Therefore $\Pi_{out}(E_\perp)$, as defined in Eq.(1.170), represents the relative probability for finding an ion with transverse energy $E_\perp$ with respect to the random case (amorphous target). As mentioned in Sect.1.3.2.1 the transverse energy conservation is valid at the half-way planes. To be more precise, $\Pi_{out}(E_\perp)$ must be calculated at the following half-way plane (at distance d/2) which changes the argument in the potential $\vec{r} \rightarrow \vec{r} + d\vec{\psi}/2$. Typical values of r ($r = |\vec{r}|$) and ψ ($\psi = |\vec{\psi}|$) are given by $\rho_\perp$ and ψ_1, respectively. For large energies in the MeV region, $d\psi_1/2$ (≈ 0.002 nm) is small compared with $\rho_\perp$ (≈ 0.01 nm). Therefore a remarkable correction of $\Pi_{out}(E_\perp)$ is only expected for small energies. This is confirmed by numerical calculations of $\Pi_{out}(E_\perp)$ as shown in Fig.1.28 for the example of 0.1 MeV H ions incident on <110> Au. The main effect is the appearance of a shoulder which is pronounced for low energies ($\lesssim$ 0.1 MeV) and negligible for large energies ($\gtrsim$ 1 MeV). This indicates that for high energy ion beam analysis the continuum approximation of $\Pi_{out}(E_\perp)$ given by Eq.(1.170) is sufficient.

1.3.2.4. Rutherford backscattering yield

1.3.2.4.1. Channeling regime

In the channeling regime (Fig.1.29) the ions which finally reach the detector enter the crystal under a small angle ψ_{in} with respect to a low index direction (1), penetrate the crystal by governed motion up to a certain depth z (2), suffer a wide angle single collision (3), and return to the surface by ungoverned motion (4). Because of the energy loss along the paths (2) and (4) the number of ions $Y(\psi_{in}, z)$ backscattered in a certain depth interval z...z+dz can be determined by measuring the energy of the detected ions as discussed in more detail in Sect.4.2.1.1 (energy-to-depth conversion). The normalized Rutherford backscattering yield is introduced by

$$\chi(\psi_{in}, z) = Y(\psi_{in}, z) \,/\, Y_r(z) \,, \qquad (1.171)$$

where Y_r is the random yield provided for large values of ψ_{in} or for amorphous targets. This quantity is determined by the relative probability (related to the random case) for large angle scattering at depth z. Notice that the impact parameter necessary for large angle scattering ($s = 10^{-4} \ldots 10^{-5}$ nm) is very small compared to the thermal vibration amplitude of the lattice atoms ($\rho_\perp \approx 10^{-2}$ nm). Therefore the physical quantity of interest is the probability of an ion at distance $\vec{r}$

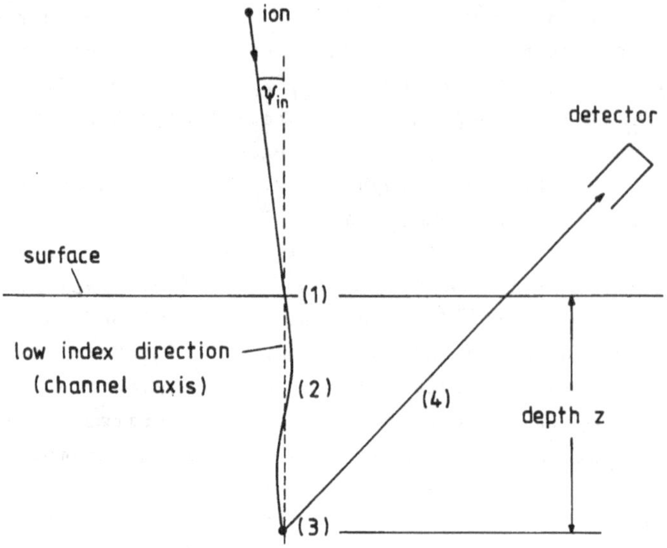

Fig.1.29. Ion path for the channeling regime

from the string and with transverse momentum $\vec{p}$ hitting a thermally vibrating lattice atom with an impact parameter $s \approx 0$. In the continuum approximation this probability is independent of $\vec{p}$ and directly given by the distribution of the thermally vibrating atoms $dP(r)$. The corresponding probability for the random case (randomly distributed target atoms) is $d(r^2)/r_o^2$. Finally the relative probability of an ion with $\vec{r}$ and $\vec{p}$ hitting a thermally vibrating lattice atom is $r_o^2\, dP(r)/d(r^2)$. Taking the average according to Eq.(1.156) we get the relative probability $\Pi_{in}(E_\perp)$ of an ion with given transverse energy $E_\perp$ for hitting a thermally vibrating lattice atom

$$\Pi_{in}(E_\perp) = \frac{r_o^2}{r_o^2 - r_{min}^2(E_\perp)} \int_{r_{min}(E_\perp)}^{r_o} dP(r) = \frac{A_o}{A(E_\perp)}\, \Pi_{out}(E_\perp) \quad . \tag{1.172}$$

In comparison with Eq.(1.170) we find that $\Pi_{in}(E_\perp)$ differs from $\Pi_{out}(E_\perp)$ only by a factor of $A_o/A(E_\perp)$ which is also valid if we use the more accurate half-way plane model. This is easily understood by the reversibility of the ion path and the fact that ions with $E_\perp$ can only start from the accessible area $A(E_\perp)$.

According to Eq.(1.164) the normalized Rutherford backscattering yield $\chi(\psi_{in}, z)$ is given by

$$\chi(\psi_{in}, z) = \int_0^\infty dE_\perp\; g(E_\perp, z; \psi_{in})\, \Pi_{in}(E_\perp) \quad . \tag{1.173}$$

For $z = 0$ the normalized yield can be calculated immediately from the initial transverse energy distribution $g(E_\perp,0;\psi_{in}) = T_{in}(E_\perp,E\psi_{in}^2)$ (see Eq.(1.165))

$$\chi(\psi_{in},0) = \int_0^\infty dE_\perp \; T_{in}(E_\perp,E\psi_{in}^2)\; \Pi_{in}(E_\perp) \quad . \qquad (1.174)$$

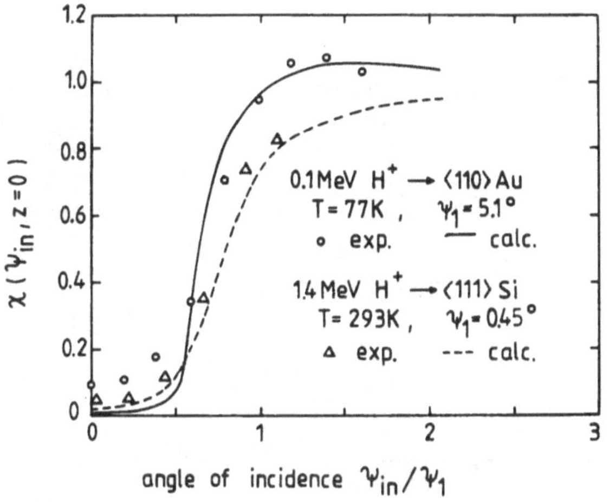

Fig.1.30. Normalized backscattering yield as a function of the angle of incidence (channeling regime). The diffusion equation results (Ref.[44]) are compared with experimental data of Refs.[45] and [46].

In Fig.1.30 the angular distributions are shown for two examples. Similar to the $E_\perp$-dependence of Π_{in} (see Eq.(1.172) and Fig.1.28) there is a shoulder in the normalized yield $\chi(\psi_{in},0)$ at ψ_{in} somewhat larger than ψ_1 for the low energy data and no shoulder for the high energy data. The minimum value at $\psi_{in} = 0$ is of the order of a few percent and for $\psi_{in} \gg \psi_1$ the yield $\chi(\psi_{in},0)$ tends to unity. The calculation of the normalized yield at depths $z > 0$ requires the corresponding transverse energy distribution $g(E_\perp,z;\psi_{in})$ which is the subject of Sect.1.3.3.

In connection with $\chi(\psi_{in},z)$ for very small depths or $z = 0$ some remarks are necessary. As mentioned in Sect.1.3.2.2. the equilibrium on the transverse energy shell is attained after a critical depth in the order of 0.1 µm which means that for smaller depths z the mean value of a physical quantity cannot be determined by the transverse energy distribution according to Eq.(1.164). In this case the full distribution in transverse phase space is required for the calculation of the statistical average of a physical quantity $f(\vec{r},\vec{p})$. Therefore the normalized yield for very small depths differs from that determined by Eq.(1.173) and tends to the yield discussed in the single scattering limit (Sect.1.3.1). This indicates that Eq.(1.173) can be used for depths $z > 0.1\,\mu m$ which is no remarkable restriction for high energy ion beam analysis because the depth intervals of interest are usually large compared to $0.1\,\mu m$. The initial angular distribution Eq.(1.174) can be understood as that received immediately after attaining the equilibrium on the transverse energy shell.

1.3.2.4.2. Blocking regime

In the blocking regime (Fig.1.31) the ions which finally reach the

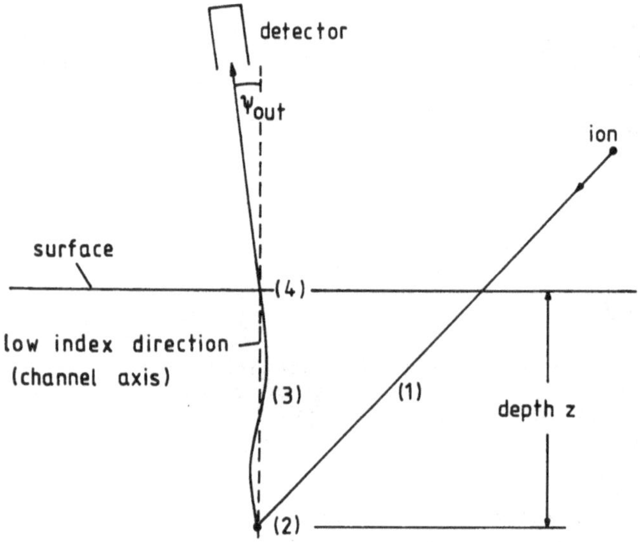

Fig.1.31. Ion path for the blocking regime

detector penetrate the crystal by ungoverned motion (1), suffer a wide angle single collision at depth z (2), return to the surface by governed motion (3), and leave the crystal at an angle ψ_{out} with respect to the low index direction (4). Assuming no change of the angle by passing the surface at position $\vec{r}$ the angular distribution for an ion with a transverse momentum $\vec{p}$ is given by $\delta(E\psi_{out}^2 - p^2/(2M_1))$. The corresponding average for a given $E_\perp$ according to Eq.(1.156) provides

$$T_{out}(E_\perp, E\psi_{out}^2) = \int_{r_{min}^2(E_\perp)}^{r_o^2} \frac{d(r^2)}{r_o^2 - r_{min}^2(E_\perp)} \; \delta(E\psi_{out}^2 - E_\perp + U(r))$$

$$= \frac{A_o}{A(E_\perp)} \; T_{in}(E_\perp, E\psi_{out}^2) \quad . \qquad (1.175)$$

The transmission probability T_{out} is normalized by $\int_0^{\infty} d(E\psi_{out}^2)\, T_{out}(E_\perp, E\psi_{out}^2) = 1$ and, except for the factor $A_o/A(E_\perp)$, equal to T_{in} because of the reversibility of the ion path through the surface. Performing the average over the transverse energy (see Eq.(1.164)) the normalized angular distribution reads

$$\chi(\psi_{out}, z) = \int_0^{\infty} dE_\perp\; g(E_\perp, z)\; T_{out}(E_\perp, E\psi_{out}^2) \quad , \tag{1.176}$$

where $g(E_\perp, z)$ now is the relative distribution of the transverse energy (normalized to the random case) with the initial distribution $g(E_\perp, 0) = \Pi_{out}(E_\perp)$ given by Eq.(1.170). Using Eqs.(1.170) and (1.175) the normalized blocking yield for $z = 0$

$$\chi(\psi_{out}, 0) = \int_0^{\infty} dE_\perp\; \Pi_{out}(E_\perp)\; T_{out}(E_\perp, E\psi_{out}^2) = \chi(\psi_{in}, 0)\Big|_{\psi_{in} = \psi_{out}} \tag{1.177}$$

is found to be the same as in the channeling case which is the result of the reversibility of the ion path (rule of reversibility) as discussed in Sect.1.3.1. The angular distribution for $z > 0$ is discussed in Sect.1.3.3.

The validity of the Lindhard concept in the case of blocking is briefly discussed here. As shown by analytical calculations (Ref.[47]) and Monte Carlo simulations (Ref.[48]) the equilibrium on the transverse energy shell is attained somewhat faster than in the channeling case. This can be understood by the fact that initially both $\vec{r}$ and $\vec{p}$ are statistically distributed in the blocking case while for channeling $\vec{p}$ is initially fixed (a lot of collisions are necessary to lose the preferred direction). The angular distributions for very small z (a few atomic layers) calculated by Monte Carlo simulation (Ref.[48]) differ remarkably from the results according to Eq.(1.176) (the dip of the distribution is not so pronounced) if $\rho_\perp > d\psi_1/2$. In this case the angular distribution discussed in Sect.1.3.1 is considerably changed by the following collisions and the integral interaction of the ion with the atoms of a string is only correct for greater depths. In summary, with respect to blocking calculations the transverse energy concept of Lindhard can be used for depth larger than a critical one in the order of some hundred atomic layers.

1.3.2.5. Ion flux in a channel

For a given transverse energy distribution $g(E_\perp, z; \psi_{in})$ (channeling regime) the normalized distribution of the flux in a channel is provided by

$$F(\vec{r},z;\psi_{in}) = \int_0^\infty dE_\perp \; g(E_\perp,z;\psi_{in}) \; F(\vec{r},E_\perp)$$
$$F(y,z;\psi_{in}) = \int_0^\infty dE_\perp \; g(E_\perp,z;\psi_{in}) \; F(y,E_\perp) \qquad (1.178)$$

where $F(\vec{r},E_\perp)$ and $F(y,E_\perp)$ are the probability distribution functions for axial and planar channeling, respectively, as they have been introduced in Sect.1.3.2.2 (Eqs.(1.155),(1.163)). Because of the strong difference between the two distributions, the planar channeling case is explicitly discussed in addition to the axial one in this section.

1.3.2.5.1. Axial channel

For simplicity let us consider $z = 0$ (or more accurately a small depth after attaining the equilibrium on the transverse energy shell) which already provides the main features of the ion flux distribution. With $g(E_\perp,0;\psi_{in}) = T_{in}(E_\perp,E\psi_{in}^2)$ (Eq.(1.165)) we find for axial channeling

$$F(\vec{r},0;\psi_{in}) = \int_{A_o} \frac{d^2\vec{r}'}{A_o} \frac{\Theta(E\psi_{in}^2+U_{tot}(\vec{r}')-U_{tot}(\vec{r}))}{A(E\psi_{in}^2+U_{tot}(\vec{r}'))} \qquad (1.179)$$

or in single string approximation ($U_{tot}(\vec{r}) = U(r)$, $U(r_o) = 0$, $A_o = \pi r_o^2$)

$$F(r,0;\psi_{in}) = \int_0^{r_m^2} \frac{d(r'^2)}{r_o^2} \frac{1}{A(E\psi_{in}^2+U(r'))} , \qquad (1.180)$$

with

$$r_m = \begin{cases} r_{min}(U(r) - E\psi_{in}^2) & \text{for} \quad r \le r_{min}\,(E\psi_{in}^2) \\ r_o & \text{for} \quad r > r_{min}\,(E\psi_{in}^2) \end{cases}$$

For $\psi_{in} = 0$ the integration can be performed analytically and provides

$$F(r,0;0) = A_o^{-1} \ln(r_o^2/(r_o^2-r^2)) , \qquad (1.181)$$

which is small for small r and increases rapidly with increasing r (infinite for $r = r_o$, i.e. in the middle of the channel). This effect is usually called flux peaking. In Fig.1.32 the numerical results for $\psi_{in} > 0$ are added. As can be seen, the flux peaking effect becomes weaker and weaker for increasing angle of incidence ψ_{in} and the flux distribution shows an edge positioned at $r = r_{min}(E\psi_{in}^2)$. Because of this edge the ion flux for a fixed position $r = r_a$ as a function of the angle

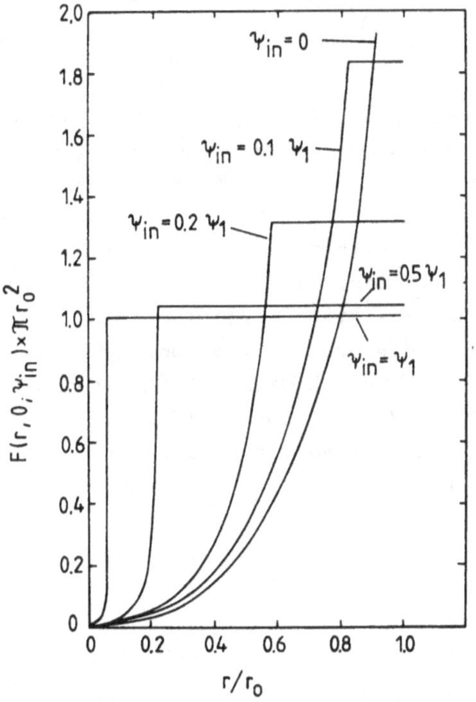

Fig.1.32. Ion flux distribution at depth $z = 0$ in single string approximation calculated using the Thomas-Fermi screening function. The results are presented for different angles of incidence ψ_{in}.

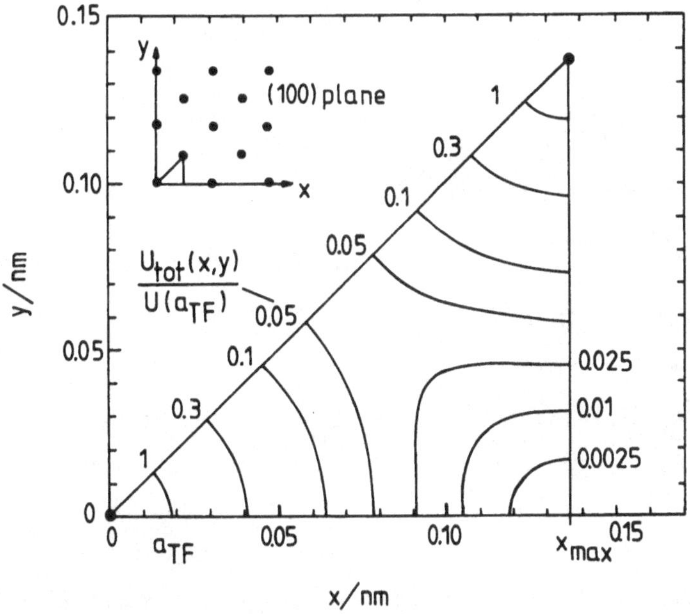

Fig.1.33. Full potential $U_{tot}(\vec{r})$ (sum of string potentials) within a symmetry element of the (100) plane of Si

of incidence ψ_{in} exhibits a weak maximum at $\psi_{in} = \psi_a$ according to

$$E\psi_a^2 = U(r_a) \ . \qquad (1.182)$$

The peak of the flux distribution for small values of ψ_{in} is situated in the middle of the channel (corresponding to very small transverse energies) and its shape is sensitive to a variation of the potential in this region. Therefore the single string approximation is not sufficient in this case and we must use Eq.(1.179) with the full potential $U_{tot}(\vec{r})$ for the calculation of the flux distribution. As an example, the total potential $U_{tot}(\vec{r})$ (with $\vec{r} = (x,y)$) for the <100> direction in silicon is given in Fig.1.33 within a symmetry element of the unit cell. The numerical results for the ion flux distribution along the x-axis $F(x,y=0,z=0;\psi_{in})$ are depicted in Fig.1.34 for different ψ_{in}. In comparison with the single string approximation, differences arise mainly in the middle of the channel for small ψ_{in}, in particular, the peak for $\psi_{in} = 0$ becomes finite.

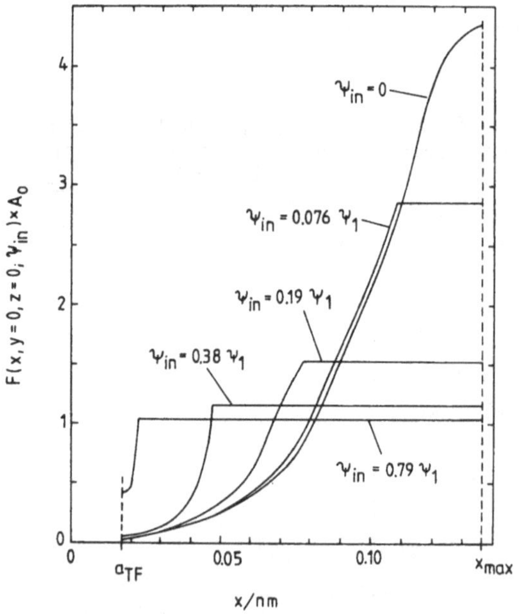

Fig.1.34. Ion flux distribution for axial channeling of He ions in <100> Si (Ref.[49]). The distribution at depth z = 0 and y = 0 is given as a function of x (see Fig.1.33) for different values of the angle of incidence ψ_{in}.

With respect to the yield of ions, backscattered from foreign atoms situated at fixed positions in the unit cell, the ion flux for a fixed position $\vec{r}_a = (x_a, y_a)$ is of interest as a function of the angle of incidence ψ_{in}. Therefore the corresponding representation of $F(x_a, y=0, z=0; \psi_{in})$ is given in Fig.1.35. As can be seen the dependence of the ion flux on ψ_{in} differs remarkably for different positions x_a. There is a strong peak for large x_a (middle of the channel) and a pronounced dip for small x_a (near the string). The latter case corresponds to the angular distribution (Eq.(1.174)) of ions backscattered from the thermally vibrating lattice atoms (but here we have a fixed position instead of a Gaussian distribution of the positions). Note that the change of the angular distribution with the position x_a is most sensitive for large values of x_a.

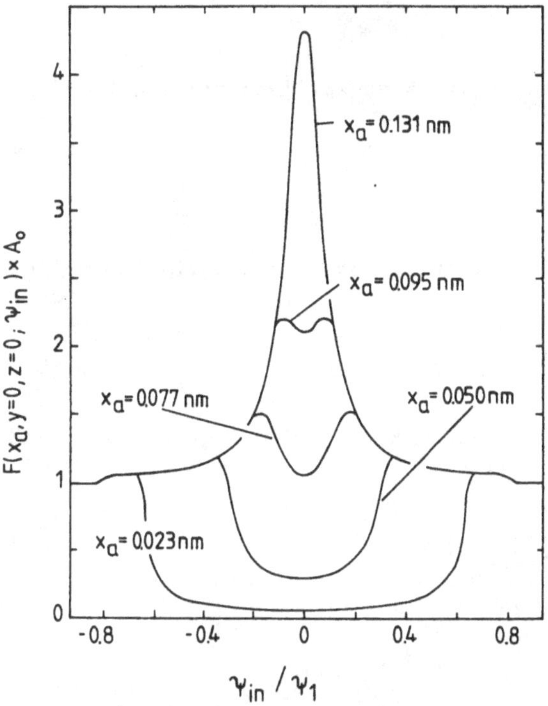

Fig.1.35. Ion flux (as given in Fig.1.34) for different fixed positions $x = x_a$, $y = 0$ in the channel as a function of the angle of incidence ψ_{in}

So far, we have considered the ion flux distribution under ideal conditions. The flux peaking effect is very sensitive to fluctuations of the angle of incidence and is rapidly suppressed with increasing depth. The influence of an ion beam divergency of 0.07^o on the flux distribution $F(x_a, y=0, z=0, \psi_{in})$ for <100> silicon is shown in Fig.1.36. In comparison with the case of an ideally collimated beam (Fig.1.35) the flux peaking for small values of ψ_{in} is reduced by a factor of about 2.

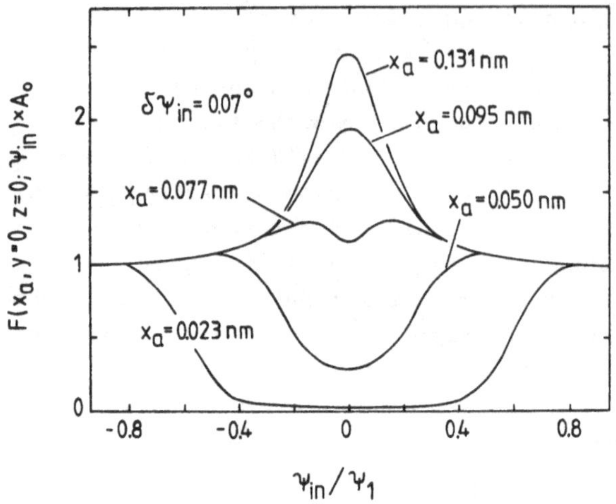

Fig.1.36. The same situation as in Fig.1.35 except that the ion beam has an angular divergence of 0.07°

1.3.2.5.2. Planar channel

With the initial transverse energy distribution computed similar to the axial case (Eq.(1.165))

$$g(E_\perp,0;\psi_{in}) = \int_0^{d_p/2} \frac{2dy'}{d_p}\,\delta(E_\perp - E\psi_{in}^2 - Y(y')) \tag{1.183}$$

(Y(y) is the plane potential introduced by Eq.(1.157)), Eqs.(1.163) and (1.178) provide for planar channeling

$$F(y,0;\psi_{in}) = \int_0^{y_m} \frac{2dy'}{d_p}\,\frac{K(E\psi_{in}^2+Y(y'))}{[E\psi_{in}^2+Y(y')-Y(y)]^{1/2}} \quad , \tag{1.184}$$

where

$$y_m = \begin{cases} y_{min}(Y(y)-E\psi_{in}^2) & \text{for } y \le y_{min}(E\psi_{in}^2) \\ d_p/2 & \text{for } y > y_{min}(E\psi_{in}^2) \end{cases} \quad .$$

In contrast to the axial case, the ion flux distribution in the planar channel has a pronounced peak at $y = y_{min}(E\psi_{in}^2)$ which is caused by the peak of $F(y,E_\perp)$ at $y = y_{min}(E_\perp)$. Fig.1.37 shows the ion flux distribution in a <110> planar channel of silicon computed by numerical integration of Eq.(1.184). For $\psi_{in} = 0$ the flux peaking in the middle of the channel is not so strong as in the axial case but with increasing ψ_{in} the peak at $y = y_{min}(E\psi_{in}^2)$ becomes more and more pronounced while the flux distribution in the axial channel has only an edge at the corresponding position.

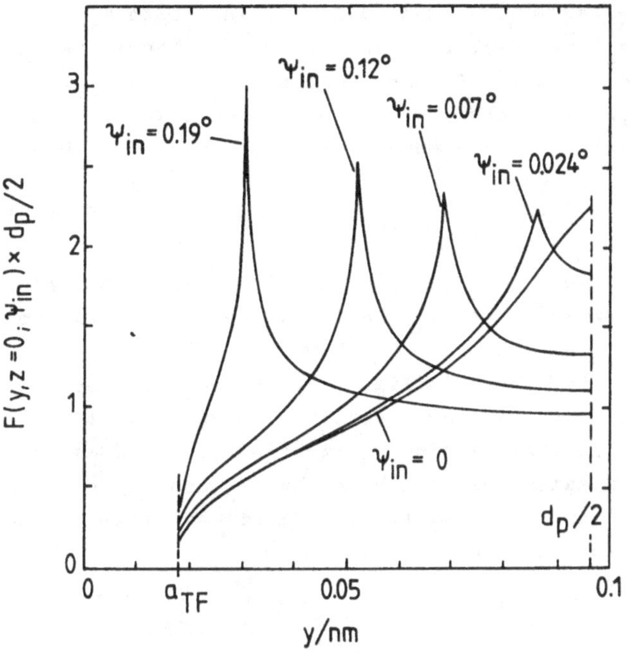

Fig.1.37. Ion flux distribution for planar channeling of He ions in <110> Si (Ref.[49]). The distribution at depth z = 0 is depicted as a function of the distance y from the plane.

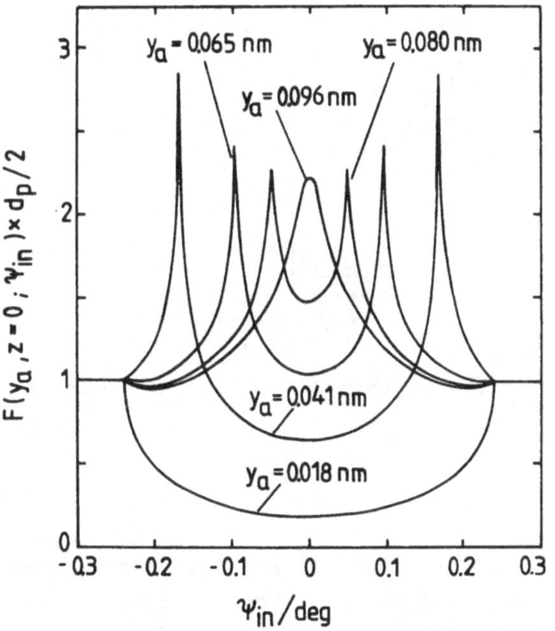

Fig.1.38. Ion flux (as given in Fig.1.37) for different fixed positions $y = y_a$ in the planar channel as a function of the angle of incidence Ψ_{in}

The dependence of $F(y_a, z=0; \psi_{in})$ on the angle of incidence ψ_{in} for different fixed values y_a is depicted in Fig.1.38. There is no substantial change in the shape of the angular distribution for different y_a, the main effect is a shift of the sharp peak. The position $\psi_{in} = \psi_a$ of the peak is immediately connected with the position y_a by

$$E\psi_a^2 = Y(y_a) . \tag{1.185}$$

As stated for the axial case, the flux peaking effect is diminished by a spread of the angle of incidence ψ_{in} and by dechanneling. For instance the influence of an ion beam divergency of 0.02° on the distribution $F(y_a, z=0; \psi_{in})$ is shown in Fig.1.39. As can be seen in comparison with Fig.1.38, the peaks are heavily reduced but yet well pronounced (measurable) and the position ψ_a of the peaks does not change, i.e. Eq.(1.185) remains valid. Note that ψ_a and y_a can be determined from a measured angular distribution most accurately for positions y_a not too close to the middle of the channel.

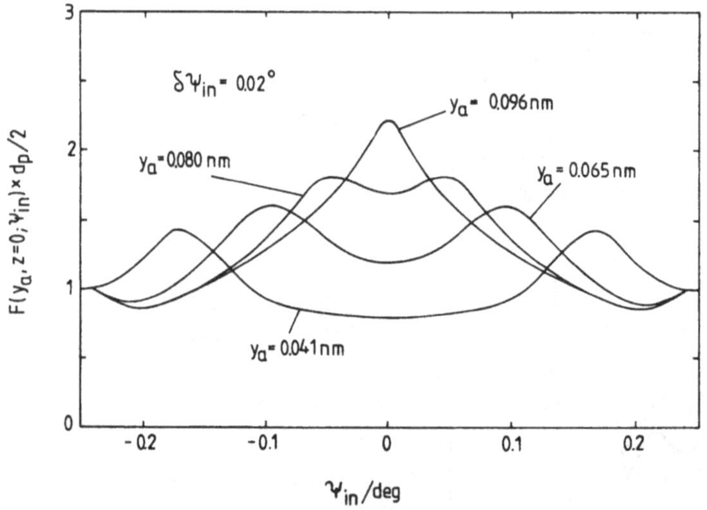

Fig.1.39. The same situation as in Fig.1.38 except that the ion beam has an angular divergence of 0.02°

1.3.3. Axial dechanneling

In the previous sections the governed ion motion in an idealized crystal lattice was considered. It is characterized by the conservation of the transverse energy for $E_\perp \leq E\psi_1^2$ and provides already the main features of the directional effects. As mentioned above for a quantitative description of the Rutherford backscattering yield the interaction process of the ions with the real crystal (electrons, thermally vibrating lattice atoms, defects) must be taken into account. Considering a single scattering event in a small depth interval δz the transverse energy of an ion with the transverse phase coordinates $\vec{r}$ and $\vec{p}$ is no longer constant but is changed by $\delta E_\perp(\vec{r},\vec{p},\beta)$, where β stands for all the additional statistical parameters with the probability $dP(\beta)$. This causes the transverse energy distribution $g(E_\perp,z)$ to change with depth z and is called axial dechanneling.

For small changes of the transverse energy ($\delta E_\perp << E_\perp$) the dechanneling can be described by a diffusion equation

$$\frac{\partial g(E_\perp,z)}{\partial z} = \frac{\partial}{\partial E_\perp}\left[A(E_\perp)\, D(E_\perp)\, \frac{\partial}{\partial E_\perp}\frac{g(E_\perp,z)}{A(E_\perp)} - g(E_\perp,z)\left(\overline{\frac{\delta E_\perp}{\delta z}}\right)_{loss}\right] \qquad (1.186)$$

derived by Bonderup et al. [50] from the master equation and by Kumakhov et al. [51] from the Fokker-Planck equation. The diffusion function $D(E_\perp)$ is determined by all contributions to the change of transverse energy $\delta E_\perp$ due to elastic scattering processes ($\delta E = 0$, $\delta E_\perp = \delta(E\psi^2+U) = E\,\delta(\psi^2)+\delta U$) and $(\overline{\delta E_\perp/\delta z})_{loss}$ is given by the energy loss contribution $\delta E_\perp = \psi^2 \delta E$.

$$D(E_\perp) = \frac{1}{2\delta z}\,\overline{\delta E_\perp^2}(E_\perp)$$

$$= \frac{1}{2}\int_A \frac{d^2\vec{r}}{A(E_\perp)} \int_0^{2\pi}\frac{d\chi}{2\pi}\int dP(\beta)\,\frac{1}{\delta z}\,\delta E_\perp^2(\vec{r}, p=\sqrt{2M_1[E_\perp - U_{tot}(\vec{r})]}, \chi, \beta) \qquad (1.187)$$

$$\left(\overline{\frac{\delta E_\perp}{\delta z}}\right)_{loss} = \frac{\delta E}{\delta z}\,\overline{\psi^2} = \frac{\delta E}{\delta z}\int_A \frac{d^2\vec{r}}{A(E_\perp)}\left(\frac{E_\perp - U_{tot}(\vec{r})}{E}\right) \quad . \qquad (1.188)$$

The averages are taken over the transverse phase space according to Eq.(1.154) and over all the statistical parameters β . By using the diffusion equation (Eq.(1.186)) Bonderup [52] derived the relation

$$\overline{\frac{\delta E_\perp}{\delta z}} = \frac{1}{A(E_\perp)}\frac{\partial}{\partial E_\perp}\left[A(E_\perp)\, D(E_\perp)\right] + \left(\overline{\frac{\delta E_\perp}{\delta z}}\right)_{loss}, \qquad (1.189)$$

which provides another representation of the diffusion function

$$D(E_\perp) = \frac{1}{A(E_\perp)}\int_0^{E_\perp} dE_\perp'\, A(E_\perp')\,\frac{1}{\delta z}\,\overline{\delta E_\perp}(E_\perp')\Big|_{\delta E=0} \qquad (1.190)$$

determined by $\overline{\delta E_\perp}$ instead of $\overline{\delta E_\perp^2}$ for the elastic scattering processes. In order to prevent the ions from diffusing to negative values of $E_\perp$ the integration constant is chosen as $D(0) = 0$. This representation is mostly preferred for the calculation of $D(E_\perp)$. Comparing the two Eqs.(1.187, 1.190) for the diffusion function we find a relation between $\overline{\delta E_\perp}$ and $\overline{\delta E_\perp^2}$ for the elastic scattering processes ($\delta E = 0$)

$$\overline{\delta E_\perp}(E_\perp) = \frac{1}{2A(E_\perp)} \frac{\partial}{\partial E_\perp} \left[A(E_\perp)\, \overline{\delta E_\perp^2}(E_\perp) \right] \quad , \tag{1.191}$$

which is understood as a consequence of the microscopic reversibility of the ion path (without energy loss) presumed for the derivation of the diffusion equation. Note that for $(\overline{\delta E_\perp/\delta z})_{loss} = 0$ the asymptotic solution of $g(E_\perp, z)$ for large z is proportional to $A(E_\perp)$.

As well known for the ion scattering on electrons and thermally displaced lattice atoms (perfect crystal) small $\delta E_\perp$ heavily dominate, for scattering on point defects small $\delta E_\perp$ are predominant but there are also a lot of large $\delta E_\perp$, and the interaction with extended defects causes large values of $\delta E_\perp$. Because of the small changes of the transverse energy for perfect crystals the diffusion model was successfully applied to the calculation of the Rutherford backscattering yield in this case (e.g. Refs.[53], [54]). In order to use the diffusion equation for point defects as well, Schiøtt et al. [53] suggested an additional absorption term for an approximative inclusion of the larger $\delta E_\perp$. Some investigations (Ref.[55]) showed that no consistent description of dechanneling due to point defects within the diffusion model can be obtained by introducing an additional absorption term. Therefore another description of dechanneling, correctly including large $\delta E_\perp$, is necessary for crystals with defects.

For this purpose, let us start from the master equation

$$\frac{\partial g(E_\perp, z)}{\partial z} = \int_0^\infty dE_\perp' \left[P(E_\perp, E_\perp')\, g(E_\perp', z) - P(E_\perp', E_\perp)\, g(E_\perp, z) \right] \tag{1.192}$$

given by Bonderup et al. [50]. The probability per depth $P(E_\perp', E_\perp)dE_\perp'$ for the transition $E_\perp \rightarrow E_\perp' \ldots E_\perp' + dE_\perp'$ is determined by the average of $\delta(E_\perp' - E_\perp - \delta E_\perp(\vec{r}, \vec{p}, \beta))$ over the transverse phase space according to Eq.(1.154) and over all the statistical parameters summarized by β

$$P(E_\perp', E_\perp) = \frac{1}{\delta z} \int_A \frac{d^2\vec{r}}{A(E_\perp)} \int_0^{2\pi} \frac{d\chi}{2\pi} \int dP(\beta)\, \delta(E_\perp' - E_\perp - \delta E_\perp) \quad ,$$

where (1.193)

$$\delta E_\perp = \delta E_\perp(\vec{r}, p = \sqrt{2M_1 [E_\perp - U_{tot}(\vec{r})]}, \chi, \beta)$$

represents the contribution from all scattering processes (also the energy loss term).

For the elastic scattering processes, the probability density $P(E'_\perp, E_\perp)$ fulfils the symmetry relation (Ref.[50])

$$P(E'_\perp, E_\perp)\, A(E_\perp) = P(E_\perp, E'_\perp)\, A(E'_\perp) \qquad \text{for} \quad \delta E = 0 \tag{1.194}$$

as a consequence of the microscopic reversibility of the ion path. In this case the asymptotic solution of the master equation (Eq.(1.192)) for large z is immediately given by

$$g_{as}(E_\perp) \sim A(E_\perp) \qquad \text{for} \quad \delta E = 0 \quad , \tag{1.195}$$

which is of course the same as found for the diffusion equation.

For crystals with defects, the dependence of the probability density $P(E'_\perp, E_\perp)$ on $E'_\perp$ for a given $E_\perp$ is qualitatively depicted in Fig.1.40.

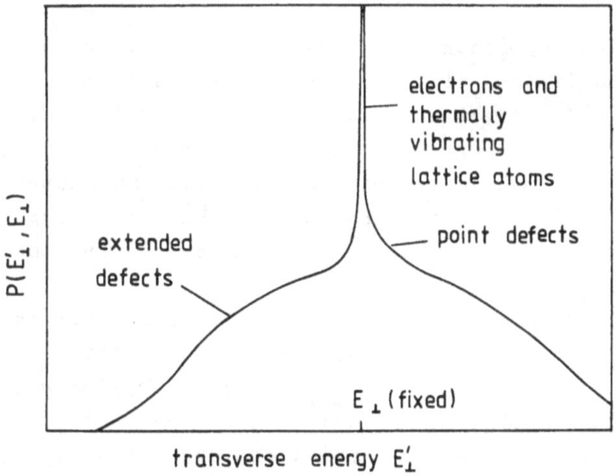

Fig.1.40. Probability density per unit depth $P(E'_\perp, E_\perp)$ for the transition from a fixed transverse energy $E_\perp$ to the final transverse energy $E'_\perp$ due to ion scattering in a crystal with different kinds of defects (qualitative representation)

There is a strong and narrow peak at $E'_\perp = E_\perp$ caused by the contributions from the interaction of the ions with the electrons and thermally vibrating lattice atoms, as well as a tail from the point defect contribution, and a smooth part introduced by the interaction with extended defects. Using this probability density, the description of all dechanneling processes by the master equation (1.192) is possible in principle but for such a complicated transverse energy dependence of $P(E'_\perp, E_\perp)$ (the widths of the peak and of the smooth distribution differ up to about 3 orders of magnitude) Eq.(1.192) cannot be solved practically. Therefore a different treatment for the dechanneling processes with small and large $\delta E_\perp$ is necessary.

For the dechanneling due to extended defects, the master equation (1.192) is a suitable description because of the smooth transverse energy dependence of the corresponding contribution to $P(E_\perp', E_\perp)$. The ion scattering on the electrons and thermally vibrating lattice atoms provides only very small $\delta E_\perp$. In this case a second order series expansion of Eq.(1.192) with respect to $\delta E_\perp$

$$\frac{\partial g(E_\perp, z)}{\partial z} = \frac{\partial}{\partial E_\perp}\left[D(E_\perp)\,\frac{\partial g(E_\perp, z)}{\partial E_\perp} - \Delta(E_\perp)\, g(E_\perp, z)\right] \tag{1.196}$$

is more useful than the master equation itself. The diffusion function $D(E_\perp)$ is already given in Eq.(1.187) and the shift term $\Delta(E_\perp)$ reads

$$\Delta(E_\perp) = \frac{1}{\delta z}\,\overline{\delta E_\perp}(E_\perp) - \frac{dD(E_\perp)}{dE_\perp} \quad . \tag{1.197}$$

Eq.(1.196) is the Fokker-Planck approach of the diffusion equation and using the relation given by Eq.(1.189) it is identical with Eq.(1.186).

The problem is that we have different equations ((1.192) and (1.196)) for the description of the dechanneling contributions from the perfect crystal and from extended defects and we cannot take into account point defects which provide both small and large changes of the transverse energy. A possibility to overcome this difficulty is described in the following (see also Ref.[56]).

The numerical solution of the master equation (1.192) requires finite intervals in $E_\perp$ in any case. Therefore let us introduce a partition of the transverse energy scale $E_{\perp i}$ (with $i = 1 \ldots i_{max}$) from the beginning. We use, for example, one of the simplest possibilities for the partition of the infinite $E_\perp$-scale into a finite even number of intervals i_{max} given by

$$E_{\perp i} = E_{\perp c}\; i/(i_{max} - i), \tag{1.198}$$

with a critical transverse energy

$$E_{\perp c} = \frac{1}{2}\, E\psi_1^2 \tag{1.199}$$

(as will be shown later the special choice of the partition hardly influences the physical quantities of interest calculated within this model). Note that half of the transverse energy intervals are situated below $E_{\perp c}$ and the other half above $E_{\perp c}$. The group i corresponds to ions with transverse energy in the interval $E_{\perp i-1} \ldots E_{\perp i}$ (see Fig.1.41) and the relative number of ions in this group is given by

$$g_i(z) = \int_{E_{\perp i-1}}^{E_{\perp i}} dE_\perp \; g(E_\perp, z) \quad , \tag{1.200}$$

with

$$\sum_i g_i(z) = \int_0^\infty dE_\perp \; g(E_\perp, z) \doteq 1 \tag{1.201}$$

(the summation always runs from $i = 1 \ldots i_{max}$). Now let us consider the change of the ion distribution $g_i(z)$ with depth z.

For dechanneling processes described by a smooth transverse energy dependence of the probability $P(E_\perp', E_\perp)$ (interaction with extended defects) dg_i/dz is immediately received by integration of the master

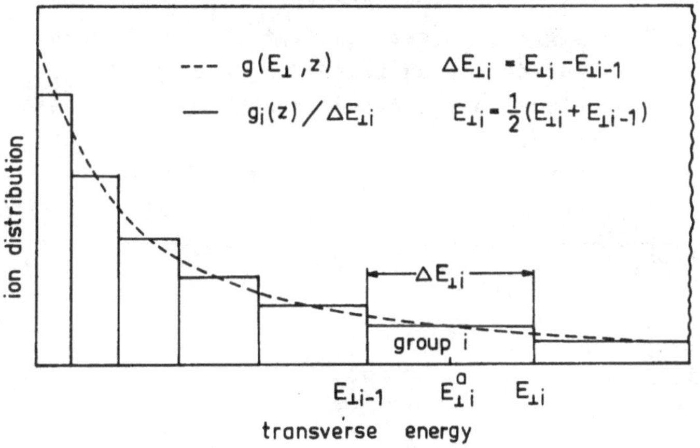

Fig.1.41. Partition of the transverse energy scale

equation (1.192) using the mean-value theorem of the integral calculus

$$\frac{dg_i(z)}{dz} = \sum_j \left[P_{ij} \; g_j(z) - P_{ji} \; g_i(z)\right] = \sum_j Q_{ij} \; g_j(z) \; , \tag{1.202}$$

with the matrix

$$Q_{ij} = P_{ij} - \delta_{ij} \sum_k P_{kj} \tag{1.203}$$

in the following referred to as dechanneling matrix, and

$$P_{ij} = \int_{E_{\perp i-1}}^{E_{\perp i}} dE_\perp \; P(E_\perp, E^a_{\perp j}) \quad , \tag{1.204}$$

where δ_{ij} is the Kronecker symbol and $E^a_{\perp j}$ is a value between $E_{\perp j-1}$ and $E_{\perp j}$. We use the mean value

$$E^a_{\perp j} = \frac{1}{2} (E_{\perp j-1} + E_{\perp j}) \quad . \tag{1.205}$$

The derivative dg_i/dz for dechanneling processes in a perfect crystal ($\delta E_\perp << E_\perp$, i.e. heavily peaked $P(E'_\perp, E_\perp)$) can be obtained by integration of the second order series expansion of the master equation (1.196)

$$\frac{dg_i(z)}{dz} = \left[D(E_\perp) \frac{\partial g(E_\perp, z)}{\partial E_\perp} - \Delta(E_\perp) \; g(E_\perp, z) \right] \Bigg|_{E_{\perp i-1}}^{E_{\perp i}} \quad . \tag{1.206}$$

Assuming a smooth dependence of $g(E_\perp, z)$ on $E_\perp$ the distribution function g and its first derivative at $E_\perp = E_{\perp i}$ are approximately given by a linear interpolation of $g(E^a_{\perp i}) = g_i/\Delta E_{\perp i}$ (see Fig.1.41). In this case we get the same type of equation for dg_i/dz as found for smooth $P(E'_\perp, E_\perp)$ (Eq.(1.202)) with the matrix elements Q_{ij} determined by the diffusion function $D(E_\perp)$ and the shift term $\Delta(E_\perp)$

$$Q_{i+1,i} = \frac{2D(E_{\perp i}) + \Delta E_{\perp i+1} \; \Delta(E_{\perp i})}{\Delta E_{\perp i} \; (\Delta E_{\perp i} + \Delta E_{\perp i+1})} \quad ,$$

$$Q_{i,i+1} = \frac{2D(E_{\perp i}) - \Delta E_{\perp i} \; \Delta(E_{\perp i})}{\Delta E_{\perp i+1} \; (\Delta E_{\perp i} + \Delta E_{\perp i+1})} \quad , \tag{1.207}$$

$$Q_{ii} = - Q_{i-1,i} - Q_{i+1,i} \quad ,$$

$$Q_{ij} = 0 \qquad \text{otherwise.}$$

Because $\delta E_\perp << E_\perp$ is fulfilled for the dechanneling processes considered, only transitions to neighbouring groups ($i \longrightarrow i \pm 1$) are possible. In this way dechanneling is described by only one equation

$$\frac{dg_i(z)}{dz} = \sum_j Q_{ij} \; g_j(z) \quad . \tag{1.208}$$

Assuming statistical independence of the different dechanneling processes, the dechanneling matrix Q_{ij} is a sum of all the individual contributions given by Eq.(1.207) for processes with $\delta E_\perp << E_\perp$ and by Eqs.(1.203, 1.204) for processes with smooth $P(E'_\perp, E_\perp)$ (large $\delta E_\perp$ dominate). Because of the unified description of processes with small and large $\delta E_\perp$, dechanneling due to point defects can also be included with two contributions to the dechanneling matrix given by Eqs.(1.203) and (1.207).

1.3.3.1. Axial dechanneling in perfect crystals

The interaction of the ions with the electrons (electronic scattering) and with the thermally vibrating lattice atoms (nuclear scattering) can be assumed to be statistically independent. Therefore the dechanneling matrix for a perfect crystal is given by the sum

$$Q_{ij} = Q_{ij}^{(e)} + Q_{ij}^{(n)} \quad , \tag{1.209}$$

where $Q_{ij}^{(e)}$ and $Q_{ij}^{(n)}$ are the contributions from electronic and nuclear scattering processes, respectively.

The interaction of the ions with the target electrons in a small depth intervall δz results in a deflection $\delta\psi$ (with azimuth angle γ) and an energy loss δE which provides

$$\delta E_{\perp}^{(e)}(r,p,\delta E,\delta\psi,\gamma) = 2E\psi\,\delta\psi\,\cos\gamma + E\,\delta\psi^2 + \psi^2\delta E \quad , \tag{1.210}$$

with $\psi = p/M_1 v$ ($\delta E, \delta\psi$ and γ are statistical parameters). The dependence of $\delta E_{\perp}^{(e)}$ on r is introduced by the probabilities $dP(\delta\psi)$ and $dP(\delta E)$ which depend on the average density

$$\rho(r) = \frac{1}{d}\int_{-\infty}^{\infty} dz\; n_e(\sqrt{r^2+z^2}\,) \tag{1.211}$$

of the electrons of a string (n_e is the density of the electrons of one atom). Because the electron mass is small compared with the ion mass, the values of $\delta\psi$ and $\delta E_{\perp}^{(e)}$ are very small. Therefore the contribution $Q_{ij}^{(e)}$ to the dechanneling matrix is determined by Eq.(1.207) with the corresponding diffusion function $D^{(e)}(E_{\perp})$ and the shift term $\Delta^{(e)}(E_{\perp})$ given by Eqs.(1.187,1.197). Using the single string approximation we obtain in the limit of $\delta z \rightarrow 0$

$$D^{(e)}(E_{\perp}) = \int_{r_{min}(E_{\perp})}^{r_o} \frac{2\pi r dr}{A(E_{\perp})}\, E\,[E_{\perp}-U(r)] \left(\frac{\delta\psi^2}{\delta z}\right) \quad , \tag{1.212}$$

$$\Delta^{(e)}(E_{\perp}) = \int_{r_{min}(E_{\perp})}^{r_o} \frac{2\pi r dr}{A(E_{\perp})} \left[E\left(\frac{\delta\psi^2}{\delta z}\right) + \frac{E_{\perp}-U(r)}{E}\left(\frac{\delta E}{\delta z}\right)\right] - \frac{dD^{(e)}(E_{\perp})}{dE_{\perp}} \quad . \tag{1.213}$$

Note that the mean energy loss $(\delta E/\delta z)$ influences only the shift term $\Delta^{(e)}(E_{\perp})$ and not the diffusion function $D^{(e)}(E_{\perp})$. For the mean angular spread $(\delta\psi^2/\delta z)$ due to multiple scattering by electrons, the expression

$$\left(\frac{\delta\psi^2}{\delta z}\right) = -\frac{Z_1^2 e^2 \hbar}{4\pi^2 \varepsilon_o M_1^2 v^4}\, \mathrm{Im}\left[\int_0^{\infty} \frac{dk}{k} \int_0^{kv} 2d\omega\, \frac{k^2-\omega^2/v^2}{\varepsilon^1(k,\omega;\rho(r))}\right] \tag{1.214}$$

derived by Lindhard [57] can be used, where $\varepsilon^1(k,\omega;\rho)$ is the longitudinal dielectric constant as a function of the absolute value of the wave vector k and frequency ω for an electron gas with the density ρ. While $D^{(e)}(E_\perp)$ increases with increasing $E_\perp$ the shift term decreases and vanishes in the limit of large $E_\perp$ except for the small loss contribution, i.e. the dechanneling due to electronic scattering is a simple diffusion process for large $E_\perp$.

A suitable depth interval δz for the ion collision with a thermally displaced lattice atom is given by the interatomic spacing d in the direction of a string. Within the depth interval $\delta z = d$ between two half-way planes (remember the fact that $E_\perp$ is defined at these planes) there is only one collision as depicted in Fig.1.42. An ion starting from the first half-way plane at $\vec{r} = (r,\varphi)$ and with a transverse momentum $\vec{p} = (p,\chi)$ will change its transverse energy due to the nuclear collision by

$$\delta E_\perp^{(n)}(r,\varphi,p,\chi,r_a) = E\,\theta^2(s) + 2E\,\vec{\psi}\vec{\theta}(\vec{s}) + U(r') - U(r)\,, \qquad (1.215)$$

where $\vec{s} = \vec{s}(r,\varphi,p,\chi,r_a)$ is the vector of the impact parameter ($s = |\vec{s}|$), $r' = r'(r,\varphi,p,\chi,r_a)$ is the distance of the ion from the string at the second half-way plane, $\vec{\theta}(\vec{s}) = \vec{s}\theta(s)/s$ is the vector corresponding to the deflection angle $\theta(s)$, $\vec{\psi} = \vec{p}/(M_1 v)$, and r_a is the distance of the thermally displaced atom from the string which is assumed to be Gaussian distributed according to Eq.(1.168).

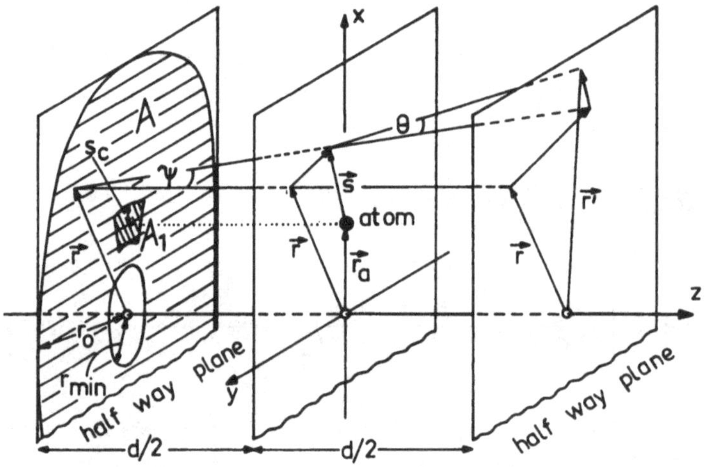

Fig.1.42. Geometry for ion scattering on a target atom displaced by a distance of r_a from the lattice site

Most of the possible ion trajectories provide small $\delta E_\perp$ ($\delta E_\perp/E_{\perp c} < 0.01$) but there are also a few trajectories with small impact parameters ($s < 0.01$ nm) which provide larger $\delta E_\perp$ ($\delta E_\perp/E_{\perp c} \leq 0.1$). Therefore a different treatment of the two kinds of ion trajectories is necessary with respect to the calculation of the dechanneling matrix. As suggested in Fig.1.42 the ion trajectories can be divided roughly into two groups according to their starting points at the first half-way plane.

Ions starting from part A_1 of the accessible area $A(E_\perp)$ may have large $\delta E_\perp$ because of the small impact parameters (the extension of A_1 is determined by a critical impact parameter s_c, the value of which is discussed later). The corresponding contribution $Q_{ij}^{(n1)}$ to the dechanneling matrix is given by Eqs.(1.203, 1.204, 1.193) with $\delta E_\perp = \delta E_\perp^{(n)}$ (Eq.(1.215)). Instead of the sixfold integration a Monte Carlo simulation method is preferred. The ions start with the mean transverse energy $E_{\perp j}^{a}$ of group j and the corresponding random numbers for the parameters r, φ, χ and r_a. The population of group i determines the dechanneling matrix contribution $Q_{ij}^{(n1)}$.

Ions starting from part $A_2 = A - A_1$ of the accessible area will have only small changes in their transverse energies. Therefore their contribution $Q_{ij}^{(n2)}$ to the dechanneling matrix is determined by Eqs.(1.207, 1.187, 1.197) with $\delta E_\perp = \delta E_\perp^{(n)}$(Eq.(1.215)) and $\delta z = d$ (for more details see Ref.[56]). In summary we find for the nuclear scattering contribution

$$Q_{ij}^{(n)} = Q_{ij}^{(n1)} + Q_{ij}^{(n2)} \tag{1.216}$$

In the channeling regime, the initial distribution $g_i(0;\psi_{in})$ for an incidence angle ψ_{in} is calculated from $g(E_\perp,0;\psi_{in})$ (Eq.(1.165)) according to Eq.(1.200)

$$g_i(0;\psi_{in}) = \int_{E_{\perp i-1}}^{E_{\perp i}} dE_\perp \; T_{in}(E_\perp, E\psi_{in}^2) \quad . \tag{1.217}$$

The numerical solution of the system of coupled differential equations (1.208) with $Q_{ij} = Q_{ij}^{(e)} + Q_{ij}^{(n)}$ provides the ion distribution $g_i(z;\psi_{in})$ at depth z. In analogy to Eq.(1.173) the normalized Rutherford backscattering yield is calculated from $g_i(z;\psi_{in})$ by

$$\chi(\psi_{in},z) = \sum_i g_i(z;\psi_{in}) \, \Pi_i \quad , \tag{1.218}$$

with the relative probability Π_i of an ion of group i hitting a thermally vibrating lattice atom determined from $\Pi_{in}(E_\perp)$ (Eq.(1.172))

$$\Pi_i = \frac{1}{\Delta E_{\perp i}} \int_{E_{\perp i-1}}^{E_{\perp i}} dE_\perp \; \frac{r_o^2}{r_o^2 - r_{min}^2(E_\perp)} \; e^{-(r_{min}^2(E_\perp)/\varrho_\perp^2)} \quad . \tag{1.219}$$

In the blocking regime, the initial transverse energy distribution $g(E_\perp,0) = \Pi_{out}(E_\perp)$ provides

$$g_i(0) = \int_{E_{\perp i-1}}^{E_{\perp i}} dE_\perp \; \Pi_{out}(E_\perp) \quad . \tag{1.220}$$

After the calculation of $g_i(z)$ the normalized Rutherford backscattering yield for the angle of emergence ψ_{out} is determined by

$$\chi(\psi_{out},z) = \sum_i g_i(z) \frac{1}{\Delta E_{\perp i}} \int_{E_{\perp i-1}}^{E_{\perp i}} dE_\perp \; T_{out}(E_\perp, E\psi_{out}^2) \quad . \tag{1.221}$$

Opposite to the channeling regime, the transverse energy distribution for blocking does not rapidly decrease with increasing $E_\perp$ but is constant for large $E_\perp$ (as can be seen in Fig.1.28). This introduces some difficulties into the description given above (e.g. $g_{i,max} \to \infty$). As shown by detailed investigations (Ref.[48]) these difficulties can be overcome by some modifications which are not discussed here.

Before considering typical results of the angular and depth dependence of the Rutherford backscattering yield, let us investigate the influence of the choice of the critical impact parameter s_c (see Fig.1.42) which determines the subdivision of the nuclear scattering processes. In order to get small $\delta E_\perp$ for all ions starting from $A_2 = A - A_1$ (which contribute to $Q_{ij}^{(n2)}$) the critical impact parameter s_c must be large enough ($s_c > d\psi_1/2$ at least). On the other hand, for a given history number (number of ion trajectories) the statistical error of the Monte Carlo simulation results of $Q_{ij}^{(n1)}$ (contribution of ions starting from area A_1) increases with increasing s_c. As shown in Ref.[56] there is an optimum value of s_c at about $s_c = 2.5\, d\psi_1$. Calculations of the channeling yield $\chi(0,z)$ for Si (Ref.[56]) and of the blocking angular distribution $\chi(\psi_{out},z)$ for Al (Ref.[48]) using different values of s_c showed that the results are independent of s_c (within the statistical error of the Monte Carlo simulation) in the range of $2d\psi_1 < s_c < 3d\psi_1$. This indicates the description of dechanneling due to nuclear collisions given above to be a consistent one. A different question is the stability of the results with respect to the choice of the $E_\perp$-partition. A strong change of $E_{\perp c}$ in Eq.(1.198) by a factor of 2 or a variation of the number of intervals $i_{max} = 10...30$ (see Ref.[48]) provides only a small influence on the calculated normalized yield ($\leq 3\%$). This is important and indicates that already a small number of intervals provides sufficiently accurate results which is very profitable for numerical calculations. Usually $i_{max} = 10$ is used. Only for blocking calculations larger i_{max} are necessary because of the large scaled transverse energy distribution.

In the following, the description of dechanneling given above is referred to as discontinuous model. All the results obtained within this model were calculated using the interatomic potential (Eq.(1.27), Ref.[4]) and the electron density (Eq.(1.211)) determined on the basis of the atomic model of Clementi and Roetti [12].

The main features of the angular distribution of the normalized Rutherford backscattering yield $\chi(\psi_{in},z)$ have already been discussed in Sect.1.3.2.4 for $z = 0$. The results for $z > 0$ obtained from Eq.(1.173) and depicted in Fig.1.43 show that the angular dip does not change in principle but becomes less and less pronounced. With increasing depth, the width of the dip decreases slightly but the yield for $\psi_{in} = 0$ increases considerably. If the dissipative processes can be neglected ($\delta E = 0$) the reversibility rule (channeling and blocking angular distributions are equal) holds also for $z > 0$. This can be proved by general considerations and also by comparing the numerical results according to Eqs.(1.218) and (1.221).

The normalized Rutherford backscattering yield at $\psi_{in} = 0$,

$$\chi_{min}(z) = \chi(0,z) \quad , \tag{1.222}$$

in the following called minimum yield for short, is an important quantity with respect to ion beam analysis. In comparison with the yield for $\psi_{in} > 0$, it provides the strongest depth dependence (as mentioned above) and is most sensitive to defects in the crystal (to be discussed later). As can be seen in Fig.1.44 the depth dependence of the minimum

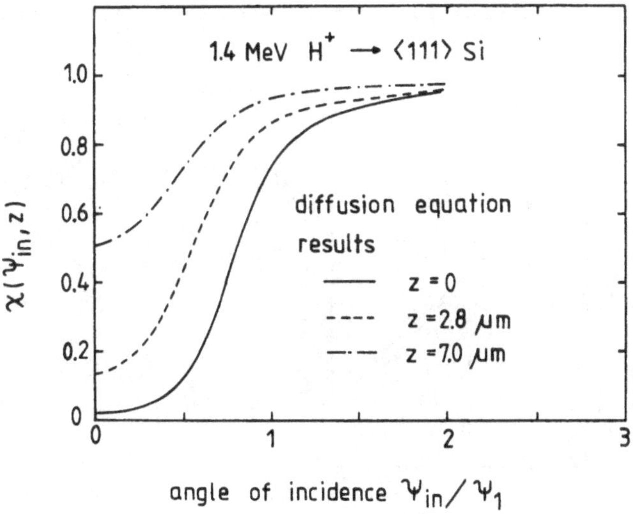

Fig.1.43. Normalized Rutherford backscattering yield $\chi(\psi_{in},z)$ as a function of the angle of incidence ψ_{in} for different depth z. The diffusion equation results are presented for 1.4 MeV H ions incident on <111> Si (Ref.[44]).

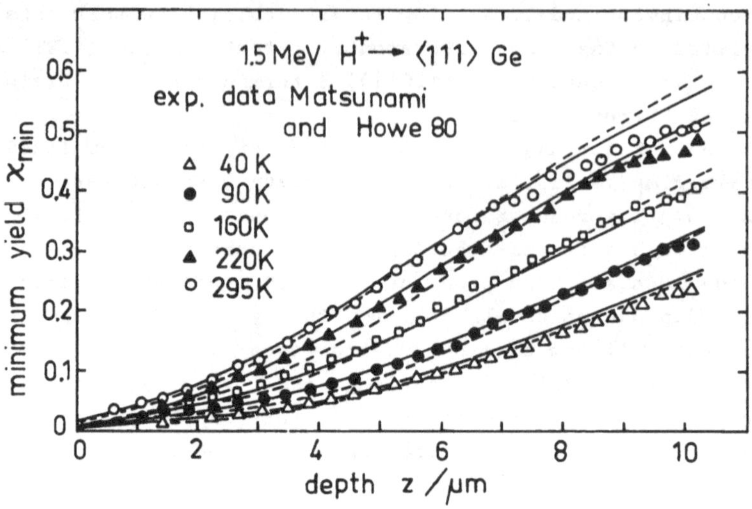

Fig.1.44. Depth dependence of the minimum yield at different temperatures for 1.5 MeV H ions incident on <111> Ge. The results obtained within the diffusion model (Ref.[54], dashed line) and within the discontinuous model (Ref.[56], solid line) are compared with experimental data of Ref.[54].

yield for different temperatures is well described by both the most elaborated diffusion model (Ref.[54]) and the discontinuous model given above. The differences between the theoretical curves are small but in contrast to the diffusion model calculations, the results from the discontinuous model are achieved without using any adjustable parameter. As was shown in Ref.[56] the contribution to the minimum yield from

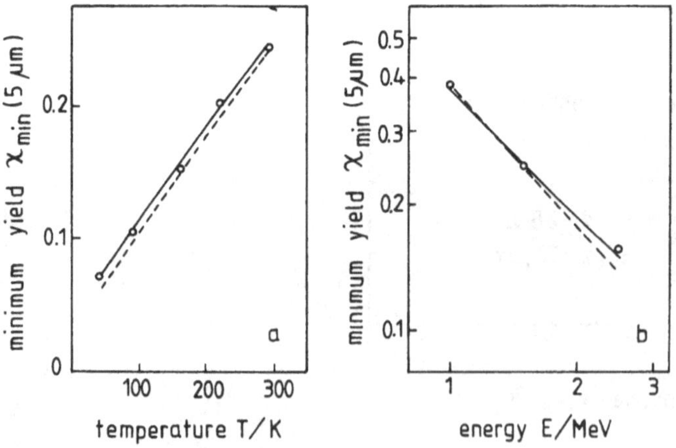

Fig.1.45. Temperature dependence and energy dependence of the minimum yield at depth z = 5 μm for 1.5 MeV H ions incident on <111> Ge. The notation corresponds to Fig.1.44.

nuclear scattering processes with larger $\delta E_{\perp}$ (all Q_{ij} with $|i-j| > 1$), which are not included in the diffusion model, is small for silicon and germanium (< 5%) but increases with increasing atomic number of the crystal (about 12% for W).

The dependence of the minimum yield on the target temperature T and the ion energy E is depicted in Fig.1.45. There is roughly a linear dependence of χ_{min} on the temperature because χ_{min} is approximately proportional to $\rho_{\perp}^2$. The E^{-1}-dependence of χ_{min} is understood by the energy dependence of multiple scattering. Both the T^1- and E^{-1}-dependence of the minimum yield as shown in Fig.1.45 are typical for perfect crystals.

1.3.3.2. Axial dechanneling due to defects

In a real crystal there are defects of different types which disturb the lattice order and influence the governed ion motion considerably. In the following let us restrict to point defects and three examples of extended defects (dislocations, stacking faults, twin lamellae) which are of particular interest with respect to defect analysis.

1.3.3.2.1. Dechanneling contributions from different defects

Dechanneling due to point defects is caused by ion scattering on uncorrelated displaced lattice atoms (or foreign atoms). For this purpose the point defects are fully characterized by the distances r_a of their position from the string (Fig.1.46a) and by their relative density $n_{pd}(z)$ as a function of the depth z (because of the large lateral extension of the beam the x- and y-dependence of n_{pd} is averaged). For large point defect densities, randomly distributed positions r_a are most probable but for small densities preferred positions are possible. In both cases the ion collisions with the defects are uncorrelated (randomly distributed r_a in one case and very large distances between the collisions in the other case). With respect to dechanneling, amorphous clusters have the same effect as the sum of all their displaced atoms and are also characterized by a mean distribution $n_{pd}(z)$, i.e. they cannot be distinguished from point defects by ion beam analysis.

Dislocations bend the channels of a crystal around the dislocation line, which disturbs the governed ion motion (see Fig.1.46b). The distortion field around a dislocation line depends on the type of the dislocation (e.g. edge dislocation, screw dislocation, dislocation loop) and on the Burgers vector $\vec{b}$. Its extension is in the order of 10 nm. In contrast to point defects, the dechanneling due to dislocations cannot

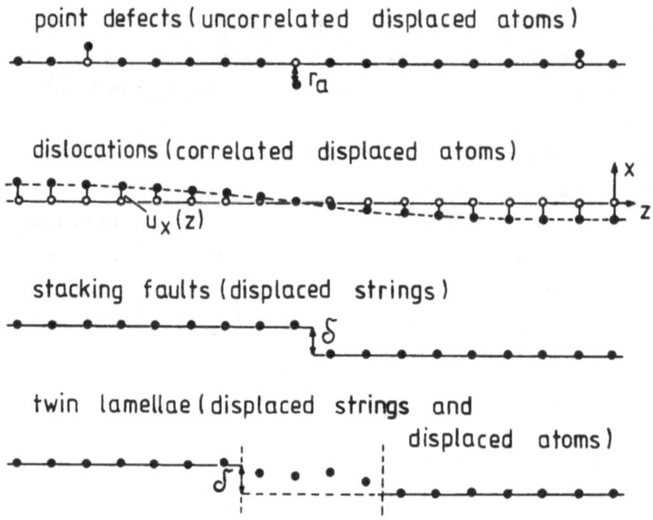

Fig.1.46. Schematic representation of different kinds of defects with respect to their influence on the channeled ion motion

be described by statistically independent single scattering events in the distorted channel because of the strong correlation between the collisions of the ions with successive displaced atoms. Therefore all the ion trajectories in the distorted region of the channel must be calculated. The dechanneling can be understood by the influence of the centrifugal force according to the curvature of the distorted channel. Usually, there is no single dislocation in a crystal but a distribution of dislocation lines characterized by their total projected length (perpendicular to the channel axis) per unit volume $n_{disl}(z)$ as a function of depth z.

For stacking faults the main contribution to dechanneling is given by the sudden displacements of the strings at the fault plane described by their displacement vector $\vec{\delta}$ (see Fig.1.46c). Assuming a statistical distribution of finite stacking defects within a macroscopic volume we may introduce the density of displaced strings $n_{dst}(z)$ which is the average number of strings becoming displaced within a unit volume at depth z. The displacement vector $\vec{\delta}$ depends on the direction $\vec{e}$ of the string with respect to the fault plane which is visualized in Fig.1.47 for a diamond-type lattice. In this case the stacking sequence of the (111) planes in the perfect crystal is AaBbCcAa... If one pair of planes is missing (e.g. Bb in Fig.1.47a) or one pair of planes is added (e.g. Bb in Fig.1.47b) we obtain an intrinsic and extrinsic stacking fault, respectively. The effect is described by a shift of one part of the crystal (open circles) with respect to the other part (full circles) by $-\vec{\delta}_o$ and $\vec{\delta}_o$, respectively ($\vec{\delta}_o = (1,1,1)\,a/3$, a being the lattice constant). In this way the vector of the displacement of strings with direction $\vec{e}$ (see Fig.1.47) is given by

$$\vec{\delta} = \vec{e} \times (\vec{\delta}_0 \times e) \quad . \tag{1.223}$$

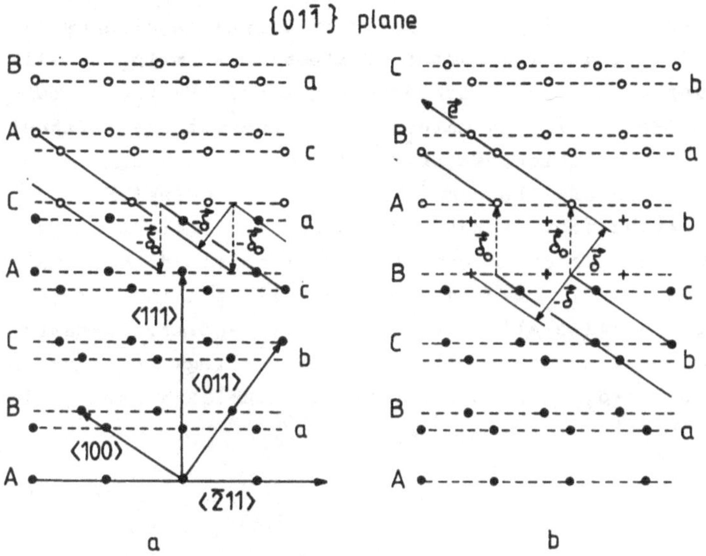

Fig.1.47. Intrinsic (a) and extrinsic (b) stacking fault in a diamond-type lattice

Note that for strings perpendicular to the fault plane there is no effect. As can be seen in Fig.1.47b for extrinsic stacking faults there are single atoms displaced by $-\vec{\delta}$ in addition to displaced strings. These atoms provide a dechanneling contribution according to point defects. If the stacking planes end within the crystal they are surrounded by dislocations of Frank type (Burgers vector $\vec{b}_F$ = (1,1,1) a/3) or Shockley type ($\vec{b}_S$ = (2,1,1) a/6) as shown in Ref.[58]. In summary, stacking faults give rise to dechanneling due to displaced strings, displaced atoms and dislocations. Because of the correlation between these defects the densities $n_{pd}(z)$ and $n_{disl}(z)$ can be assumed to be approximately proportional to the density of displaced strings $n_{dst}(z)$.

Twin lamellae are twinned layers of small thickness d_{lam} in the order of nm. For a diamond type lattice it corresponds to some (111) planes turned around their normal axis by 180°. Within the twinned layer the atoms form a new set of strings with a different crystallographic direction. Because these strings are very short (only a few atoms) and deformed by the influence of the two interfaces the penetrating ions experience nearly uncorrelated collisions. Therefore, with respect to dechanneling, the atoms within the twin lamella can be treated as point defects with roughly randomly distributed positions r_a (see Fig.1.46d) and a relative concentration

$$n_{pd}(z) = (d_{lam}/d)\, n_{st}(z)/N \quad , \tag{1.224}$$

where n_{st} is the number of strings crossing the twin lamellae per unit volume, N is the number of crystal atoms per unit volume as already has been introduced and d_{lam}/d is the number of atoms in a twin lamella belonging to one string (d is the interatomic spacing). Behind a twin lamella, the strings of the untwinned material continue with or without displacement depending on the thickness of the lamella or its equivalent number of pairs of (111) planes. For 3n (n = 0,1,2 ...) pairs of (111) planes there is no displacement of the strings and for 3n+1 and 3n+2 pairs, the strings are displaced according to extrinsic and intrinsic stacking faults, respectively. Because of the fluctuation of the thickness of the twin lamellae all three cases will appear. Assuming equal probability for these cases, only 2/3 of the strings crossing a twin lamella become displaced, i.e. the density of displaced strings is given by

$$n_{dst}(z) = \frac{2}{3}\, n_{st}(z) \quad . \tag{1.225}$$

Thus, with respect to dechanneling, a distribution of twin lamellae with an average thickness d_{lam} approximately can be described by a density distribution $n_{dst}(z)$ of displaced strings and randomly distributed point defects with a relative density

$$n_{pd}(z) = \frac{3}{2}\, (d_{lam}/d)\, n_{dst}(z)/N \quad . \tag{1.226}$$

1.3.3.2.2. Calculation of the dechanneling matrix

For the defects considered above there are three different dechanneling processes: the interaction of the ions with uncorrelated displaced atoms (point defects), with correlated displaced atoms (dislocations) and with displaced strings. Assuming statistical independence of these processes and electronic and nuclear scattering, the dechanneling matrix of the perfect crystal (Eq.(1.209)) is added to by three further contributions determined in the following.

(i) Point defects

The point defect contribution of the dechanneling matrix $Q_{ij}^{(pd)}$ is calculated by the same procedure as used for the nuclear scattering contribution $Q_{ij}^{(n)}$ (Eq.(1.216)), there is only a difference in the distribution of the positions r_a (see Fig.1.42).

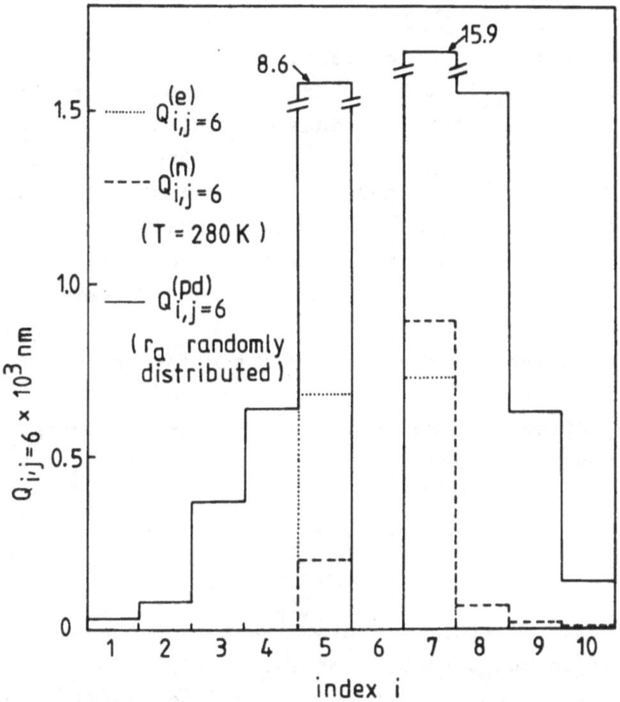

Fig.1.48. Elements of the dechanneling matrices for electronic scattering, nuclear scattering and scattering on randomly distributed displaced atoms (point defects). The matrix elements for fixed j = 6 and different i are depicted as a histogram. The values correspond to 1.0 MeV H ions incident on <111> Si.

Different point defects are represented by displaced atoms with different distributions of r_a. There are two limiting cases, r_a fixed (δ-distributed) and uniformly distributed r_a (within the unit area A_o), which are also of practical relevance. In comparison to the thermal vibration with mean amplitude $\rho_\perp$ in the order of 0.01 nm, the displacement r_a for point defects is much larger ($\leq$ 0.1 nm). Therefore, on one hand the additional thermal vibration of a point defect can be neglected and on the other hand scattering processes with large $\delta E_\perp$ are more probable. For comparison, the dechanneling matrices $Q^{(e)}_{ij}$, $Q^{(n)}_{ij}$, $Q^{(pd)}_{ij}$ for j = 6 are given in Fig.1.48. As can be seen point defects provide considerable contributions to the dechanneling matrix also for $|i-j| > 1$ (for more details see Ref.[59]). Since not all atoms are displaced, the contribution to the dechanneling matrix from point defects with a relative density $n_{pd}(z)$ is given by

$$Q_{ij}\Big|_{pd} = n_{pd}(z)\, Q^{(pd)}_{ij} \quad . \tag{1.227}$$

(ii) Dislocations

We will now discuss the contribution to the dechanneling matrix from dislocations with a density $n_{disl}(z)$. Let us first consider the transition probability $\tilde{P}_{ij}^{(disl)}$ for ions moving in a channel distorted by a dislocation of given type and Burgers vector $\vec{b}$. To this end, the change of the transverse energy $\delta E_{\perp}^{(disl)}$ during the ion motion in the distorted part of the channel must be determined. First, this is possible by Monte Carlo computer simulations which are very time consuming and provide results only for a given parameter set. Therefore they are not suitable for defect analysis because a lot of calculations are necessary in this case. A second possibility is the integration of the classical equations of motion. This way the ion motion in a distorted channel can be described as the motion of ions in a straight channel with an additional inertia force (Refs.[60,61,62])

$$- M_1 \ddot{u}_x(z(t)) = - M_1 v^2 \frac{d^2 u_x(z)}{dz^2} , \qquad (1.228)$$

where $u_x(z)$ is the displacement of the string atoms as a function of the depth z (the x-axis is chosen in the direction of the displacement). For edge and screw dislocations, $u_x(z)$ is given in Ref.[63]. If the potential $V_{ch}(x,y)$ of the channel (sum of the potentials of the surrounding strings) can be approximated by a harmonic potential with different frequencies ω_x and ω_y for the two directions in the transverse plane, the equations of motion

$$M_1 \ddot{x} = - M_1 \omega_x^2 x - M_1 \ddot{u}_x \qquad (1.229)$$

$$M_1 \ddot{y} = - M_1 \omega_y^2 y \qquad (1.230)$$

can be solved analytically. As can be seen in Fig.1.49 the main effect

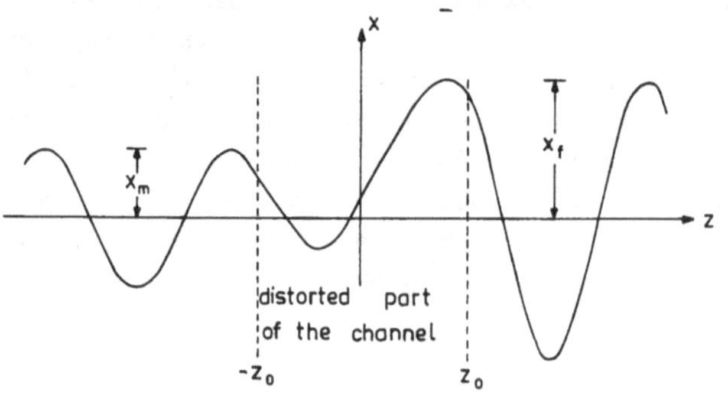

Fig.1.49. x-component of the ion trajectory in an axial channel distorted in the range of $z = -z_0 \ldots z_0$

of the distorted part of the channel $(-z_o \ldots z_o)$ on the ion motion is the change of the amplitude of the harmonic oscillation in x-direction (the oscillation in y-direction remains unchanged). With the analytical solution of the final amplitude x_f given in Ref.[64], the change of transverse energy $\delta E_\perp^{(disl)}$ reads

$$\begin{aligned}\delta E_\perp^{(disl)}(x_m, \varphi_x, x_d) &= \frac{1}{2} M_1 \omega_x^2 \, (x_f^2 - x_m^2) \\ &= \frac{1}{2} M_1 \omega_x^2 \, (x_d^2 + 2 x_m x_d \cos\varphi_x) \quad ,\end{aligned} \tag{1.231}$$

where x_m and φ_x are the amplitude and phase of the initial harmonic oscillation (for $z < -z_o$) and x_d is determined by the distortion field $u_x(z)$ of the dislocation. For an edge dislocation we find (see Ref.[63])

$$x_{d,edge}^2 = \left(x_d^{(s)}\right)^2 \cos^2\zeta + \left(x_d^{(as)}\right)^2 \sin^2\zeta \tag{1.232}$$

with

$$x_d^{(s)} = \frac{b}{2}\left[\frac{1-2\nu}{2(1-\nu)} + \frac{1}{2(1-\nu)} \frac{R\omega_x}{v}\right] e^{-\frac{R\omega_x}{v}} \tag{1.233}$$

$$x_d^{(as)} = \frac{b}{2}\left[1 + \frac{1}{2(1-\nu)} \frac{R\omega_x}{v}\right] e^{-\frac{R\omega_x}{v}} \quad . \tag{1.234}$$

As can be seen in Fig.1.50 ζ is the angle between the Burgers vector $\vec{b}$ (parallel to x'-axis) and the direction of the channel (z-axis), R is the minimum distance between the dislocation line (parallel to y- and y'-axis) and the channel, b is the absolute value of the Burgers vector $\vec{b}$, and ν is Poisson's ratio.

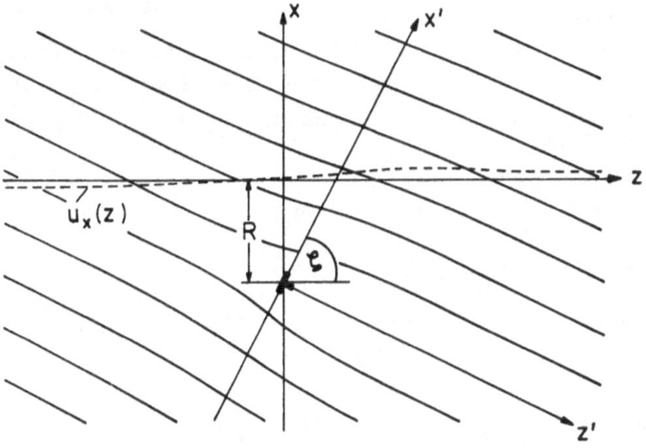

Fig.1.50. Distortion of an axial channel by an edge dislocation

Usually $x_d^{(s)}$ is much smaller than $x_d^{(as)}$ ($\nu = 0.42$ for Si), i.e. the edge dislocation is most effective with respect to dechanneling for $\zeta = \pi/2$. For screw dislocations we get

$$x_{d,screw} = \frac{b}{2} e^{-\frac{R\omega_x}{v}} \tag{1.235}$$

and for dislocation loops we refer to Ref.[63].

For different initial phases φ_x $(0...2\pi)$ there is a widespread distribution of $\delta E_\perp^{(disl)}$ (Eq.(1.231)) up to large $\delta E_\perp^{(disl)}$ comparable with $E_\perp$. Therefore the transition probability $\tilde{P}_{ij}^{(disl)}$ corresponding to the ion motion in one channel must be calculated according to Eqs.(1.203, 1.204, 1.193) with $\delta E_\perp = \delta E_\perp^{(disl)}$ (Eq.(1.231)). The integration can be performed analytically and provides

$$\tilde{P}_{ij}^{(disl)}(R/v) = f_{ij}(R/v) - f_{i-1,j}(R/v) \quad , \tag{1.236}$$

with

$$f_{ij}(R/v) = \begin{cases} 0 & \text{for } \zeta_{ij} \le -1 \\ 1 - \frac{1}{\pi}[\arccos\zeta_{ij} - \zeta_{ij}(1-\zeta_{ij}^2)^{1/2}] & \text{for } |\zeta_{ij}| \le 1 \\ 1 & \text{for } \zeta_{ij} \ge 1 \end{cases} \tag{1.237}$$

and

$$\zeta_{ij} = \frac{E_{\perp i} - E_{\perp j}^a - \frac{1}{2} M_1\omega_x^2\, x_d^2(R/v)}{[2M_1\omega_x^2\, x_d^2(R/v)\, E_{\perp j}^a]^{1/2}} \tag{1.238}$$

(note that the initial values x_m and φ_x are connected with the x-components of the transverse phase space by $x = x_m \cos\varphi_x$ and $p_x = x_m\omega_x \sin\varphi_x$, for more details see Ref.[63]).

If the distortion fields of the dislocation lines with a total projected length L in the area A_c of the crystal do not overlap, the probability per depth for finding a channel in the distance R...R + dR is given by

$$2L\, dR/(\delta z\, A_c) = 2\, n_{disl}\, dR \tag{1.239}$$

(the factor 2 is introduced because there are channels on both sides of the dislocation line). Finally, the contribution of the dechanneling matrix due to dislocations is given according to Eq.(1.203) by

$$\begin{aligned} Q_{ij}|_{disl} &= 2n_{disl}(z) \int_0^R dR \left[\tilde{P}_{ij}^{(disl)}(R/v) - \delta_{ij}\sum_k \tilde{P}_{kj}^{(disl)}(R/v)\right] \\ &= n_{disl}(z)\, Q_{ij}^{(disl)} \quad , \end{aligned} \tag{1.240}$$

with

$$Q_{ij}^{(disl)} = 2\, w_{ij}\, v \int_0^\infty d\tau\, \tilde{P}_{ij}^{(disl)}(\tau) \qquad \text{for} \quad i \neq j$$

and (1.241)

$$Q_{jj}^{(disl)} = - \sum_{i(\neq j)} Q_{ij}^{(disl)} \quad ,$$

where $\tau = R/v$. The correction factor $w_{ij} \leq 1$ as introduced and discussed in Ref.[63] roughly takes into account the following effect. For large values of x and y, the channel potential $V_{ch}(x,y)$ has a valley between two neighbouring strings surrounding a channel. This provides an opportunity for the ions to escape from the channel which decreases the dechanneling matrix calculated within the harmonic approximation of the channel potential and is expressed by the factor w_{ij}.

As can be seen, all elements of the matrix $Q_{ij}^{(disl)}$ have the same energy dependence given by $E^{1/2}$ which is the consequence of the R/v-dependence of $\tilde{P}_{ij}^{(disl)}$ and of the statistical independence of the contributions from the individual dislocation lines. For dislocation densities $n_{disl} > 10^{11}$ cm^{-2} the distortion fields of the individual dislocation lines overlap, i.e. the statistical independence of their contributions and Eq.(1.240) are not valid. In this case the dechanneling matrix is no longer proportional to n_{disl} and $E^{1/2}$ (Ref.[65]).

(iii) Displaced strings

The contribution of the dechanneling matrix from a distribution $n_{dst}(z)$ of displaced strings can easily be calculated. Fig.1.51 shows the

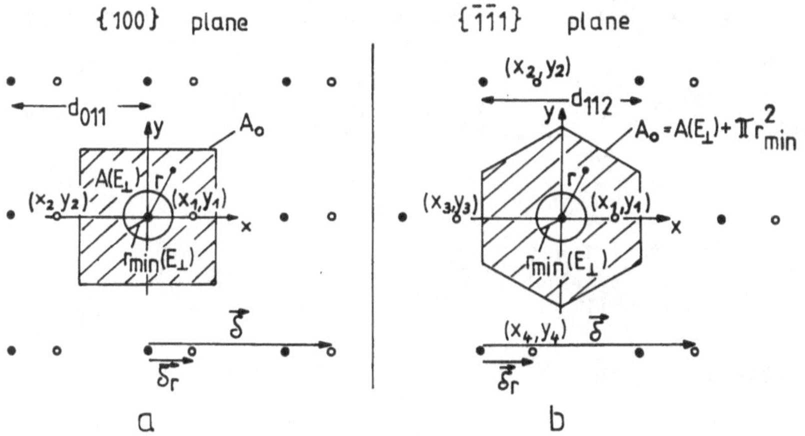

Fig.1.51. Original strings (•) and displaced strings (o) in <100> direction (a) and <111> direction (b) of a diamond-type lattice. The accessible area $A(E_\perp)$ is given by the hatched area.

original and displaced strings for the <100> and <111> direction in silicon. Because of the symmetry, $\vec{\delta}$ can be replaced by the reduced displacement vector $\vec{\delta}_r$ with the values $\delta_r = d_{011}/3$ and $\delta_r = d_{112}/3$ for the <100> and <111> directions, respectively. The displacement $\vec{\delta}_r$ of a string causes a sudden change of the potential energy while the angle ψ remains unchanged. Therefore the corresponding change of transverse energy

$$\delta E_\perp^{(dst)}(x,y) = \sum_{n=1}^{n_{max}} U(\sqrt{(x-x_n)^2+(y-y_n)^2}) - U(\sqrt{x^2+y^2}) \tag{1.242}$$

is only a function of the transverse space coordinates (x,y), where (x_n, y_n) are the positions of the displaced strings. Because of the short range of U(r), $n_{max} = 2$ and $n_{max} = 4$ are sufficient for <100> and <111> directions, respectively. The transition probability $\tilde{P}_{ij}^{(dst)}$ due to one displaced string per unit cell A_o in the transverse plane is given according to Eqs.(1.204) and (1.193) because there is a smooth and widespread distribution of $\delta E_\perp^{(dst)}$.

$$\tilde{P}_{ij}^{(dst)} = \int_{E_{\perp i-1}}^{E_{\perp i}} dE_\perp' \int_{A(E_{\perp j}^a)} \frac{dx\ dy}{A(E_{\perp j}^a)} \delta(E_\perp' - E_{\perp j}^a - \delta E_\perp^{(dst)}(x,y)) \tag{1.243}$$

The integration over the accessible area $A(E_{\perp j}^a)$ (hatched area in Fig.1.51, single string approximation is not possible in this case) is preferentially performed by Monte Carlo computer calculations (for more details see Ref.[66]). According to Eq.(1.203) the contribution to the dechanneling matrix from a distribution $n_{dst}(z)$ of displaced strings is given by the probability $\tilde{P}_{ij}^{(dst)}$ of one displaced string per unit cell A_o multiplied by the average number of displacements $A_o n_{dst}$ per unit depth within an area A_o

$$Q_{ij}\Big|_{dst} = A_o n_{dst}(z)\ [\tilde{P}_{ij}^{(dst)} - \delta_{ij} \sum_k \tilde{P}_{kj}^{(dst)}] = n_{dst}(z)\ Q_{ij}^{(dst)} , \tag{1.244}$$

with

$$Q_{ij}^{(dst)} = A_o\ \tilde{P}_{ij}^{(dst)} \qquad \text{for} \quad i \neq j$$

and (1.245)

$$Q_{jj}^{(dst)} = - \sum_{i(\neq j)} Q_{ij}^{(dst)} \quad .$$

As can be seen from Eqs.(1.242, 1.243, 1.245) the matrix $Q_{ij}^{(dst)}$ does not depend on the energy E of the ions.

(iv) Real crystal

Finally, the dechanneling matrix for a real crystal, taking into account point defects, dislocations, stacking faults and twin lamellae, is the sum of all contributions given by Eqs.(1.209, 1.227, 1.240, 1.244). Then, the master equation (1.208) reads

$$\frac{dg_i(z)}{dz} = \sum_j Q_{ij}\, g_j(z) = \sum_j \Big\{ Q_{ij}^{(e)} + [1-n_{pd}(z)]\, Q_{ij}^{(n)} + n_{pd}(z)\, Q_{ij}^{(pd)} + n_{disl}(z)\, Q_{ij}^{(disl)} + n_{dst}(z)\, Q_{ij}^{(dst)} \Big\} g_j(z) \tag{1.246}$$

1.3.3.2.3. Rutherford backscattering yield in channeling regime

Ions will be backscattered from point defects, from the head atoms of displaced strings (direct scattering part), and from the remaining thermally vibrating atoms (dechanneling part). The direct backscattering part from a dislocation line can be neglected with respect to the dechanneling contribution of a dislocation. In analogy to Eq.(1.218), the normalized Rutherford backscattering yield for real crystals in channeling regime reads

$$\chi(\psi_{in}, z) = \sum_i g_i(z;\psi_{in})\, \tilde{\Pi}_i(z) \quad , \tag{1.247}$$

with

$$\tilde{\Pi}_i(z) = [1-n_{pd}(z)-n_{dst}(z)/N]\, \Pi_i + n_{pd}(z)\, \Pi_i^{(pd)}(r_a) + [n_{dst}(z)/N]\, \Pi_i^{(pd)}(\delta_r) \quad . \tag{1.248}$$

The probability Π_i of an ion in the group i hitting a thermally vibrating lattice atom is already given by Eq.(1.219) and the corresponding probability $\Pi_i^{(pd)}(r_a)$ of an ion hitting a lattice atom displaced by a distance r_a from the string is found analogously. It follows for fixed r_a

$$\Pi_i^{(pd)}(r_a) = \frac{1}{\Delta E_{\perp i}} \int_{E_{\perp i-1}}^{E_{\perp i}} dE_\perp \, \frac{r_o^2}{r_o^2 - r_{min}^2(E_\perp)} \quad \Theta(E_\perp - U(r_a)) \tag{1.249}$$

and for randomly distributed r_a

$$\Pi_i^{(pd)} = 1 \quad . \tag{1.250}$$

The notation "randomly distributed r_a" is used for the case, where the displaced atoms are uniformly distributed within the unit cell A_o of the transverse space. The head atoms of the displaced strings are treated as point defects (with respect to $\Pi_i^{(pd)}$) with $r_a = \delta_r$ and a relative density $n_{dst}(z)/N$.

In contrast to perfect crystals, the accuracy of the theoretical description of dechanneling due to defects can be directly checked only for such cases where the depth distributions and the other parameters necessary for the characterization of the defects are well known from other investigations. An amorphous layer of thickness d_{am} on top of a perfect crystal can be understood as a very extreme distribution of point defects with $n_{pd}(z) = 1$ for $z < d_{am}$ and $n_{pd}(z) = 0$ otherwise, and with displacements r_a randomly distributed in the unit area A_o. For this extreme case (multiple scattering limit) the description of the dechanneling contribution due to point defects can be tested. Fig.1.52 shows the minimum yield $\chi_{min}(z) = \chi(0,z)$ for 1.4 MeV He ions incident on <111> silicon with an amorphous surface layer of about 0.5 µm. From the good agreement of the calculated minimum yield (for $z > d_{am}$) with the experimental data for this extreme situation we expect the dechanneling to be well described also for other point defect distributions.

The description of dechanneling due to dislocations can be tested for a simple and well known distribution of dislocations as for instance found in <100> Si crystals implanted with 10^{16} As^+ cm^{-2} of 80 keV and annealed in a dry oxygen atmosphere at 1220 K for 960 min (Ref.[67]). The TEM measurements (Ref.[67]) show a regular two-dimensional misfit dislocation network of pure edge dislocations in a (100) plane at fixed depth (z = 0.175 µm, measured by RBS). In this case the integral density of the dislocations $N_{disl} = \int_0^\infty dz\, n_{disl}(z)$ can be

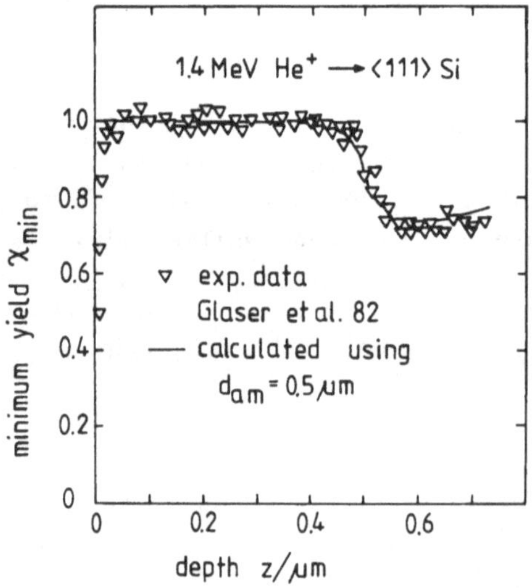

Fig.1.52. Depth dependence of the minimum yield of 1.4 MeV He ions incident on <111> Si with an amorphous layer of thickness d_{am} = 0.5 µm. The results of the calculation within the discontinuous model (Ref.[59]) is compared with the experimental data of Ref.[68].

determined relatively accurately from the TEM micrograph and provides $N_{disl} = 6.2*10^4\ cm^{-1}$. The difference of the RBS minimum yield of the distorted and virgin crystal calculated with these data for the distribution of dislocations is depicted in Fig.1.53. As can be seen there is good agreement with the experimental results (Ref.[67]).

With respect to stacking faults, a direct test of the description of dechanneling is not possible because no example of stacking faults with the density known from other measurements is available.

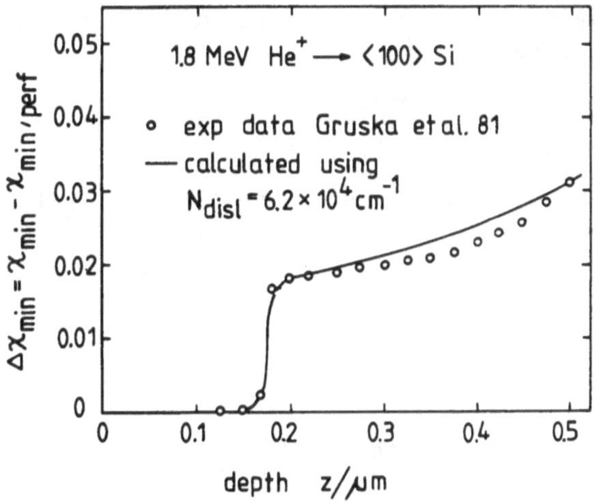

Fig.1.53. Depth dependence of the difference of the minimum yield of 1.8 MeV He ions incident on <100> Si with edge dislocations and the corresponding minimum yield for perfect <100> Si. The dislocations have been produced by implantation of 10^{16} $As^+\ cm^{-2}$ of 80 keV at room temperature and annealing in a dry oxygen atmosphere at 1220 K for 960 min. The results obtained within the discontinuous model (Ref.[63]) are compared with experimental data of Ref.[67].

1.3.4. Two-beam approximation

Especially for the application to defect analysis, Bøgh [69] developed a model of dechanneling which can be considered as a limit of the discontinuous model using only two groups of ions ($i_{max} = 2$). Therefore, in the following, this model is referred to as two-beam approximation. The relative parts of ions of group 1 $g_1(z) = 1-\chi_r(z)$ and group 2 $g_2(z) = \chi_r(z)$ are called aligned and random beam, respectively. Neglecting the probability for the transition from random to aligned beam, the dechanneling matrix is given by only one parameter, the probability per unit depth P for the transition from aligned to random beam ($Q_{21} = P$, $Q_{12} = 0$, $Q_{11} = -P$, $Q_{22} = 0$), where $P = P_{perf} + P_{def}$ is a sum of the statistically independent contributions from the perfect crystal and the defects. In this way the master equation (1.246) reduces to

$$\frac{d\chi_r(z)}{dz} = [P_{perf}(z) + P_{def}(z)]\,[1-\chi_r(z)] \qquad (1.251)$$

and can be solved immediately

$$\chi_r(z) = 1 - [1-\chi_r(0)]\, e^{-\int_0^z dz'[P_{perf}(z') + P_{def}(z')]} \quad . \qquad (1.252)$$

With the simplified probability Π_i for hitting a thermally vibrating lattice atom ($\Pi_1 = 0$, $\Pi_2 = 1$, see Eq.(1.219)), the minimum yield for a perfect crystal Eq.(1.218) is given by

$$\chi_{min,perf}(z) = \chi_{r,perf}(z) = 1 - [1-\chi_r(0)]\, e^{-\int_0^z dz' P_{perf}(z')} \quad . \qquad (1.253)$$

The advantage of this model is that the unknown probability per unit depth P_{perf} in Eq.(1.252) can be replaced by the minimum yield of the perfect crystal (Ref.[69]) which is available from the experiment in most cases. This provides

$$\chi_r(z) = \chi_{min,perf}(z) + [1-\chi_{min,perf}(z)][1-e^{-\gamma(z)}] \quad . \qquad (1.254)$$

The integral over the probability P_{def} per depth

$$\gamma(z) = \int_0^z dz' P_{def}(z') \qquad (1.255)$$

is usually called dechanneling parameter or dechanneling probability and $\tilde{P} = 1-\exp(-\gamma(z))$ can be understood as the total probability for the transition of an ion from the aligned beam to the random beam, in the following called dechanneling function.

(i) Point defects

For point defects with randomly distributed positions within the lattice cell ($\Pi_1^{(pd)} = \Pi_2^{(pd)} = 1$, see Eq.(1.250)) the minimum yield according to Eq.(1.247) is given by

$$\chi_{min}(z) = [1-\chi_r(z)]\, n_{pd}(z) + \chi_r(z) \quad . \qquad (1.256)$$

This means that ions of the random beam can be backscattered from all regular and displaced atoms of the target while ions of the aligned beam are backscattered only from the displaced atoms. The dechanneling parameter $\gamma(z)$ as calculated from the probability for single scattering is valid only for very small integral densities $\int_0^z dz'\, n_{pd}(z')$ of point defects. In most cases of interest, multiple scattering processes must be taken into account (Refs.[70, 71]). Therefore the probability $\tilde{P}$ of the transition of an ion from the aligned beam to the random beam is determined directly from the multiple scattering angular distribution $P(\varphi,z)\,2\pi\varphi\,d\varphi$, as has been introduced in Sect.1.2.5. In this way the dechanneling function $\tilde{P}$ is equal to the relative part of ions which become deflected to angles (after traversing a thickness z) larger than a critical angle ψ_c for channeling. Using the universal angular distribution (Eq.(1.133)) we find for axial channeling

$$1-e^{-\gamma(z)} = \int_{\psi_c}^{\infty} 2\pi\varphi\,d\varphi\,P(\varphi,z) = \tilde{P}(\tilde{\psi}_c,\tau) \quad , \tag{1.257}$$

with

$$\tilde{\psi}_c = \frac{Ea_{TF}}{2Z_1Z_2}\,\frac{4\pi\varepsilon_0}{e^2}\,\psi_c = c_\varphi\,\psi_c \tag{1.258}$$

and

$$\tau = \pi a_{TF}^2 \int_0^z dz' n_{pd}(z') = \pi a_{TF}^2\, N_{pd}(z) = c_z\, N_{pd}(z) \quad . \tag{1.259}$$

The probability $\tilde{P}$ is a universal function of the parameters $\tilde{\psi}_c$ and τ, and is depicted in Fig.1.54 and tabulated in Appendix 9. The depth dependence is introduced by the integral density $N_{pd}(z)$ of the point defects.

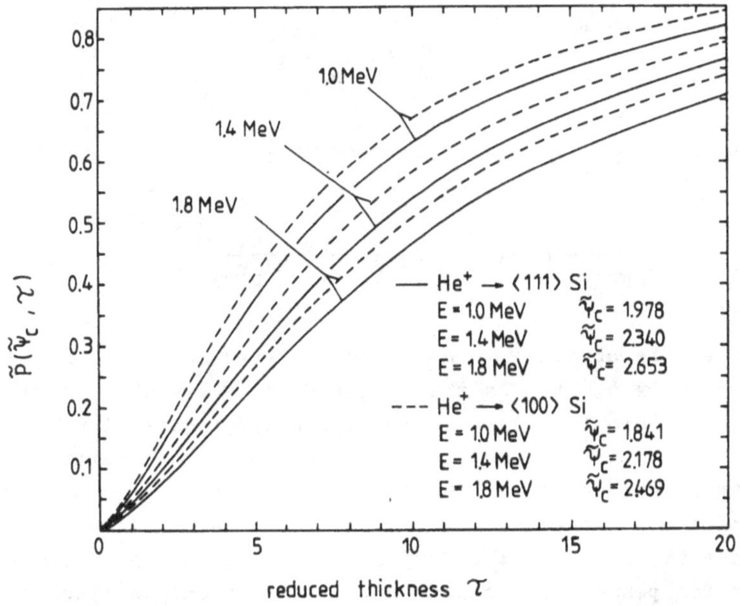

Fig.1.54. Dechanneling function obtained from the multiple scattering distribution of Meyer [38] for some values of the reduced critical angle ψ_c corresponding to He ions of different energies incident on <111> and <100> Si ($\psi_c = \psi_1/\sqrt{2}$ is used)

(ii) Extended defects

Neglecting the very small direct backscattering contribution of stacking faults and dislocations, the minimum yield for crystals with such extended defects is (according to Eq.(1.247)) given by the random beam

$$\chi_{min}(z) = \chi_r(z) = 1 - [1-\chi_{min,perf}(z)]\, e^{-\gamma(z)} \tag{1.260}$$

with $\gamma(z)$ being the sum of the dechanneling probabilities of the different extended defects.

For <u>dislocations</u> with a density $n_{disl} \leq 10^{11}\ cm^{-2}$ (the distortion fields do not overlap) the dechanneling probability is proportional to the integral density N_{disl} of dislocations

$$\gamma_{disl}(z) = \lambda \int_0^z dz' n_{disl}(z') = \lambda\, N_{disl}(z) \quad , \tag{1.261}$$

with λ usually being called dechanneling diameter. Using a static description of dechanneling (equilibrium between centrifugal force in the curved channel and the channel force) Quéré [60] found the dechanneling diameter to be proportional to $E^{1/2}$

$$\lambda = \lambda_o\ (E/MeV)^{1/2} \quad , \tag{1.262}$$

with

$$\lambda_{o,ax} = \lambda_{o,ax,Q} = \left(\frac{bda_{TF}\ MeV\ 4\pi\,\varepsilon_o}{\beta_{ax} Z_1 Z_2\ e^2}\right)^{1/2} = 191.7 \left(\frac{bda_{TF}}{\beta_{ax} Z_1 Z_2 a_o}\right)^{1/2} \tag{1.263}$$

for axial channeling and

$$\lambda_{o,pl} = \lambda_{o,pl,Q} = \left(\frac{b\ MeV\ 4\ \ _o}{\beta_{pl} Z_1 Z_2 Nd_p e^2}\right)^{1/2} = 191.7 \left(\frac{b}{\beta_{pl} Z_1 Z_2 Nd_p a_o}\right)^{1/2} \tag{1.264}$$

for planar channeling. The absolute value of the Burgers vector $\vec{b}$ of the dislocation is denoted by b, a_{TF} is the Thomas-Fermi screening length (Eq.(1.29)), d is the interatomic spacing for a low index direction, d_{pl} is the spacing between low index planes, β_{ax} is a geometrical parameter for axial channeling which depends on the type of the dislocation and on the direction $\vec{e}$ of the channel with respect to the direction $\vec{l}$ of the dislocation and the direction of the Burgers vector $\vec{b}$, and β_{pl} is a geometrical factor for planar channeling which depends on the type of the dislocation and on the direction of the normal $\vec{n}$ of the planar channel with respect to the directions of the vectors $\vec{b}$ and $\vec{l}$. For several different cases the values of β_{ax} and β_{pl} are presented in Table 1.1.

TABLE 1.1

Parameters β_{ax} and β_{pl} for different types of dislocations after Quéré [60].

type of dislocation	β_{ax}	β_{pl}
screw dislocation ($\vec{b} \parallel \vec{l}$)		
$\zeta = \sphericalangle(\vec{b},\vec{e})$, $\zeta_{pl} = \sphericalangle(\vec{b},\vec{n})$	$3.91/\sin^3\zeta$	$3.93/[\cos\zeta_{pl}(1-0.36\sin^3\zeta_{pl})^2]$
$\vec{b} \perp \vec{e}$	3.91	
$\vec{b} \parallel \vec{n}$		3.93
average over all directions of $\vec{l}$	12.5	8.6
edge dislocation ($\vec{b} \perp \vec{l}$)		
$\vec{b} \perp \vec{e}$, $\vec{l} \perp \vec{e}$	1.45	
$\vec{b} \parallel \vec{e}$, $\vec{l} \perp \vec{e}$	1.41	
average over all directions of $\vec{b}$ and $\vec{l}$	4.57*	
$\vec{b} \parallel \vec{n}$, $\vec{e} \perp \vec{l}$		1.45**
$\vec{b} \perp \vec{n}$, $\vec{e} \perp \vec{l}$		1.41**

* value of Quéré assuming the average over all directions of $\vec{l}$ to provide the same factor as for screw dislocations

** values from Ref.[67]

Because only a part of the ion trajectories is influenced by the distortion of the strings or planes, the dechanneling diameter is smaller than the values given by the formulae of Quéré (Eqs.(1.263) and (1.264)). This effect was investigated by van Vliet [72] using computer simulation. For large ion energies ($E > Z_1 Z_2 e^2 d/(4\pi\varepsilon_o a_{TF}^2)$) the results of van Vliet can be expressed by

$$\lambda_{o,ax} = Nd\ r_{crit} b \sin\zeta\ \lambda_{o,ax,Q} \qquad (1.265)$$

for axial channeling (Ref.[72]) and

$$\lambda_{o,pl} = \frac{b \cos\zeta_{pl}}{2(d_p - 2y_{crit})}\ \lambda_{o,pl,Q} \qquad (1.266)$$

for planar channeling (Ref.[67]), where $\zeta = \sphericalangle(\vec{b},\vec{e})$, $\zeta_{pl} = \sphericalangle(\vec{b},\vec{n})$, and the critical distances r_{crit} and y_{crit} can be approximated by

$r_{crit} = (r^2_{min}(E_{\perp c}) + a^2_{TF})^{1/2} \approx 1.2a_{TF}$ and $y_{crit} \approx a_{TF}$, respectively. For lower energies van Vliet found the dechanneling diameter to be independent of the ion energy.

The dechanneling due to stacking faults is mainly caused by displaced strings (see Sect.1.3.3.2.1) and displaced planes for axial and planar channeling, respectively.

In the axial case, the dechanneling probability of ions being scattered on $n_{dst}(z)$ displaced strings per unit volume is given by

$$\gamma_{dst}(z) = \sigma_{dst} \int_0^z dz' n_{dst}(z') = \sigma_{dst} N_{dst}(z) \quad , \tag{1.267}$$

where σ_{dst} is the dechanneling cross section of one displaced string. Considering a fault plane as a new interior surface as suggested by Campisano et al. [73] the dechanneling cross section of one displaced string can be determined by the minimum yield of the corresponding perfect crystal at the surface.

$$\sigma_{dst} = A_o \chi_{min,perf}(0) \quad . \tag{1.268}$$

Within the continuum approximation (Ref.[74]) we obtain

$$\sigma_{dst} = \pi (\rho_\perp^2 + a^2_{TF}) \quad , \tag{1.269}$$

with $\rho_\perp$ being the mean thermal vibration amplitude as introduced in Eq.(1.169). A fit of the Monte Carlo computer simulation results (Ref.[75]) provides

$$\sigma_{dst} = \pi (3\rho_\perp^2 + 0.2\, a^2_{TF}) \quad . \tag{1.270}$$

The best way is to use the measured value of $\chi_{min,perf}(0)$ directly.

In the planar case, the dechanneling probability of ions being scattered on displaced planes reads

$$\gamma_{dpl}(z) = S_{dpl} \int_0^z dz'\, l_{dpl}(z') = S_{dpl} L_{dpl}(z) \quad , \tag{1.271}$$

where $L_{dpl}(z)$ is the total projected edge length (on the plane normal to the direction of incidence) of all displaced planes per unit volume at depth z and S_{dpl} is the dechanneling cross section per unit projected edge length of a displaced plane. A similar consideration as performed for the axial case (Eq.(1.268)) provides

$$S_{dpl} = d_p \chi_{min,perf}(0) \quad . \tag{1.272}$$

Using the continuum approximation for the minimum yield at the surface, the dechanneling cross section per unit projected length of a displaced plane is given by

$$S_{dpl} = 2(a_{TF} + \rho_{\perp}/\sqrt{2}) \quad . \tag{1.273}$$

1.3.5. Defect analysis

As described in Sect.1.3.3, the Rutherford backscattering minimum yield can be calculated straightforwardly if all parameters of the defect distributions are known. Conversely, the determination of the defect distribution from the measured minimum yield is only possible by iterative procedures and, in general, additional information (e.g. about the type of the defect) is necessary.

1.3.5.1. Analysis of point defects

(i) Two-beam approximation

This approximation can be applied only for point defects with randomly distributed positions r_a. In this case Eq.(1.256) provides

$$n_{pd}(z) = \frac{\chi_{min}(z) - \chi_r(z)}{1 - \chi_r(z)} \quad , \tag{1.274}$$

with

$$\chi_r(z) = \chi_{min,perf}(z) + [1-\chi_{min,perf}(z)]\, \tilde{P}(\tilde{\psi}_c, \tau(N_{pd})) \tag{1.275}$$

and $\tau = \pi\, a_{TF}^2\, N_{pd}(z)$ being proportional to the integral density of defects $N_{pd}(z) = \int_0^z dz' n_{pd}(z')$. If both the minimum yield for the crystal with defects and for the perfect crystal can be measured (e.g. after and before irradiation in the case of radiation induced point defects) and $\tilde{P}(\tilde{\psi}_c, \tau)$ is known (see Appendix 9) the relative density of point defects $n_{pd}(z)$ can be calculated iteratively using Eqs.(1.274) and (1.275). Starting from $z = 0$ with $N_{pd}(0) = 0$ we find $n_{pd}(0)$ from Eq.(1.274). For the next step, $N_{pd}(\Delta z)$ is approximated by $N_{pd}(0) + n_{pd}(0)\Delta z$ which provides $\chi_r(\Delta z)$ and $n_{pd}(\Delta z)$ and so on. In this way, $n_{pd}(z)$ can be calculated step by step.

(ii) Discontinuous model

For analysis of crystals containing only point defects with known positions r_a, Eq.(1.247) can be rearranged

$$n_{pd}(z) = \frac{\chi_{min}(z) - \sum_i g_i(z)\, \Pi_i}{\sum_i g_i(z)\, [\Pi_i^{(pd)}(r_a) - \Pi_i]} \quad . \tag{1.276}$$

This is no explicit expression for $n_{pd}(z)$ because $g_i(z)$ also depends on the relative point defect density according to the master equation (from Eq.(1.246))

$$\frac{dg_i(z)}{dz} = \sum_j \left\{ Q_{ij}^{(e)} + [1-n_{pd}(z)]\, Q_{ij}^{(n)} + n_{pd}(z)\, Q_{ij}^{(pd)} \right\} g_j(z) \quad , \qquad (1.277)$$

but Eqs.(1.276) and (1.277) can be solved iteratively in a similar way as described above within the two-beam approximation. The point defect density $n_{pd}(0)$ for $z = 0$ can be calculated immediately because the initial distribution $g_i(0)$ is independent of n_{pd} and given by Eq.(1.217). With $n_{pd}(0)$ we find g_i for a small depth interval Δz from a second order series expansion

$$g_i(\Delta z) = g_i(0) + \left.\frac{dg_i(z)}{dz}\right|_{z=0} \Delta z + \frac{1}{2} \left.\frac{d^2 g_i(z)}{dz^2}\right|_{z=0} \Delta z^2 \qquad (1.278)$$

where the coefficients are given by Eq.(1.277) and its first derivative. With $g_i(\Delta z)$ Eq.(1.276) provides $n_{pd}(\Delta z)$. By successive continuation of this procedure we find the relative point defect density $n_{pd}(z)$ for all depths $z = n\,\Delta z$ $(n = 0,1,2...)$.

1.3.5.1.1. Heavily damaged crystals

If the Rutherford backscattering minimum yield for a damaged crystal shows a pronounced profile corresponding to the damage profile, in the following such crystals are referred to as heavily damaged crystals. In this case the direct backscattering part of the minimum yield dominates the dechanneling part and the displaced atoms may be assumed to be randomly distributed within the lattice cell because preferred positions are improbable. The relative point defect density $n_{pd}(z)$ can readily be calculated from the measured minimum yield $\chi_{min}(z)$ using the two-beam approximation (Eqs.(1.274, 1.275)) with $\tilde{P}$ being determined by multiple scattering (Eqs.(1.257, 1.258, 1.259)) or using the discontinuous model (Eqs.(1.276, 1.277)) with Π_i for randomly distributed r_a given by Eq.(1.250). Within the two-beam approximation, uncertainties are introduced by uncertainties of the critical angle for channeling ψ_c used in the dechanneling function $\tilde{P}(\tilde{\psi}_c,\tau)$. For limited defect distributions (such as those caused by implantation) this problem can usually be overcome by adjusting ψ_c in this way, so that $n_{pd}(z) = 0$ is provided for depths much larger than the range of the defects. The two-beam approximation results obtained in this way agree with the relative point defect concentration $n_{pd}(z)$ as provided by the discontinuous model (without any adjustable parameter). Examples are presented in Sect.5.2.1.1.

1.3.5.1.2. Weakly damaged crystals

Damaged crystals which provide a Rutherford backscattering minimum yield weakly and continuously increasing with increasing depth, with the enhancement of the minimum yield being of the same order of magnitude as the minimum yield of the perfect crystal, are referred to in the following as weakly damaged crystals. In this case the minimum yield is mainly determined by its dechanneling part and preferred positions of the displaced atoms within the lattice cell are possible. Therefore, besides $n_{pd}(z)$, the position r_a of the displaced atoms within the lattice cell (or its distribution) must be determined, too. For this purpose the measured minimum yield $\chi_{min}(z)$ as a function of depth z no longer provides enough information. Additional information may be available, for example, from the temperature dependence or from the energy dependence of the minimum yield.

(i) Temperature dependence

Let us consider crystals with point defects (displaced atoms) with only one fixed position r_a within the lattice cell. As was shown by theoretical investigations, for a given defect density $n_{pd}(z)$ the absolute value of the minimum yield and its temperature dependence are considerably different for different positions r_a of the displaced atoms. Moreover, a variation of the density $n_{pd}(z)$ for a given value of r_a changes the absolute value of the minimum yield but hardly influences the temperature dependence of χ_{min}. Therefore the temperature dependence of the minimum yield for a crystal with point defects is characteristic for the position r_a of these defects rather than for their density n_{pd}. This provides the possibility of determining both the density $n_{pd}(z)$ and the position r_a from the measured temperature dependence of the minimum yield $\chi_{min}(z,T)$ for different temperatures T. The procedure is described in the following.

In the first step we start from the measured minimum yield $\chi_{min}(z,T_f)$ at a fixed temperature T_f. Assuming a set of different values r_a the corresponding point defect densities $n_{pd}(z)$ are calculated from $\chi_{min}(z,T_f)$ according to the procedure given by Eqs.(1.276) and (1.277). This is demonstrated in Fig.1.55 for point defects in silicon generated by irradiation with He ions and subsequent thermal annealing. Fig.1.55a shows the depth dependence of the minimum yield $\chi_{min}(z,T_f)$ measured at a temperature T_f = 160 K for 1 MeV H ions incident on the <111> direction of the irradiated silicon. The corresponding minimum yield $\chi_{min,perf}(z,T_f)$ belonging to the perfect crystal is also depicted for

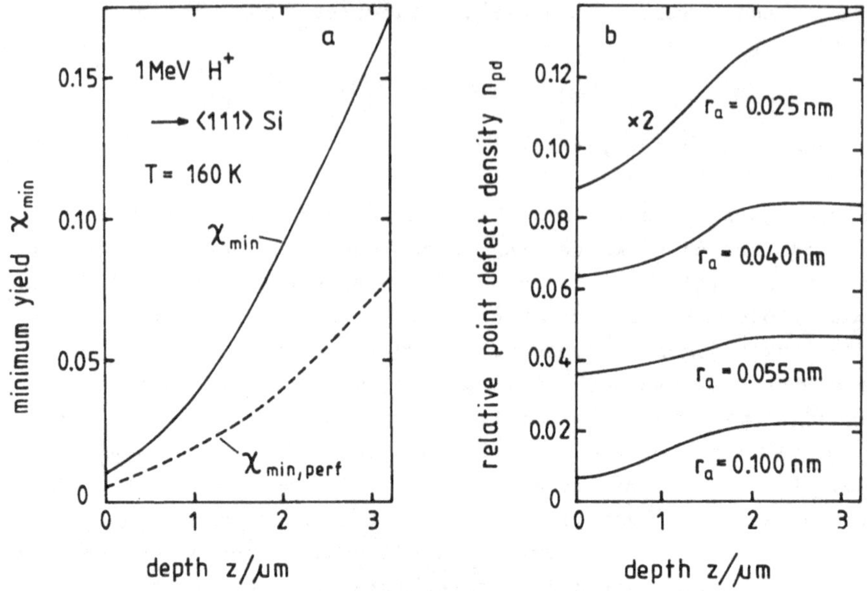

Fig.1.55. (a) Depth dependence of the minimum yield of 1.0 MeV H ions incident on perfect <111> Si at a temperature T = 160 K (dashed line) and the corresponding curve for <111> Si with point defects (solid line). (b) Relative density of point defects calculated from the minimum yield (a) for different positions r_a of the displaced atoms (discontinuous model)

comparison. As shown in Fig.1.55b the point defect densities $n_{pd}(z)$ calculated from $\chi_{min}(z,T_f)$ differ heavily for different positions r_a. Without any additional information it is not possible to find the positions r_a and the density $n_{pd}(z)$ of the present defects.

In the second step, the temperature dependence of the minimum yield is calculated for different positions r_a using the corresponding densities $n_{pd}(z)$ (Fig.1.55b) as have been determined from $\chi_{min}(z,T_f)$. The influence of the defects on the temperature dependence is more distinctly seen for the difference

$\Delta\chi_{min}(z,T) = \chi_{min}(z,T) - \chi_{min,perf}(z,T)$ of the minimum yields for the crystal with defects and for the perfect crystal. Therefore the temperature dependence of $\Delta\chi_{min}(z_f,T)$ at a fixed depth $z_f = 2\,\mu m$ is depicted in Fig.1.56a for different positions r_a of the point defects. According to the procedure of determining the defect densities, the difference of the minimum yields $\Delta\chi_{min}(z_f,T_f)$ at $T = T_f$ is equal for all values of r_a. As can be seen, the temperature dependence of $\Delta\chi_{min}$ varies strongly with r_a. There is even a change from increasing $\Delta\chi_{min}$ for large r_a to decreasing $\Delta\chi_{min}$ for small r_a. The data are shown in a different manner in Fig.1.56b. As a measure of the increase of $\Delta\chi_{min}$ with increasing temperature, the second difference

$\Delta^2\chi_{min} = \Delta\chi_{min}(z_f,280\text{ K}) - \Delta\chi_{min}(z_f,40\text{ K})$ for a fixed depth z_f is

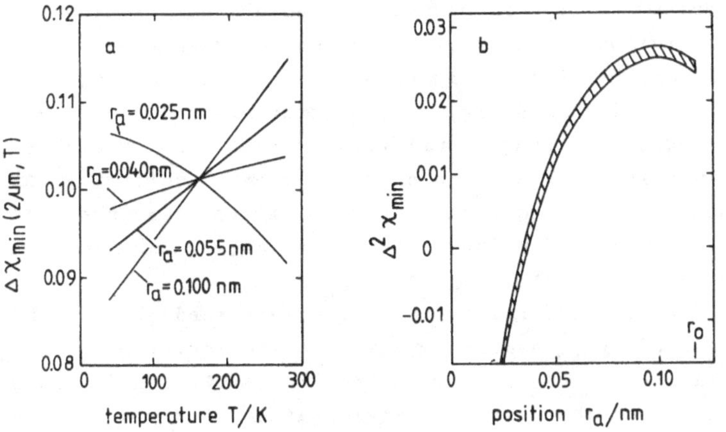

Fig.1.56. (a) Temperature dependence of the difference of the minimum yield $\Delta\chi_{min} = \chi_{min} - \chi_{min,perf}$ at fixed depth $z = 2\,\mu m$ for different values of r_a. The results are obtained using the point defect densities shown in Fig.1.55b. (b) Second difference of the minimum yield (see text) as a function of the position r_a of the displaced atoms

depicted as a function of the position r_a. There is a strong dependence of $\Delta^2\chi_{min}$ on the position for small r_a up to about 0.07 nm and a weak dependence for large r_a.

The different temperature dependence of $\Delta\chi_{min}$ for different r_a may be understood in the following way. In a perfect crystal, dechanneling is caused by the combined action of electronic and nuclear scattering. This may be simplified by two steps in the increase of the transverse energy $E_\perp$. Initially, most of the ions have a very small transverse energy (see Fig.1.26) which is increased by electronic scattering up to a medium value in the order of $E_{\perp c} = \frac{1}{2} E\psi_1^2$. For $E_\perp \geq E_{\perp c}$ nuclear scattering becomes effective and is responsible for a further increase of $E_\perp$. Including point defects with larger r_a (located near the middle of the channel) there is an additional dechanneling contribution which is most effective for ions with small transverse energies $E_\perp < E_{\perp c}$ similar to the electronic contribution. Therefore more ions with medium transverse energies are available for nuclear scattering. Because of the resulting enhanced influence of the nuclear scattering, the temperature dependence of the minimum yield is increased by such defects with respect to the perfect crystal. The result is a positive temperature dependence of $\Delta\chi_{min}$ ($\Delta^2\chi_{min} > 0$) as shown in Fig.1.56a. In the case of point defects with small r_a (displaced atoms are near the strings of atoms) the situation is changed. The displacement of the atoms is on the order of the thermal displacement and therefore the dechanneling contribution of such defects is more comparable to that of nuclear scattering. In this case the increase of the transverse energy up to medium values $E_{\perp c}$ is mainly due to electronic scattering (as for the perfect crystal) while the further enhancement of $E_\perp$ is caused by both the temperature dependent nuclear scattering and scattering on weakly displaced atoms which is nearly independent of the temperature. With respect to the temperature dependence of dechanneling, the scattering by defects and the nuclear scattering are no longer amplifying but rival processes. Therefore the dechanneling contribution due to nuclear scattering is relatively suppressed. The consequence is a temperature dependence of dechanneling weaker than that for the perfect crystal, which means a negative temperature dependence of $\Delta\chi_{min}$ ($\Delta^2\chi_{min} < 0$).

As can be seen in Fig.1.56 and has already been mentioned above, the temperature dependence of $\Delta\chi_{min}$ depends greatly on the position r_a for small and medium values of r_a ($r_a \leq 0.07$ nm for <111> silicon). This effect offers a sensitive method for the determination of r_a as far as the displaced atoms are not positioned near the middle of the channel. To this end, the minimum yield for the perfect crystal and for the crystal with defects must be measured at different temperatures. By comparison of the resulting $\Delta^2\chi_{min}$ with the calculated $\Delta^2\chi_{min}$ as a function of r_a (Fig.1.56b) we find the value of r_a. For the position r_a known, the corresponding point defect density $n_{pd}(z)$ can be determined immediately. Examples are given in Sect.5.3.3.2.

The main idea of the procedure described above is the following: There are a lot of point defect distributions given by different densities $n_{pd}(z)$ and positions r_a (Fig.1.55) which provide the same minimum yield measured at a fixed temperature T_f but only one of them also provides the measured temperature dependence.

(ii) Energy dependence

As was shown by calculations for 0.5 - 2 MeV H ions incident on <111> and <110> silicon, the difference of the minimum yields $\Delta\chi_{min}$ decreases with increasing energy. Fitting by a power of E, the energy dependence varies between $E^{-0.65}$ and $E^{-0.95}$ for different energy regions and different values of n_{pd} and r_a. In contrast to the temperature dependence, the energy dependence of $\Delta\chi_{min}$ is not characteristic of the position r_a of the displaced atoms.

In order to estimate the variation of the energy dependence due to the variation of the density n_{pd}, let us consider point defects which are randomly distributed within the lattice cell. Within the two-beam approximation the dechanneling due to such defects is described by multiple scattering. The energy dependence of the dechanneling function varies between E^{-1} fcr small n_{pd} (single scattering limit) and $E^{1/2}$ for larger n_{pd} (plural scattering limit). This estimation indicates that the energy dependence of $\Delta\chi_{min}$ (approximately equal to that of the dechanneling function) is considerably influenced by the value of n_{pd}. The full variation of the calculated energy dependence of $\Delta\chi_{min}$ ($E^{-0.65}$... $E^{-0.95}$) can be achieved by a variation of the values of r_a as well as by a variation of the density n_{pd}. Therefore the energy dependence of the minimum yield provides no useful information about the position of point defects within the lattice cell.

1.3.5.2. Analysis of crystals with one kind of extended defect

In contrast to the situation for point defects, the direct backscattering contribution of extended defects to the RBS minimum yield is mostly negligible with respect to its dechanneling contribution. Therefore the first derivative of the minimum yield is important for the determination of the defect densities rather than its absolute value. In this section, crystals with only one kind of extended defects predominating are considered.

(i) Two-beam approximation

According to Eq.(1.260) the dechanneling probability for crystals with extended defect can be determined immediately from the measured minimum yield $\chi_{min}(z)$ and the corresponding $\chi_{min,perf}(z)$ measured for the perfect crystal

$$\gamma(z) = \ln\left(\frac{1-\chi_{min,perf}(z)}{1-\chi_{min}(z)}\right) . \tag{1.279}$$

If the crystal contains predominantly only one kind of extended defect, the corresponding defect density can be determined from the dechanneling probability $\gamma(z)$. For <u>dislocations</u> Eq.(1.261) provides

$$n_{disl}(z) = \frac{1}{\lambda}\frac{d\gamma(z)}{dz} \tag{1.280}$$

and in the case of <u>displaced strings</u> (stacking faults, see Eq.(1.267)) we find

$$n_{dst}(z) = \frac{1}{\sigma_{dst}}\frac{d\gamma(z)}{dz} \tag{1.281}$$

(ii) Discontinuous model

After rearranging the first derivative of Eq.(1.247) we find for crystals containing only <u>dislocations</u>

$$n_{disl}(z) = \frac{\frac{d\chi_{min}(z)}{dz} - \sum\limits_{i,j} n_i \left[Q_{ij}^{(e)} + Q_{ij}^{(n)}\right] g_j(z)}{\sum\limits_{i,j} n_i\, Q_{ij}^{(disl)}\, g_j(z)} , \tag{1.282}$$

with $g_i(z)$ determined by

$$\frac{dg_i(z)}{dz} = \sum_j \left\{ Q_{ij}^{(e)} + Q_{ij}^{(n)} + n_{disl}(z)\, Q_{ij}^{(disl)} \right\} g_j(z) \quad , \tag{1.283}$$

according to Eq.(1.246). For the kind of dislocation and its Burgers vector known (e.g. from transmission electron microscopy) the dechanneling matrix $Q_{ij}^{(disl)}$ can be calculated using Eq.(1.241) and the density of dislocations $n_{disl}(z)$ can be determined iteratively by Eqs.(1.282) and (1.283). The initial distribution $g_i(0)$ is known and provides $n_{disl}(0)$. With $n_{disl}(0)$ we find $g_i(\Delta z)$ at the next depth step from Eq.(1.283) and $n_{disl}(\Delta z)$ according to Eq.(1.282). Continuing this procedure we get the dislocation density $n_{disl}(z)$ for all depth z.

If the dechanneling due to stacking faults is only determined by the contribution from the <u>displaced strings</u>, their density can be calculated in an analogous way as the dislocation density. From Eq.(1.247) we find

$$n_{dst}(z) = \frac{\dfrac{d\chi_{min}(z)}{dz} - \sum\limits_{i,j} \Pi_i \left[Q_{ij}^{(e)} + Q_{ij}^{(n)} \right] g_j(z)}{\sum\limits_{i,j} \Pi_i \, Q_{ij}^{(dst)} \, g_j(z)} \quad , \tag{1.284}$$

with g_i given by

$$\frac{dg_i(z)}{dz} = \sum_j \left\{ Q_{ij}^{(e)} + Q_{ij}^{(n)} + n_{dst}(z)\, Q_{ij}^{(dst)} \right\} g_j(z) \quad . \tag{1.285}$$

Examples for analysis of extended defects within the two-beam approximation and the discontinuous model are given in Sect.5.2.2.

1.3.5.3. Analysis of crystals with complex defect structures

Mostly real crystals or damaged crystals do not contain only one kind of defect. Let us consider a crystal with different defect structures (e.g. point defects, dislocations, stacking faults, twin lamellae). With respect to dechanneling, these defect structures are described by the density $n_{pd}(z)$ and position r_a (or a distribution of r_a) of the point defects, by the dislocation density $n_{disl}(z)$, by the density of displaced strings $n_{dst}(z)$ and by some geometrical parameters of the extended defects (e.g. directions of the Burgers vector and the dislocation line with respect to the channel axis, mean size of the stacking planes, thickness of the twin lamellae, see also Sect.1.3.3.2.1). Information about the presence of dislocations, stacking faults and twin lamellae and about their geometrical parameters can be obtained for instance from TEM measurements. For known geometrical parameters of the extended defects, the RBS minimum yield

$\chi_{min}(z)$ can be calculated for given $n_{pd}(z)$, $n_{disl}(z)$, $n_{dst}(z)$, and r_a but conversely the defect densities and the position of point defects r_a cannot be determined only from the depth dependence of the minimum yield. There is no general procedure of defect analysis by dechanneling for crystals with such complex defect structures, but in some special cases all the defect parameters can be determined iteratively using additional experimental data. Two situations are discussed in the following.

(i) First, let us consider a crystal containing all kinds of defects mentioned above, however, the dechanneling contribution of one of them is dominating and all the defect densities are approximately proportional to each other $n_{pd}(z) \sim n_{disl}(z) \sim n_{dst}(z)$, which is true for instance for stacking faults and twin lamellae as discussed in Sect.1.3.3.2.1. In this case the following procedure is used. Neglecting all the other defects, the density of the dominating defects is calculated in first approximation from the measured $\chi_{min}(z)$ using Eqs.(1.276), (1.282) or (1.284), respectively. If the factors between the different defect densities are known, they provide the other defect densities in first approximation which are used for determining the second approximation of the density of the dominating defects.

In the case of <u>point defects predominating</u> their density $n_{pd}(z)$ is calculated iteratively for given distributions $n_{disl}(z)$ and $n_{dst}(z)$ by

$$n_{pd}(z) = \frac{\chi_{min}(z) - \sum_i g_i(z)\, \Pi_i}{\sum_i g_i(z)\, [\Pi_i^{(pd)}(r_a) - \Pi_i]} \tag{1.286}$$

(derived from Eq.(1.247), the direct backscattering part from the head atoms of displaced strings is neglected) and by the master equation (1.246)

$$\frac{dg_i(z)}{dz} = \sum_j \Big\{ Q_{ij}^{(e)} + [1-n_{pd}(z)]\, Q_{ij}^{(n)} + n_{pd}(z)\, Q_{ij}^{(pd)} + n_{disl}(z)\, Q_{ij}^{(disl)} + n_{dst}(z)\, Q_{ij}^{(dst)} \Big\}\, g_j(z) \quad . \tag{1.287}$$

If <u>displaced strings</u> are the <u>dominating</u> defects, their density n_{dst} can be determined iteratively (analogously to Eq.(1.284)) for given $n_{pd}(z)$ and $n_{disl}(z)$ by

$$n_{dst}(z) = \frac{\frac{d\chi_{min}(z)}{dz} - \sum_i g_i(z)\, \frac{d\tilde{\Pi}_i(z)}{dz} - \sum_{i,j} \tilde{\Pi}_i(z)\, \tilde{Q}_{ij}(z)\, g_j(z)}{\sum_{i,j} \tilde{\Pi}_i(z)\, Q_{ij}^{(dst)}\, g_j(z)} \tag{1.288}$$

with

$$\tilde{\Pi}_i(z) = \Pi_i + n_{pd}(z)\,[\Pi_i^{(pd)}(r_a) - \Pi_i] \qquad (1.289)$$

(the direct backscattering part from the head atoms of the displaced strings is neglected) and

$$\tilde{Q}_{ij}(z) = Q_{ij}^{(e)} + [1-n_{pd}(z)]\,Q_{ij}^{(n)} + n_{pd}(z)\,Q_{ij}^{(pd)} + n_{disl}(z)\,Q_{ij}^{(disl)}. \qquad (1.290)$$

Similar equations hold for the analysis of crystals with dislocations being the dominating defects. They are provided from Eqs.(1.288, 1.289, 1.290) by exchanging the indices "dst" and "disl". The depth dependence of $g_i(z)$ is given by Eq.(1.287) in all cases.

If the ratios of the defect densities are not known from other measurements, some information can be obtained from the measured energy dependence of the minimum yield because there is a different energy dependence of the minimum yield for different kinds of defects (point defects: $E^{-1/2}...E^{-1}$, dislocations: $E^{1/2}$, displaced strings: E^{0}). This is possible if the dechanneling contributions from the different kinds of defects do not differ too much, otherwise the energy dependence introduced by the dominating defects is hardly changed. If point defects dominate, their position r_a can be determined from the temperature dependence of the minimum yield as described in Sect.1.3.5.1.2.

(ii) Now a second example of a complex defect structure is discussed. Let us consider a crystal with one kind of extended defect and point defects of high densities ($n_{pd} \approx 0.1...1$) in a depth interval small compared to the extension in depth of the extended defects. In this case the first approximation of the point defect density $n_{pd}(z)$ is immediately obtained by the pronounced direct backscattering peak (subtracting the dechanneling part of the minimum yield in a simple way). Using this distribution $n_{pd}(z)$ the density of extended defects can be calculated from the measured minimum yield $\chi_{min}(z)$ by the procedure described above. The resulting density of extended defects can be used for the determination of the second approximation of $n_{pd}(z)$ according to Eqs.(1.286) and (1.287) and so on. The defect analysis for different ion energies is useful, in that it may provide consistent results and fewer uncertainties.

It must be emphasized that the analysis of crystals with complex defect structures cannot be performed by straightforward procedures and requires some experience in this field. An example is given in Sect.5.2.2.4.

Chapter 2

EXPERIMENTAL EQUIPMENT

A. Witzmann

This chapter provides an introduction into the main components of the equipment for backscattering analysis. The typical arrangement, as shown schematically in Fig.2.1, consists of devices for ion beam generation and handling (ion source, accelerator, focusing system, magnet, collimator), of the target system (sample holder, current integrator, target chamber), and of the detecting and analyzing system (detector, preamplifier, spectroscopic amplifier, multichannel analyzer).

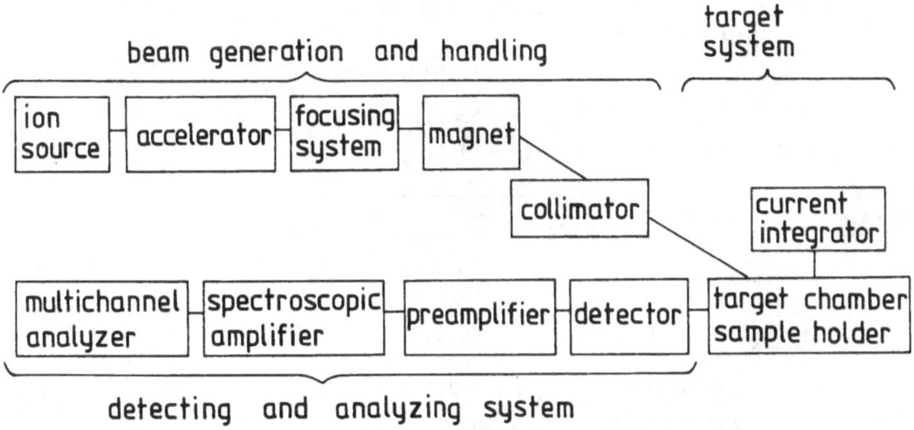

Fig.2.1. Block diagram of the experimental equipment for RBS

An ion beam generated in the ion source and accelerated to an energy in the MeV region impinges on the target after passing devices for mass separation and focusing. The backscattered ions are detected and processed by linear and digital electronic devices, thus providing information about the target by means of their energy spectrum. In the following, the components represented in Fig.2.1 will be discussed in detail and a survey of their mode of operation will be given.

2.1. ION BEAM GENERATION AND HANDLING

2.1.1. Ion source

The ion source has to provide an ion current in the range of 1 μA to 100 μA with an energy width of several 10 eV. Generally, radio-frequency (rf) gas discharge ion sources or for special applications low voltage arc discharge ion sources are applied (Ref.[76]). In both cases the ionization is achieved by collisions of the gas atoms with high-energy electrons.

2.1.1.1. Radio-frequency source

In radio-frequency (rf) sources (see Fig.2.2) an oscillation of the electrons within a gas tube is initiated by applying either an alternating electric field (linear rf discharge) or an alternating magnetic field (ring rf discharge).

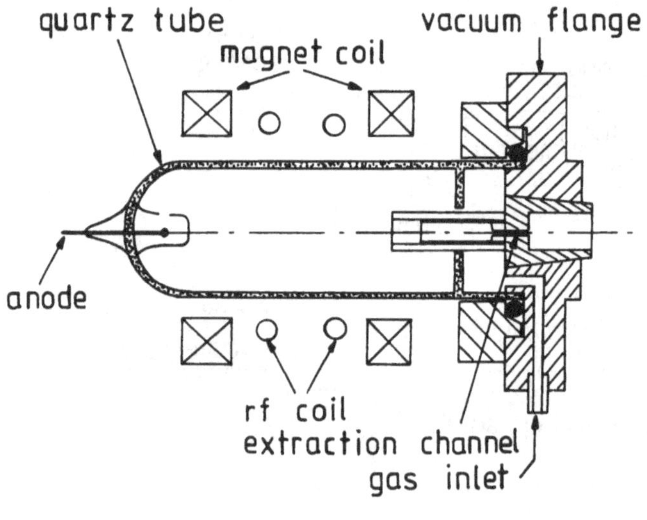

Fig.2.2. Schematic diagram of an rf discharge ion source

The kinetic energy of free electrons is increased by the applied field within the time between two collisions with gas atoms. Therefore the maximum energy of the electrons is determined by the magnitude and frequency of the accelerating field and by the mean free path length of the electrons which is given by the gas pressure. If the energy of the electrons exceeds the ionization energy of the gas atoms, electrons can be released from the gas atoms. By choosing the process parameters in such a way that the generation of electrons exceeds the loss of electrons, mostly due to ejection from the discharge column, a self-consistent gas discharge is obtained. The ions generated within the discharge plasma are extracted by an axial electric field through the exit channel.

Additional application of a static axial magnetic field (magnetic coil in Fig.2.2) increases the plasma density and therefore increases the ion current at constant rf power. Maximum ion currents are obtained using inhomogeneous fields.

The advantages of rf sources are simple design and a high life time exceeding 1000 h.

2.1.1.2. Low voltage arc discharge source

The plasma is generated by a high dc discharge current > 0.1 A flowing through a gas at a pressure in the range from 0.1 to 100 Pa (arc discharge). A stable discharge even at low voltages is achieved by introducing thermionic electrons into the discharge volume (heating filament). An increase of the plasma density is gained by a mechanical narrowing of the discharge tube using an intermediate electrode with a diameter of about 5 mm. In the case of the duoplasmatron (see Fig.2.3)

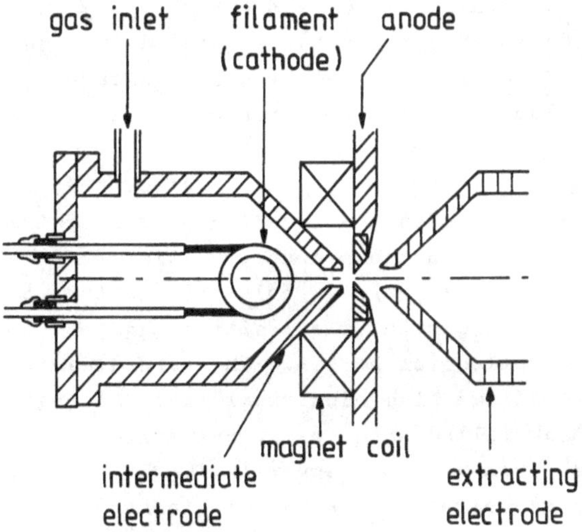

Fig.2.3. Schematic diagram of a duoplasmatron

the plasma is collimated twofold by the additional application of a magnetic field. Usually, the intermediate electrode and the anode are used as magnetic pole pieces. The lifetime of a duoplasmatron is in the order of several 100 h at ion currents up to 100 mA.

2.1.1.3. Ion sources for special applications

In tandem accelerators (see Sect.2.1.2) negatively charged ions are used. Such ions are generated within the high density plasma of a duoplasmatron. Moreover they can be gained from charge transfer processes usually due to the interaction of positively charged ions with alkali metal vapour. Another efficient source for negatively charged ions is based on surface ionization of atoms by sputtering from a cathode that contains the desired elements.

Multiply charged ions may be extracted from high-density plasma sources too. But the generation of multiply charged ions by stripping of electrons when ions pass through a gas or foil target (e.g. C, Ni, Al_2O_3) is more common because of its application in tandem accelerators (see next section).

2.1.2. Accelerator

The ions of the ion source are usually accelerated by an electrostatic accelerator to energies of several MeV. The accelerators differ in their methods of high-voltage generation. In the Crockroft-Walton generator, a voltage cascade (Greinacher circuit) is applied to the multiplication and rectification of an initially alternating voltage of several 100 kV. Voltages up to about 2 MV and currents of about 10 mA are obtained.

The principle of the van de Graaff generator is represented in Fig.2.4. The high-voltage terminal of the van de Graaff generator gains a high positive potential with respect to the ground potential by the accumulation of positive charges. The moving belt consisting of insulating material serves for the transport of the positive charges to the high-voltage terminal via the collection screen. To avoid unwanted electrical discharge and breakdown at the high electrical field strength the generator is mounted in a tank filled with electronegative gases (dry N_2, CO_2, SF_6, $CClF_3$) at a pressure of several 10^5 Pa. Thus, voltages exceeding 10 MV can be generated. In order to control the generator voltage and to stabilize the voltage (fluctuations arise, for example, from charge fluctuations on the belt) a discharge is created between the high-voltage terminal and a corona triode. By controlling the discharge current the generator voltage can be kept constant.

A gradated field distribution within the evacuated accelerator tube is obtained by a column of ring electrodes at equidistant potential levels fixed by a resistor chain. The electrodes are designed in this way to achieve focusing during the acceleration.

The Pelletron generator differs from the van de Graaff generator only in the charging system. In the Pelletron generator the charging belt is replaced by a chain of metal pellets connected by insulating links.

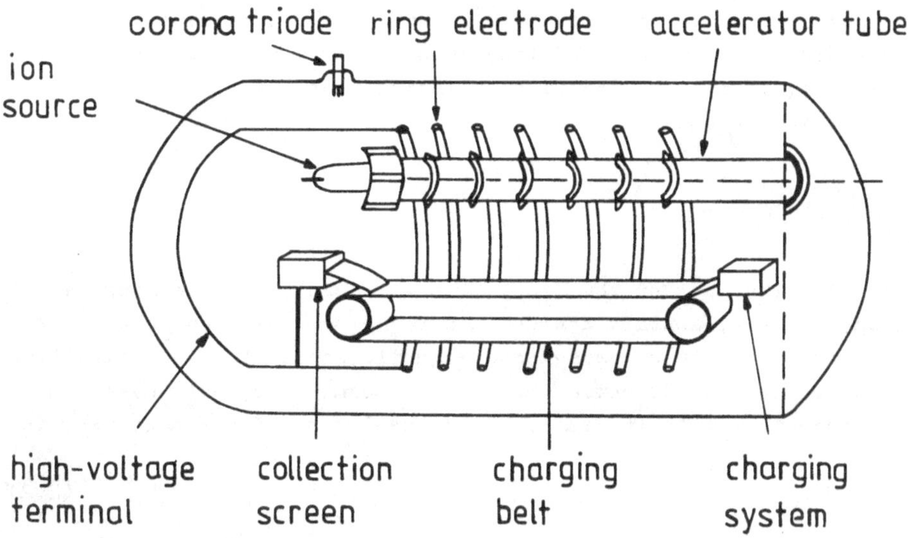

Fig.2.4. Scheme of a van de Graaff accelerator

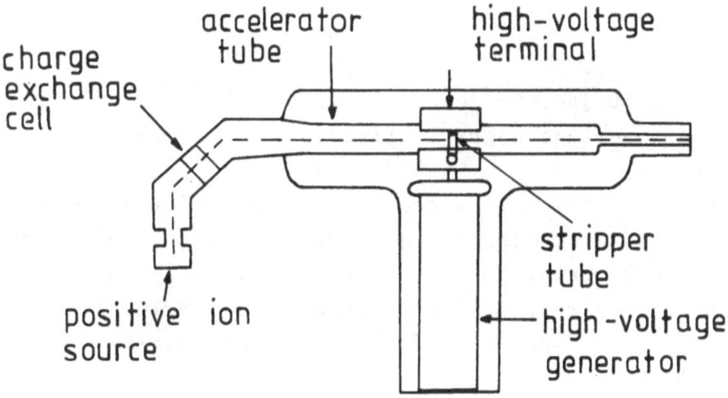

Fig.2.5. Scheme of a tandem accelerator

In a tandem accelerator (see Fig.2.5) the high voltage is used twice. First, negatively charged ions are accelerated towards the positive terminal potential. After passing a stripper target, the ions are positively charged and therefore repelled from the terminal potential.

Thus, for example, the generation of 3 MeV He^{2+} ions requires only a terminal voltage of 1 MV.

2.1.3. Mass separation

For the analysis of solids with high-energy ions, a monoenergetic ion beam containing only one kind of ions is necessary. The selection of the ion species required is performed by a magnetic analyzer (Ref.[77]).

After the acceleration to the kinetic energy

$$\frac{M_1}{2} v^2 = i\ e\ V \quad , \tag{2.1}$$

where M_1 is the mass of the ion, v the velocity, i the degree of ionization, e the elementary charge, and V the acceleration voltage, the ions are deflected by a homogeneous magnetic field with its direction perpendicular to the direction of the ion beam. The ions move on a circular trajectory with the radius r by the Lorentz force according to

$$M_1\ v = i\ e\ B\ r \tag{2.2}$$

with B being the magnetic induction. Using Eq.(2.1) the radius of the beam in the magnetic field is given by

$$r = \left(\frac{2\ V}{B^2\ i\ e/\ M_1}\right)^{1/2} \quad . \tag{2.3}$$

Hence, the radius of the trajectory depends only on the ratio i/M_1 at constant voltage and at constant magnetic field. The arrangement is shown in Fig.2.6.

The monoenergetic beam focused on the object slit is deflected by the magnet on its image plane. According to Eq.(2.3) ions with different ratios i/M_1 strike different positions of the focal plane. The ions that will be used for further treatment must pass through the fixed slit in the image plane. By varying the magnetic field the ions desired for the investigation can be selected.

A mass spectrum of the ion beam may be gained by sweeping the magnetic field over an appropriate range while registering the ions passing through the image slit. A typical spectrum of an ion beam generated in a source working with a H-He mixture is represented in Fig.2.7.

The mass separation by the system described above fails in two cases. First, ions with equal ratio i/M_1 move on identical trajectories as

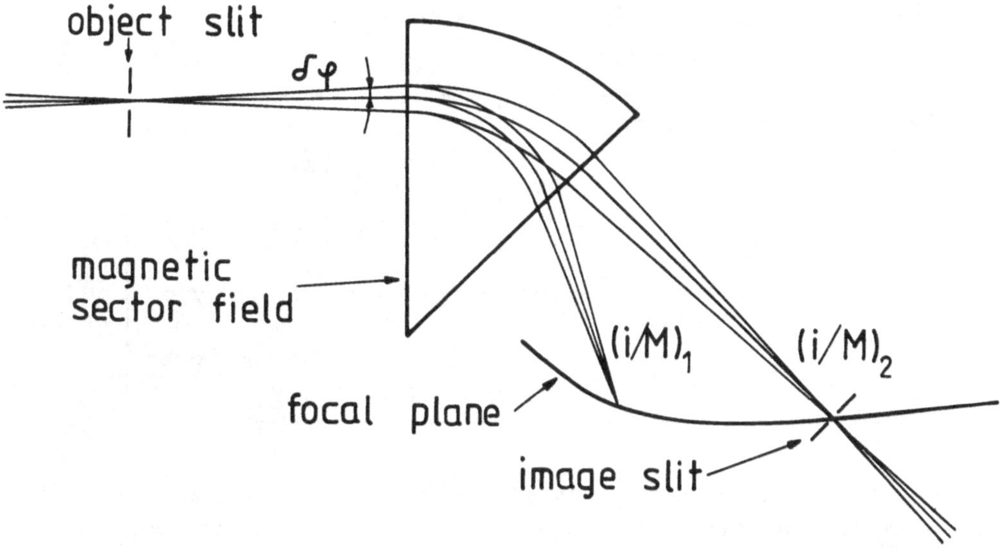

Fig.2.6. Principle of the mass separation by a magnetic sector field for an incident beam with the angular divergence δφ

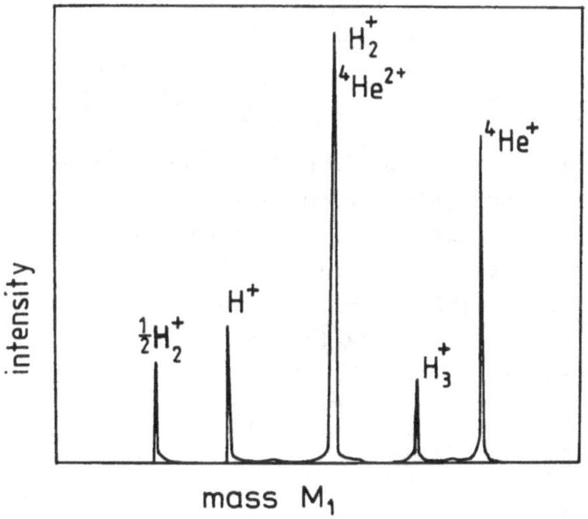

Fig.2.7. Mass spectrum of an ion beam from a H-He ion source

follows from Eq.(2.3) and are not separated by the magnet (see Table 2.1). Second, stripping processes occuring after acceleration but before mass separation cause a change in the ratio i/M_1 of the ions. Using Eq.(2.2) with the ratio $(i/M_1)'$ for the ion after the stripping process and connecting it with Eq.(2.1), identical trajectories are obtained for ions with $(i/M_1)' \sqrt{M_1/i}$ = constant (see Table 2.1). This so

TABLE 2.1

Examples for beam contamination

accelerated ion	deflected ion	$(i/M_1)'\sqrt{M_1/i}$	
$^4He^{2+}$	$^4He^{2+}$	$1/\sqrt{2}$	no separation
H_2^+	H_2^+	$1/\sqrt{2}$	
$^4He^+$	$^4He^+$	1/2	no separation
$^{16}O^+$	$^{16}O^{2+}$	1/2	

called beam contamination causes a background signal in the backscattering spectrum. According to the scattering cross section, the background signal is proportional to the atomic number squared and therefore most troublesome for beam contaminations with high atomic numbers. The reduction of beam contamination may be obtained by improving the vacuum conditions which diminishes the interaction between the beam and the residual gas atoms.

2.1.4. Beam handling

The task of the beam handling system is to guide the beam along the beam line with all its elements, providing optimum beam position within the line and the focusing of the beam on the target. Electrostatic and magnetic fields and limiting apertures are applied to this end. The specially formed fields act as ion-optical lenses, influencing the path of charged particles similar to optical elements in the case of light. The main elements of the beam handling system will be discussed for the system represented in Fig.2.8.

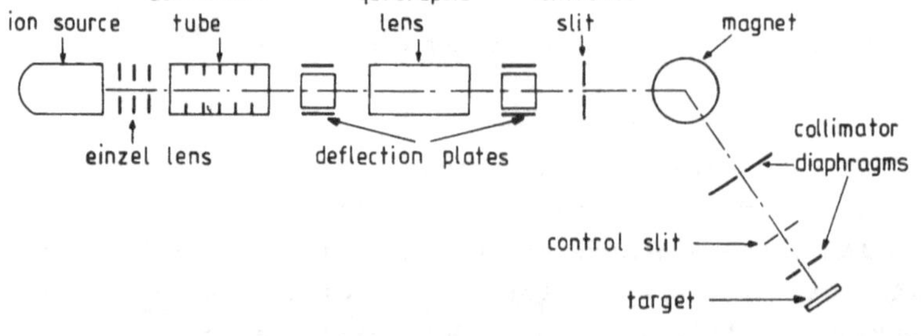

Fig.2.8. Schematic representation of the beam handling system

First, the ions pass a system of annular electrodes, the einzel lens of the source, and the electrodes of the accelerator tube acting as a system of electrostatic lenses. The focus may be varied by changing the potential applied to the electrodes of the einzel lens. But, usually the field of the einzel lens is optimized to guarantee a stable extraction of ions from the ion source. Then an additional quadrupole lens can be used for focusing the ion beam on the entrance slit of the magnet.

By ensuring a central and axial pass of the beam through the entrance slit, an automatic control system diminishes beam current losses and optimizes the path of the beam through the magnet. Information about the actual beam position can be gained by comparing the ion currents falling on two opposite edges of a slit. If the two currents are equal, the beam passes the slit centrally. The difference between the two currents provides information about the degree and the direction of displacement of the beam by the magnitude and sign of the difference current. In the system shown in Fig.2.8 the entrance slit of the magnet acts as a sensor for beam position and the control slit in connection with the entrance slit of the collimator acts as a sensor for angular alignment. The feedback of the difference current on the deflection plates stabilizes the beam at the optimum path.

Before striking the target, the beam passes the collimator which is a system of diaphragms for shaping the beam, limiting the angular divergence and protecting the target from scattered particles. Generally, a system of three diaphragms as shown in Fig.2.9 is used. The ion beam is shaped by the collimator diaphragms. Ions scattered at the edges of the collimator diaphragms cause a background signal when striking the target outside the beam spot or falling upon the sample holder. These scattered ions are suppressed by the antiscattering diaphragm. In order to avoid scattering at the edge of the antiscattering diaphragm, its diameter should be somewhat greater than the diameter of the beam.

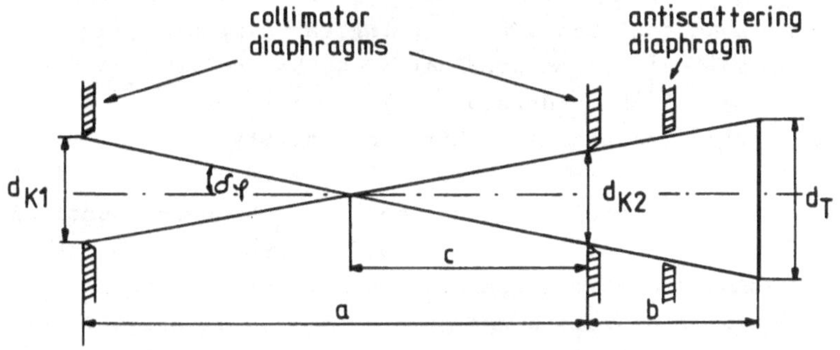

Fig.2.9. Schematic diagram of a collimator

From simple considerations, the following relations are obtained for the diameter of the beam spot on the target d_T and the angular divergence $\delta\varphi$

$$d_T = d_{K2} + 2\ b\ \tan\delta\varphi \tag{2.4}$$

$$\delta\varphi = \arctan\left(\frac{d_{K1} + d_{K2}}{2\ a}\right) \tag{2.5}$$

where d_{K1} and d_{K2} are the diameters of the collimator diaphragms, a is the distance between the collimator diaphragms, and b is the distance between the exit collimator diaphragm and the target as represented in Fig.2.9.

For channeling experiments, the angular divergence should be below 0.05°. The spot size which can be applied is limited by the structure of the target. At low spot diameters, the loss of beam current due to its absorption in the diaphragms must be compensated by an efficient detector system (see Sect.2.4.1.2) or by the prefocusing of the beam.

Devices for monitoring the beam, for instance quartz plates for ocular observation or Faraday cups (see Sect.2.2.2) for current measurement facilitate the adjustment of the beam line.

2.1.5. Energy stabilization and calibration

In order to obtain an energy spread of the ion beam less than 1 keV the generator voltage must be stabilized. This can be done by feedback circuits which stabilize the generator voltage, controlling the input voltage in the case of Crockroft-Walton generators or controlling the discharge current to the corona triode and the charging of the belt in the case of van de Graaff generators, respectively. The input signal of the feedback circuit, which is proportional to the deviation of the generator voltage from the preset value may be gained from a generating voltmeter or from the beam position after passing the analyzing magnet.

Using a generating voltmeter, an accuracy of 0.05% can be obtained; however, after opening of the generator tank the calibration of the voltmeter has to be repeated because of its sensitivity to changes in the geometry.

The beam position on the exit slit of the magnet, which is identical with the entrance slit of the collimator, depends only on the energy of the ions at a constant magnetic field strength as can be seen from Eq.(2.3). An increased or decreased energy of the ions causes the beam to be more or less deflected by the magnet. The difference current of the opposite edges of the slit which depends on the beam position therefore may be used as a feedback signal to control the generator voltage.

Since a change in the deflection of the ion beam is also caused by a variation of the magnetic field the energy width of the beam depends on the stability of the magnetic field, and the geometric arrangement. By means of highly stabilized magnets (e.g. by a Hall detector or by NMR permitting a stability in the order of 10^{-5}) and small slits less than 1 mm in the object and focal plane of the magnet, the energy spread of the beam may be kept below 300 eV. Such a value is necessary for high resolution measurements. For routine measurements, an energy spread of 1 to 2 keV is sufficiently low.

Generally, the energy calibration of an accelerator is performed using nuclear reactions with sharp resonances or endothermic nuclear reactions, measuring the threshold energy (see Sect.3.3). Secondary standards such as the generating voltmeter or the magnet are calibrated at several energies, allowing interpolation to other energy values.

For example the calibration of the magnet by the $^{27}Al(p,\gamma)^{28}Si$ reaction at 992 keV is represented in Fig.2.10. A significant increase of the γ-yield occurs if the energy of the incident protons reaches the resonance energy at 992 keV.

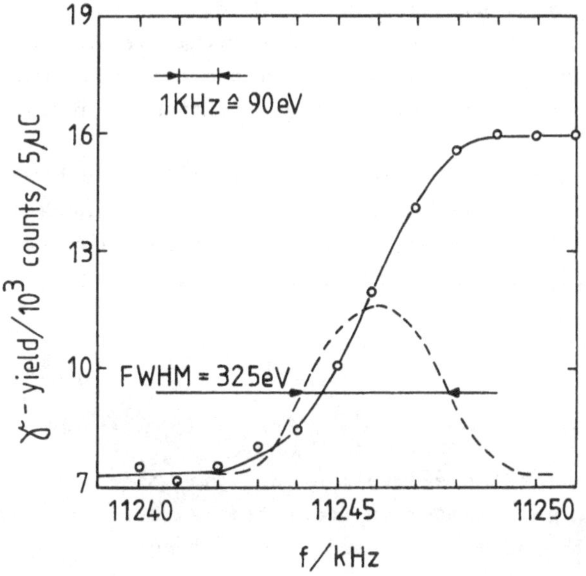

Fig.2.10. The γ-yield as a function of the NMR frequency for the $^{27}Al(p,\gamma)^{28}Si$ reaction at 992 keV (full line). The dashed line represents the graphically derived curve.

For higher energies, the resonance condition is fulfilled for atoms within the target. The γ-yield of a thick target is therefore a step function convoluted with the energy distribution of the incident beam. Thus, the energy distribution of the ion beam may be obtained from the first derivative of the γ-yield (see Fig.2.10). The resulting peak, which is identical with the yield from a thin target, allows the calculation of the NMR frequency at the peak maximum. Knowing the slope of the calibration curve from measurements at several energies, the energy width of the beam can be calculated.

The calibration of a generating voltmeter for protons is sufficient since it is independent of the kind of ions needed. However, the calibration of the magnet must be performed for every kind of ion that is used, because of its dependence on the mass and charge state of the ion (see Eq.(2.3)). An indirect calibration is necessary if there are no suitable reactions in the energy range between 0.5 MeV and 3 MeV. For example, three methods of the calibration for He ions will be given.

(i) Knowing the field distribution of the magnet, the trajectory of the ions can be calculated and the dependence of the ion energy selected on the magnetic field strength can be determined.

(ii) After calibrating for protons, a target with a known composition is irradiated with protons and He ions. From the knowledge of the proton energy and the kinematics of the scattering process, the energy of the ions backscattered can be calculated and is used for the calibration of the detecting system. By means of this calibrated detecting system, the energy of He ions backscattered may be determined and the energy of the incident He particles can be calculated, giving a calibration point for the magnetic field.

(iii) An ^{241}Am source of α particles (energy 5.477 MeV and 5.435 MeV) is applied to the calibration of the detecting system. Having checked the linearity of the detecting system, the calibration can be extrapolated to lower energies for the calibration of the magnetic field.

2.1.6. Vacuum system

The vacuum system is designed for the evacuation of the accelerator tube, the beam line, and the target chamber in order to:

- diminish the interaction of the beam with residual gas atoms (suppression of ion losses by scattering, suppression of secondary electrons, neutrons),
- avoid electrical breakdown,
- prevent surface contamination of the target, and
- provide the vacuum conditions for other methods of surface analysis and target treatment.

The pumps for the parts of the vacuum system have to meet different requirements. The pumps for the tube of the accelerator must have a high pump speed for the compensation of the gas load from the ion source. Usually, turbomolecular pumps or diffusion pumps with a pump speed of more than 500 l/s are applied. Ion pumps or cryo pumps are not useful because of their low pump speed for hydrogen and rare gases (Ref.[78]).

The pumps for the beam line must guarantee an operation pressure of about 10^{-5} Pa. Generally, the beam line is seldom exposed to air and therefore pumps with a low pumping speed are sufficient. To avoid electrical breakdown, a security circuit is applied that automatically switches off the high voltage if the pressure exceeds 10^{-3} Pa.

The target chamber is separated from the beam line by a gate valve to allow separate pumping. The pumps for the target chamber should have a high pump speed to allow a fast change of the targets. Usually, turbomolecular pumps or diffusion pumps are applied to reach operation pressures of about 10^{-5} Pa. In order to obtain UHV conditions, ion-getter pumps are preferred. Then the target chamber must be preheatable to about 500 K and the application of metal gasket rings is necessary.

A sorption trap or baffle should be applied to prevent forepump oil from entering the high-vacuum system. Diffusion pumps should be liquid N_2-trapped.

A local improvement of the vacuum in the target region down to about 10^{-9} Pa may be obtained by a cryoshield with small apertures for the ion beam of about 10^{-3} sr cooled down to about 20 K.

2.2. TARGET SYSTEM

The target system involves the sample holder, devices for beam measurement, supports for the detector system and additional devices, such as a laser for alignment, a cryoshield, devices for in-situ investigation etc. all within the target chamber.

2.2.1. Sample holders

The design of the sample holder is determined by its specific experimental purpose. Generally, multiple-sample holders are applied to routine measurements and special sample holders are used for channeling experiments, experiments at high or low temperatures, and at different geometrical configurations (e.g. glancing angle incidence).

Multiple-sample holders allow a very fast performance of routine measurements since the evacuation of the target chamber must be done only once for a set of samples. In the example shown in Fig.2.11a, the sample to be investigated may be selected by rotating the sample holder.

The sample holder allows the positioning of the samples in two directions perpendicular to the beam axis. Usually multiple-sample holders provide only restricted possibilities for channeling measurements.

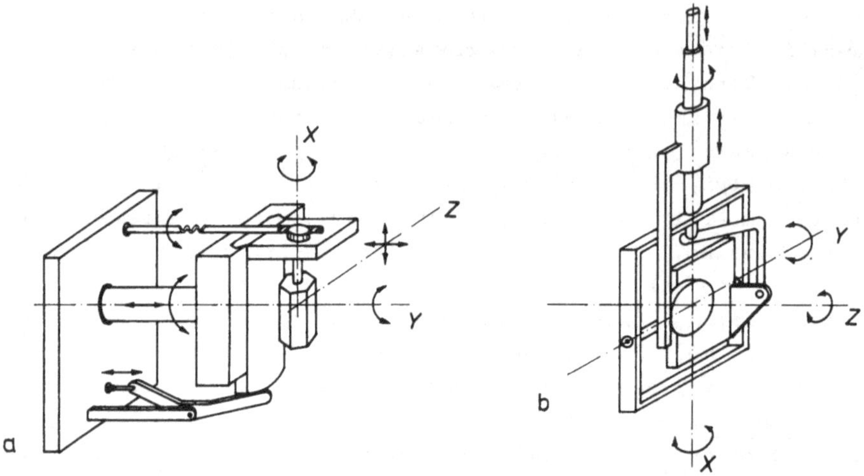

Fig.2.11. Scheme of a multiple-sample holder (a) and of a three-axis goniometer (b)

The most important group of <u>special sample holders</u> are two-axis or three-axis goniometers for channeling measurements. When tilting the axes an angular accuracy of less then 0.05° is required for high precision channeling experiments. In order to prevent surface contamination of the target, precision goniometers are capable of operating at UHV conditions. The scheme of a three-axis UHV goniometer is represented in Fig.2.11b. The sample may be rotated around the z-axis in the range of -10° to $+90^\circ$ with an accuracy of 3'; around the x-axis in the range of -180° to $+180^\circ$, and around the y-axis in the range of -10° to $+10^\circ$, both with an accuracy of 6'.

The fitting of stepping motors to the goniometer allows the computer-controlled performance of channeling experiments.

Other special sample holders have been designed for operation at temperatures in the range of 4 K to about 1000 K, cooling the target with liquid helium or nitrogen or heating the target by means of a filament, a halogen lamp, a laser or an electron beam. In every case, an influence on other devices within the target chamber (e.g. the very sensitive detector) must be avoided.

2.2.2. Beam current measurement

For a quantitative analysis of the sample, the measurement of the number of incident particles is necessary. This measurement is usually performed by integrating the beam current. With the knowledge of the degree of ionization, the number of projectile atoms can easily be calculated. However, the beam is not homogeneous, due to the interaction of the beam with the residual gas atoms and with apertures and due to beam contamination (see Sect.2.1.3). The interaction of the beam with the residual gas atoms depends on the pressure in the vacuum system and on the energy and the mass of the ions. For light projectiles with an energy of more than about 2 MeV at a pressure below about 10^{-5} Pa, the scattering processes may be neglected, but at lower energy and at higher pressure, scattering effects have to be taken into account.

Generally, insulated Faraday cups are used for the measurement of the ion current along the beam line. The Faraday cup is a metallic tube with only a small entrance aperture for the ion beam. The charges collected by the Faraday cup are measured by a current integrator.

A small entrance aperture is necessary to prevent the loss of secondary electrons which are generated by the ion beam striking the wall of the tube. Otherwise, the escape of electrons would create additional positive charges indicating an ion current greater than the real current. In order to perform the current measurement during the backscattering investigation, the target fixed in a Faraday cup is used as a current detector. Since the application of a small entrance aperture requires restrictions in the geometric arrangement (detector, sample etc.) plates and grids with a negative potential are usually applied for the suppression of the secondary electrons.

The described method fails in the case where the samples are insulators. In this case, arrangements similar to that shown in Fig.2.12 can be used.

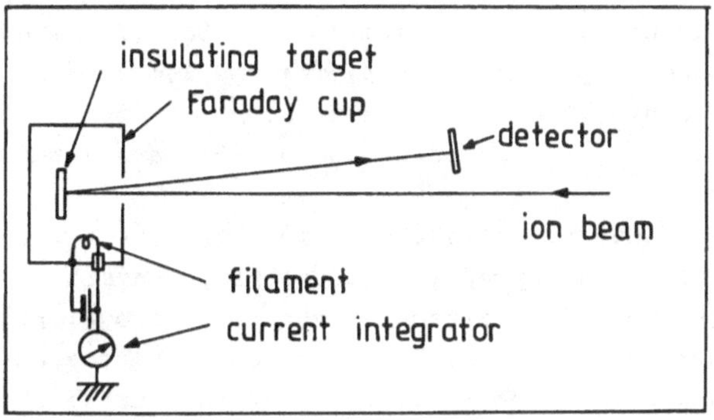

Fig.2.12. Schematic representation of current measurement by charge compensation

The sample and a filament heated by an isolated power source are placed within a Faraday cup. The electrons emitted by the filament compensate for the positive charges within the target which are introduced by the ion beam. The magnitude of the compensation current is equal to the ion current and may be measured by a current integrator.

An indirect method which may be applied for any target material is represented in Fig.2.13. A motor-driven rotating vane is used to chop the ion beam. Thus, a constant percentage of the beam is backscattered by the vane on the beam detector. The rotating vane is made of material consisting of light elements covered with a thin layer of a material with a large atomic number. Thus, the ions backscattered from the thin top layer appear as a separated peak with a high yield in the spectrum detected by the beam detector (see Fig.2.13b).

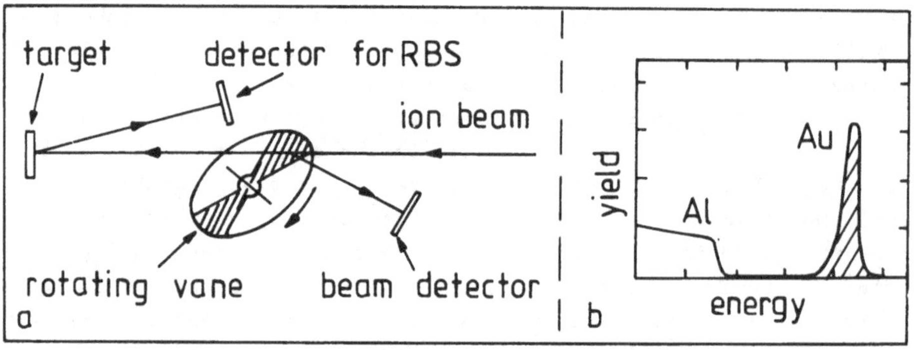

Fig.2.13. Principle of current measurement by a rotating vane (a) and the energy spectrum of ions backscattered from an Al vane covered with a thin layer of Au (b)

Their total yield is proportional to the number of ions striking the target with a proportionality factor which can be calculated knowing the geometric arrangement. In this way precise measurements are possible since neutral atoms or electrons within the beam do not falsify the result. The calibration may also be performed comparing the ion current at the place of the target measured by means of a Faraday cup to the yield detected by the beam detector, but in this case neutral atoms and electrons influence the result.

A block diagram of a current integrator is shown in Fig.2.14. The charge collected on the target is transferred to a capacitor, converting the charge signal in a voltage one. The voltage increases with a slope proportional to the input current. If the voltage reaches a preset value, the comparator causes a high current discharge of the feedback capacitor providing an output pulse and the charging of the capacitor starts again.

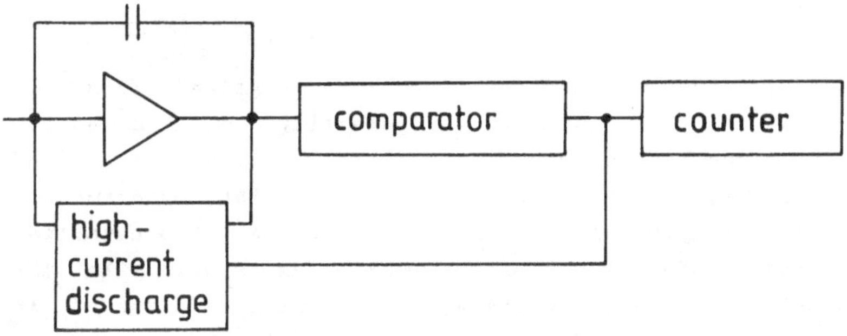

Fig.2.14. Block diagram of a current integrator operating by current-to-frequency conversion

The output frequency of the comparator, which is proportional to the input current of this current-to-frequency converter, is used for beam current measurement. With this type of current integrator, ion currents down to the range of 0.1 nA can be measured.

2.3. SAFETY CONSIDERATIONS

The operation of ion accelerators is accompanied by the generation of bremsstrahlung radiation induced by high-energy electrons and of neutrons from nuclear reactions. The electric field of the accelerator tube causes the acceleration of electrons released by the ionization of residual gas atoms and of secondary electrons from electrodes and apertures towards the positive high-voltage terminal. These high-energy electrons generate bremsstrahlung radiation when impinging on the wall of the tube or on the ion source. The X-ray flux can be diminished by decreasing the pressure within the accelerator tube, thus reducing the interaction of the ion beam with residual gas atoms. In order to guarantee a mean free path length of the ions of more than 10 m the pressure should be kept below $2*10^{-3}$ Pa. An additional decrease of the X-ray flux can be achieved by preventing the electrons from traversing the entire acceleration voltage. For this purpose, specially formed acceleration fields or traverse magnetic fields are applied to deflect the electrons on the electrodes of the accelerator tube.

Special safety precautions have to be taken during the acceleration of deuterons since the reaction $D(d,n)^3He$ releases a high flux of neutrons. The reaction occurs through self-targeting at diaphragms and at walls of the beam line.

In every case, a continuous inspection of the gamma and the neutron flux is required to ensure that the exposure of the personnel does not exceed the limits established by law.

2.4. DETECTING SYSTEM

In the case of RBS, the experimental information is gained from the energy of the backscattered particles. Generally, a high energy resolution and a large backscattering yield are required. It is of essential importance to fit the detecting system to the experimental requirements because energy resolution and high counting rates establish different directions in the design of the detecting system. In the following, the major components of the equipment will be discussed in more detail and general aspects of their application will be presented.

2.4.1. Detectors

In backscattering spectrometry, two main principles are applied to the measurement of the energy of the particles; the counting of particles after having passed a magnetic or electrostatic analyzer and the utilization of energy-dispersive semiconductor detectors.

2.4.1.1. Magnetic and electrostatic detectors

A homogeneous magnetic sector field (see Sect.2.1.3) or the electric field of a cylindrical capacitor are used as energy-dispersive elements for the backscattered particles. For example the principle of an electrostatic detector is shown in Fig.2.15. The cylindrical capacitor consists of a pair of sectors of a cylinder with different radii.

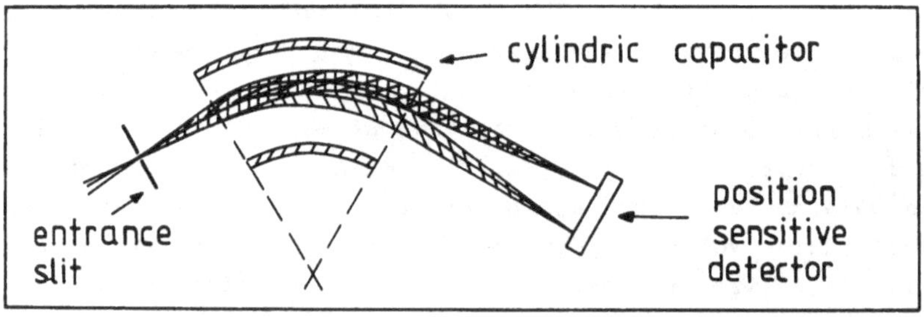

Fig.2.15. Principle of an electrostatic detector

A collimated beam of backscattered particles is deflected by the electric field and focused on the focal plane. Particles of different energies appear at different places on the focal plane. Usually, the energy of the particles to be counted is selected by the variation of the electric field using a fixed detector slit. The energy resolution of the system is given by the resolution of the cylindrical capacitor and the size of the entrance aperture of the counting detector. Magnetic and electrostatic detector systems, respectively, permit an energy resolution lower than 1 keV for MeV He^+ particles. The increase of the energy resolution (for example by diminishing the entrance aperture) is, however, obviously accompanied by a decrease of the counting rate and therefore of the statistic accuracy. Moreover, the successive counting of particles of different energies causes the irradiation of the target with a high dose of ions at a high flux to reduce the measuring time. Thus, surface contamination and radiation damage can lessen the accuracy of the results significantly. The method can be improved by using a position-sensitive detector to count the particles within a certain range of energy in one run.

2.4.1.2. Semiconductor detectors

Semiconductor detectors are silicon or germanium diodes of the p-n or p-i-n type or surface barrier diodes operated in the reverse bias mode (Ref.[79]). Ionizing radiation penetrating the depletion layer releases charge carriers (electrons and holes) which are separated by the electric field and collected at the elctrodes. The carriers generated provide information about the energy of the ionizing radiation and the time of the occurrence.

<u>Surface barrier detectors</u> are used for both particle counting and energy analysis of particles. Usually, surface barrier detectors for particle spectrometry are formed by the deposition of a thin gold front electrode on high resistivity n-Si as shown in Fig.2.16a. An incident particle creates free electron-hole pairs within the depletion layer by transferring an energy of 3.6 eV for each pair in Si at room temperature.

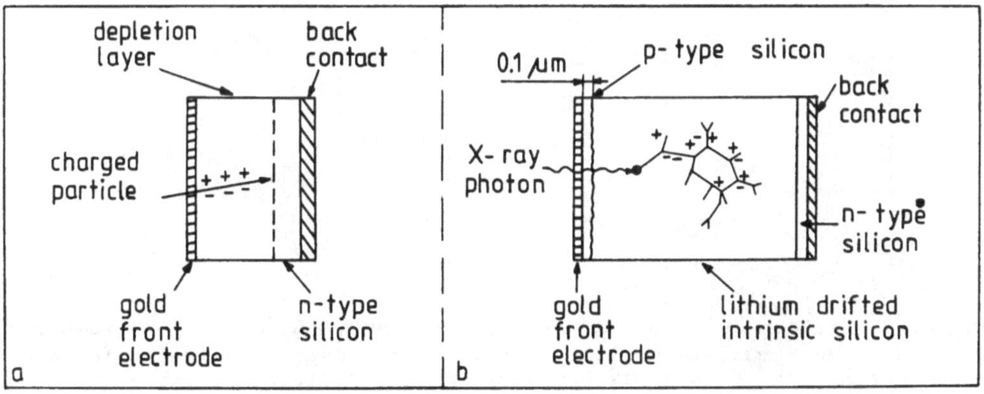

Fig.2.16. Typical cross section of a surface barrier detector (a) and of a Si(Li) X-ray detector (b) (Ref.[80])

This value is independent of the particle energy over a wide range of particle types and energies. Therefore, the detector charge pulse is proportional to the energy of the particle, provided that the sensitive depth exceeds the range of the particle in silicon and that the electric field is sufficiently large to separate the carriers before recombination. The correlation between the H and He ion energy, respectively, and the minimum depletion layer depth is given in the central part of Fig.2.17. A straight line which intersects the center vertical line at the required depletion depth gives useful combinations of resistivity and detector bias.

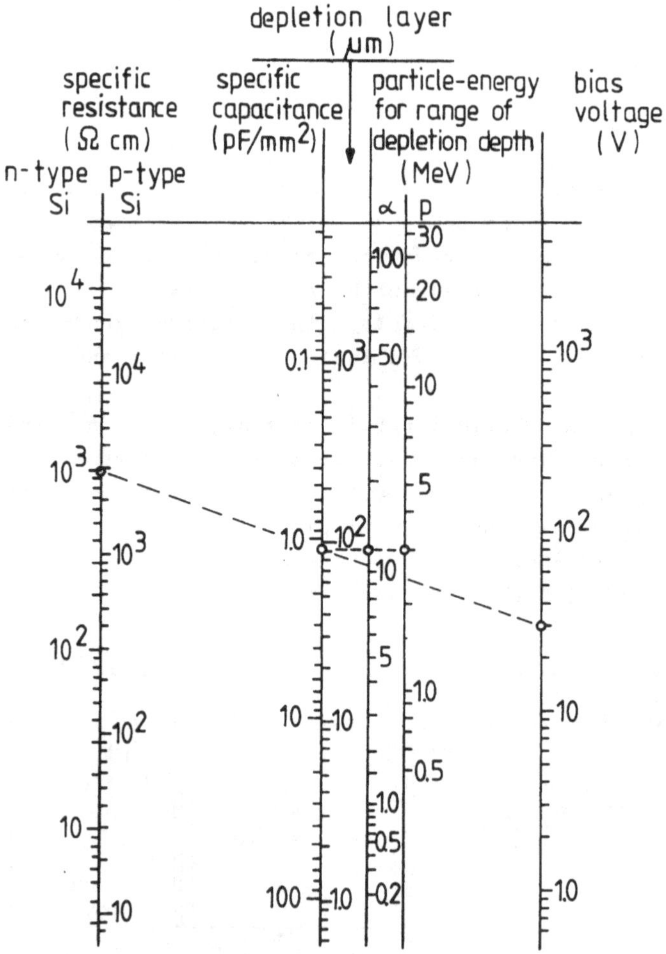

Fig.2.17. Silicon detector parameters nomogram (Ref.[80]). For example a voltage of 30 V must be applied to a 10 Ωcm n-type Si detector to stop an 11 MeV He ion or a 3 MeV H ion within the depletion depth.

The energy resolution of the detecting system is limited by the noise contribution of the detector and the electronic system. Detector noise originates from straggling processes of the incoming particles in the front electrode layer, the roughness of the layer, the statistical spread in the number of electron-hole pairs released in the depletion layer and from the recombination processes. The detector noise is therefore proportional to the active volume. Besides this contribution the detector generates noise in the preamplifier which is proportional to the capacitance and the reverse leakage current of the detector. The capacitance of the detector is given by the thickness of the depletion layer (see Fig.2.17) and the detector area. Minimum noise is gained by choosing a high quality detector (low detector noise and low reverse current) with a sensitive area as small as possible. The area of the detector influences the detection geometry (counting rate) and the noise (detector and preamplifier noise). The dichotomy between energy resolution and statistics can be seen again.

By increasing the depletion layer, the noise due to the capacitance of the diode decreases but on the other hand the larger active volume gives rise to an increase in the noise. Minimum noise is achieved by equalizing these two contributions. Using small detectors, an energy resolution of less than 15 keV for MeV He ions is attainable.

<u>p-i-n detectors</u> (see Fig.2.16b) have a large sensitive volume and are therefore especially capable of detecting photons in the range from keV to some MeV. They are applied to X-ray and γ-ray spectrometry. The active volume of a Si(Li) detector is given by the intrinsic Si layer formed by a lithium drift into p-Si.

The detectors usually operate at the temperature of liquid nitrogen in order to diminish the leakage current (determining the noise of the system) and to avoid an undesirable diffusion of the lithium atoms. For protection against contamination, the detectors are kept in vacuum. The entrance window for the photons is usually made of low Z material. Very often a thin beryllium foil is used. However, the window causes a strong absorption of X-rays with an energy below about 1 keV and determines the lowest detectable energy. The increase of the absorption length for high-energy photons provides an upper energy limit which depends on the thickness of the active layer. These effects are displayed in Fig.2.18. An increased thickness of the Be window causes the shift of the lowest detectable energy to higher values, whereas an increased thickness of the active volume permits the detection of photons with higher energy.

As in the case of surface barrier detectors, the energy resolution of p-i-n detectors strongly depends on the detector area. The resolution ranges from less than 150 eV for detectors with a diameter of 4 mm to about 300 eV for detectors with 16 mm diameter.

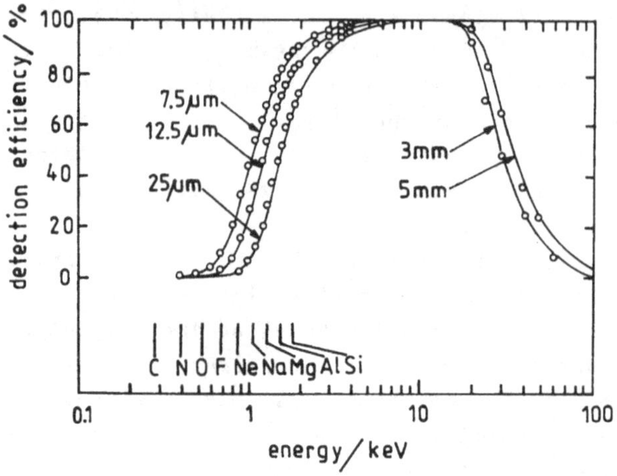

Fig.2.18. Detection efficiency as a function of the X-ray energy. The parameters are the Be window thickness (left edge) and the Si(Li)-crystal thickness (right edge) (Ref.[80]). The position of the K-lines of light elements is indicated.

2.4.2. Preamplifier

The output signal of the detector is amplified by a charge-sensitive preamplifier that is specially designed to preserve maximum signal-to-noise ratio. Generally, charge sensitive amplifiers are used in spectroscopy (Fig.2.19). The charges from the detector are integrated into the feedback capacity of the amplifier. The rise time of the signal is equal to the duration of the detector current pulse in the ideal case (< 10 ns). The decay time of the pulse given by the time constant R_fC_f is in the order of 100 µs. Therefore the output signal of the preamplifier is approximately given by a step function. A pile up of the output signals that exceeds the linear region of the preamplifier and causes incorrect processing must be avoided by decreasing the counting rate, the decay time constant, and the amplification, respectively.

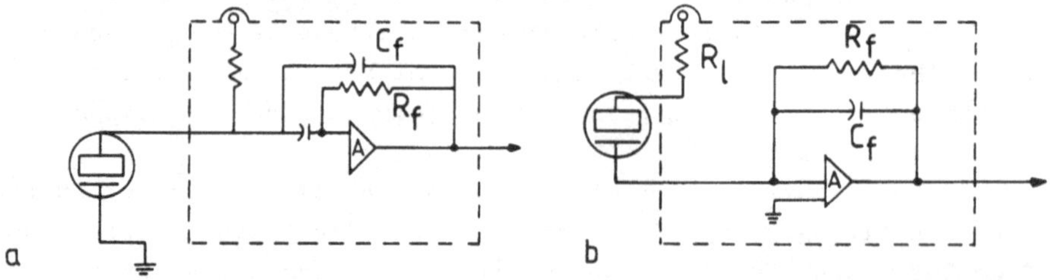

Fig.2.19. The ac-coupled (a) and dc-coupled (b) bias circuits of a charge-sensitive preamplifier

Two typical bias circuits for charge-sensitive preamplifiers are represented in Fig.2.19. The dc-coupled preamplifier allows high-resolution measurements but only relatively low rates. The ac-coupled preamplifier permits high signal rates but the coupling capacity causes a distortion of the pulse shape. However, the ac-coupled preamplifier establishes a good compromise between high rates and high resolution.

Sources of noise are the leakage current of the detector, which is usually the dominating contribution, the input FET, the resistance connected to the input and the capacitances of the detector and of the cables. Hence, much attention must be focused on the design of the input configuration of the preamplifier. The preamplifier should be selected to provide minimum noise for the external capacitance required by the experiment.

2.4.3. Spectroscopic amplifier

The preamplifier output signal is shaped and filtered by the spectroscopic amplifier in order to improve the signal-to-noise ratio and to shorten the pulse length. The shortening of the nearly step function pulses is necessary to prevent an overlap of successive pulses (pile-up effect) at high counting rates. At the output of the spectroscopic amplifier, the pulses have a constant shape but an amplitude which is proportional to the height of the input step pulse. Two methods are employed for pulse shaping.

2.4.3.1. CR-RC pulse shaping

During amplification, the pulse passes one or several passive or active CR-RC networks (Fig.2.20a). The CR-high-pass filter causes a significant

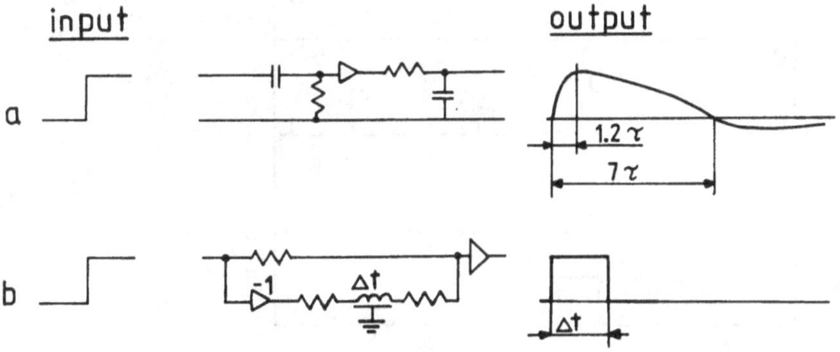

Fig.2.20. Pulse shaping by a CR-RC network (a) and by a single-delay-line (b), respectively

decrease of the decay time of the pulse whereas the rise time of the pulse is increased by the RC-low-pass filter. The shaping method by CR and RC filters is connected with a bandwidth limitation of the signal. Because of the much larger bandwidth of the noise in comparison to the signal, this bandwidth limitation attenuates the noise significantly more than the signal, which yields an improvement of the signal-to-noise ratio. A maximum signal-to-noise ratio is gained for nearly equal time constants τ of the filters. For surface barrier detectors the optimum time constant is about 0.5 μs. The undershoot of the baseline behind the pulse (see Fig.2.20) limits the maximum pulse rate unless the baseline is restored by special electronic circuits.

2.4.3.2. Delay-line pulse shaping

In this case, the output pulse is the sum of the input pulse and the inverted and delayed pulse. This method yields rectangular pulses without baseline undershoot (see Fig.2.20b). It is therefore preferable for high counting rate experiments and pulse widths less than 1 μs. However, the signal-to-noise ratio is poorer than for CR-RC pulse shaping because of the missing bandwidth limitations.

2.4.3.3. Pile-up effects

The probability that two output pulses pile up increases with increasing counting rate. At very high counting rates the pile-up signal (Fig.2.21b) which appears in the high energy part of the spectrum cannot be neglected. In this case the application of a pile-up branch circuit is useful.

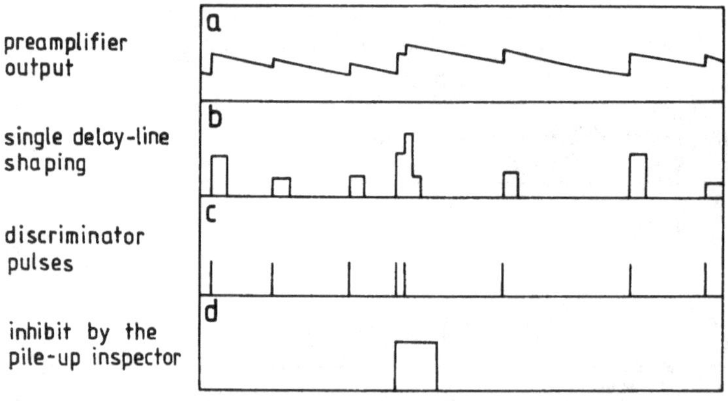

Fig.2.21. Schematic of the pulse processing

Fast electronic circuits (timing filter amplifier, 100-MHz discriminator) detect the leading edges of the current pulses and deliver short logical pulses to the pile-up inspector (Fig.2.21c).

Every input pulse switches a trigger with a presettable holding time (inspection time) within the pile-up inspector. If a second pulse arrives within this inspection time, the pile-up inspector generates an inhibit pulse which prevents the pile-up pulse from further processing by shutting a linear gate (Fig.2.21d).

2.4.4. Multichannel analyzer

Multichannel analyzers (MCA) are multifunctional devices that permit data acquisition, storage, display and interpretation. Generally, MCA's are connected with a computer or are special computers themselves. Typical functional elements of an MCA are displayed in Fig.2.22. The input signals pass through a single-channel-analyzer (SCA) or an analog-to-digital converter (ADC) to the central unit that organizes the transfer of the data to the memory in several operation modes. The minimum output configuration of the MCA should include a display, an x-y plotter, a hard-copy printer and a storage device.

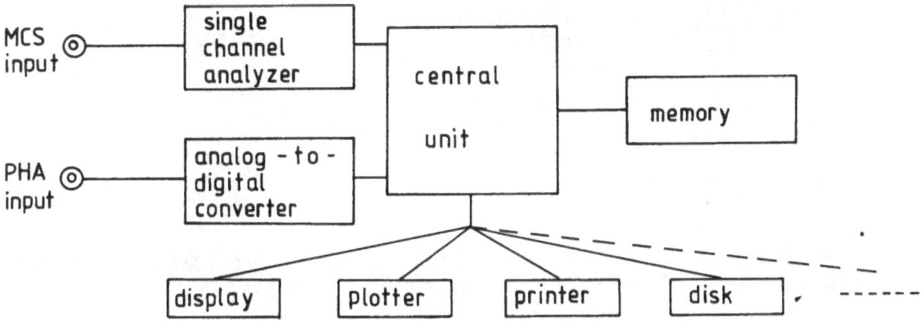

Fig.2.22. Block diagram of a multichannel analyzer (MCA)

MCA's are characterized by their ability to work in either of two distinct data analysis modes; the pulse-height-analysis (PHA) mode and the multichannel scaling (MCS) mode. In both the PHA mode and the MCS mode an input signal causes the incrementation or decrementation of one register in the memory which corresponds to a channel of the spectrum. The MCS mode and the PHA mode differ in the determination of the register number.

Generally, the MCA is used in PHA mode for spectroscopic applications Every input pulse causes the incrementation of a register the number of which is proportional to the amplitude of the pulse. The resulting spectrum shows the number of detected events versus the pulse height (i.e., the energy of the backscattered particle).

In MCS mode, the output signals of the single-channel-analyzer are counted for a given dwell time into one register. At the completion of the dwell time the register number is increased by one and the signals are counted into the next channel. The result is a time histogram of the input signal where each channel represents a sequential time interval. The dwell time can be set by an internal clock or by an external logical signal. The MCS mode is used, for instance, for channeling dip measurements.

The accuracy of a spectrum accumulated in PHA mode is limited by the linearity of the ADC that digitizes the amplitudes of the input pulses. High linearity is achieved using the method of amplitude-to-time conversion. The conversion principle is discussed by means of the simplified block diagram in Fig.2.23. The input pulse passes a linear gate and charges the charge capacitor to the voltage of the input pulse.

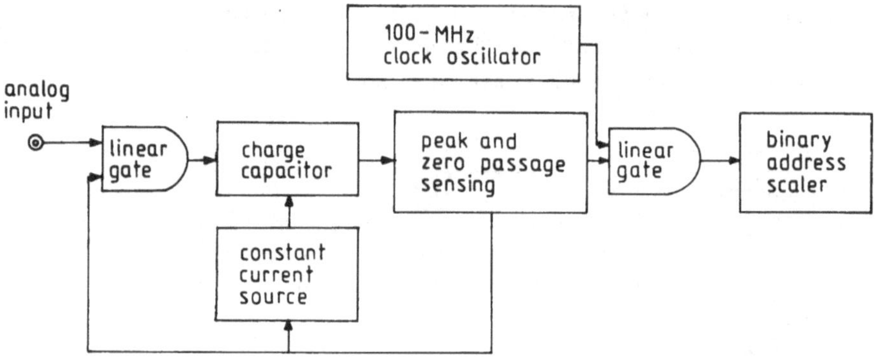

Fig.2.23.Block diagram of an analog-to-digital converter (ADC) operating by the method of amplitude-to-time conversion

When the peak of the pulse is detected by the peak sensing circuit, the charge capacitor is discharged by a constant current. The time that is required to discharge the capacitor to the baseline is proportional to the amplitude of the input pulse. During this time, a gate is open for counting clock pulses of a 100-MHz oscillator by the binary address scaler. The binary number represents the digitized value of the input pulse height and is used as the address of a register in the MCA-memory.

The ADC operating by the method of successive approximation permits a high-speed conversion but it requires a greater expense to guarantee the desired linearity. The integral and differential nonlinearity of an ADC for high-resolution spectrometry should be less than 1% of full scale, over the top 99% of the full-scale setting. The integral nonlinearity characterizes the maximum deviation of any address from its ideal position described by a straight-line plot of the address versus the input pulse amplitude. Differential nonlinearity describes the uniformity of address widths over the entire number of addresses in the ADC.

Beginning from the time of peak detection, the input gate prevents the acceptance of additional input pulses during the processing time (dead-time) of the MCA. The dead-time must be compensated for by an increased measuring time. A simple correction factor is given by 1/(1 - dead-time fraction) with the dead-time fraction being the ratio of the dead-time to the measuring time which is usually displayed by the MCA (Ref.[80]). Using a current integrator for controlling the measuring time, the dead-time may be compensated for by suppressing the current integration when a second pulse arrives during the dead-time of signal processing.

Significant advantages arise from the application of computer-aided systems. They allow data accumulation and processing or output at the same time. A software library containing programs for data handling, spectrum interpretation, service functions, and the necessary data of the elements is very suitable for fast spectrum interpretation. For example it is very useful to have programs for energy calibration, dose calculation, smoothing of the spectrum, energy-to-depth conversion and computation of concentration profiles.

Chapter 3

GENERAL EXPERIMENTAL METHODS

G. Götz and A. Dittmar

3.1. RUTHERFORD BACKSCATTERING SPECTROMETRY (RBS)

3.1.1. Scattering geometry

A schematic drawing of the geometrical arrangement for RBS is shown in Fig.3.1. A well-collimated monoenergetic ion beam (energy E_o, ion mass M_1, atomic number Z_1) impinges on a target which contains atoms of the mass M_2 and the atomic number Z_2. If there is more than one kind of atom M_2, this will be denoted by $M_2^{(i)}$ ($i = 1,2,3,\ldots$). The detector registers ions backscattered at an angle θ (in the laboratory system) in the solid angle $\Delta\Omega$ given by the detector geometry. In the case of RBS the scattering angle θ is always greater than 90°.

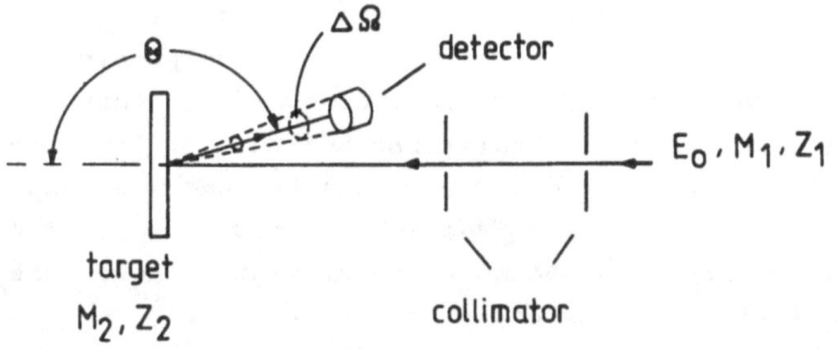

Fig.3.1. Geometrical arrangement for RBS measurements

3.1.2. Kinematic factor

According to Sect.1.1.1, an elastic scattering process between a moving ion (M_1) and a target atom (M_2) initially at rest yields an energy transfer from the moving ion to the target atom. The energy E of the ion after the collision is reduced, which can be described by the kinematic factor K

$$E = KE_o. \tag{3.1}$$

Corresponding to Eq.(1.11) in Sect.1.1.1.2, the kinematic factor K in the laboratory system is given by

$$K = \left\{ \frac{\cos\theta + [(M_2/M_1)^2 - \sin^2\theta]^{1/2}}{1 + M_2/M_1} \right\}^2 . \tag{3.2}$$

The kinematic factor depends only on the scattering angle θ and on the ratio of the masses M_1/M_2. The dependence of the kinematic factor on the scattering angle for different ratios of M_1/M_2 is shown in Fig.3.2. It can be seen that for any mass ratio, K has its lowest values at $\theta = 180^\circ$. From Eq.(3.2) follows

$$K(\theta = 180^\circ) = \left(\frac{M_2 - M_1}{M_1 + M_2} \right)^2 . \tag{3.3}$$

Considering RBS, the highest values for K are always obtained at $\theta = 90^\circ$

$$K(\theta = 90^\circ) = \frac{M_2 - M_1}{M_1 + M_2} . \tag{3.4}$$

Low values of K mean that there is a large difference between the energy of the incident and the scattered ion.

If we want to distinguish between two types of target atoms then the energy difference of the ions scattered at the atom $M_2^{(1)}$ and at the atom $M_2^{(2)}$, respectively, should be as large as possible. For that the scattering angle $\theta = 180^\circ$ is favoured. In practice, it is only possible to place the detector at angles $\theta < 180^\circ$, usually at $\theta \approx 170^\circ$, otherwise the path of the ions will be obstructed. Scattering angles near 180° are possible using annular detectors. From Fig.3.2 it can be concluded that small deviations of the scattering angle from 180° do not change the energy of the scattered ions very much, because the kinematic factor has only a weak dependence on the scattering angle in this region.

In practice, mostly H or He ions are used for RBS. In Fig.3.3 the kinematic factor versus the mass of the target atoms is shown for these ions considering the two limits of the scattering angle, $\theta = 90^\circ$ and $\theta = 180^\circ$. Because of the strong dependence of K on M_2, the kind of atoms (M_2) in a target can be determined from Eq.(3.2) measuring the energy E_2 of the scattered ions, provided that the mass M_1 and the energy E_0 of the incident ions are known. If there is more than one type of atom in the target, the mass composition of the target material can be measured by RBS. For H and He ions the kinematic factor K is represented in the Appendices 2 and 3 as a function of the scattering angle θ and the mass M_2 of the target atoms.

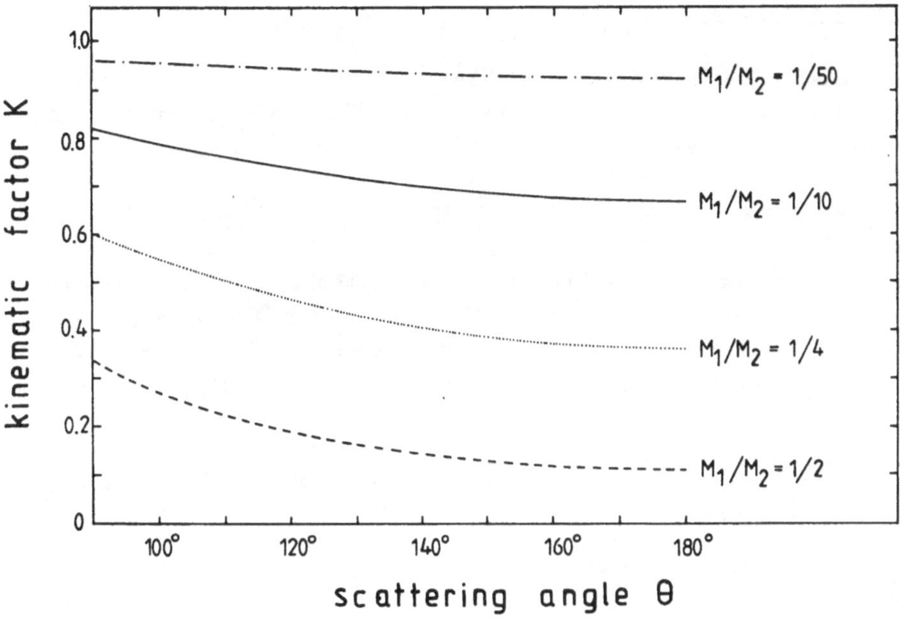

Fig.3.2. Kinematic factor K as a function of scattering angle θ for different mass ratios M_1/M_2 = 1/2, 1/4, 1/10, 1/50

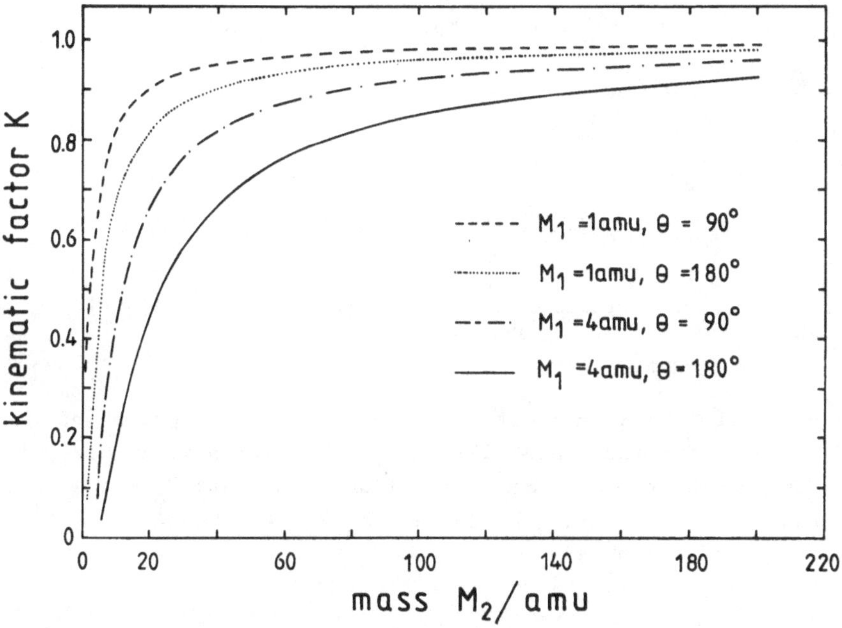

Fig.3.3. Kinematic factor K as a function of the mass of target atoms M_2 for H and He ions and the scattering angles θ = 90°, 180°

3.1.3. Mass Resolution

To distinguish between two kinds of atoms, which differ in their masses by δM_2, the energy change ΔE_2 must be greater than the energy resolution δE_d of the detecting system. The limit is given by

$$\delta E_d = \Delta E_2 = E_o \, (dK/dM_2) \, \delta M_2 \quad . \tag{3.5}$$

As already mentioned, the highest value of the energy difference is given at the scattering angle 180°. For this angle dK/dM_2 is plotted as a function of the mass M_2 of the target atoms for H and He ions in Fig.3.4 (left-hand scale).

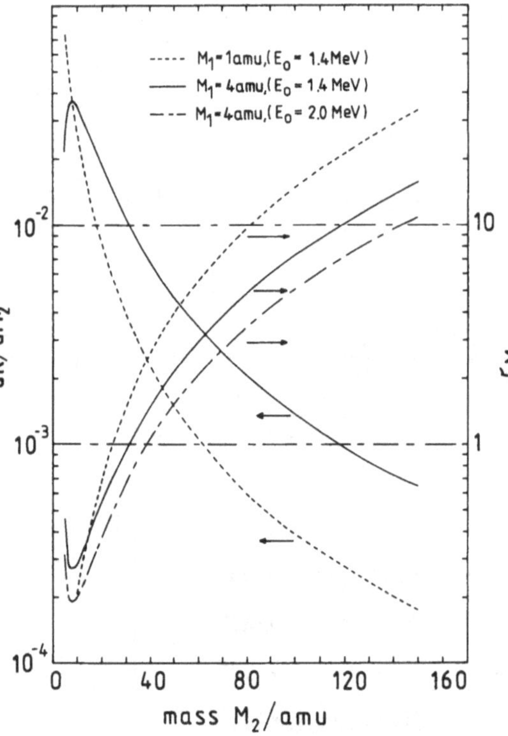

Fig.3.4. Mass resolution as a function of the mass of target atoms M_2 for H and He ions; left-hand scale: dK/dM_2, right-hand scale: absolute values δM_2 for usual experimental conditions (H ions: E_o = 1.4 MeV, energy resolution δE_d = 8 keV; He ions: E_o = 1.4 MeV and 2.0 MeV, energy resolution δE_d = 14 keV)

Of most interest is the absolute value δM_2 which can be resolved by the detecting system. This is represented in Fig.3.4 (right-hand scale) for conditions often used: energy of the incident particles E_0 = 1.4 MeV for H ions, E_0 = 1.4 MeV and 2.0 MeV for He ions, energy resolution (surface barrier detectors, see Sect.2.4.1) δE_d = 8 keV (FWHM) for H ions, δE_d = 14 keV for He ions. It can be seen that the resolution for neighbouring masses, (i.e. δM_2 = 1) is only possible up to target masses of about M_2 = 25 amu for 1.4 MeV H ions and about M_2 = 32 amu for 1.4 MeV He ions and M_2 = 38 amu for 2.0 MeV He ions. If we have to separate target masses which differ by δM_2 = 10, then this can be done by 1.4 MeV H beams up to about M_2 = 82 amu, by 1.4 MeV He beams up to about M_2 = 118 amu and by 2.0 MeV He beams up to about M_2 = 143 amu. Generally, the mass resolution deteriorates for higher target masses. From Fig.3.4 it can be concluded that He beams usually provide higher mass resolution than H beams. But for target atoms with very low mass values (M_2 < 15 amu) H beams yield a somewhat better resolution than He beams. In all cases an increase of the ion energy E_0 gives a better mass resolution (compare the curves for 1.4 MeV and 2.0 MeV He beams in Fig.3.4).

For further improvement of the mass resolution, analyzers with better energy resolution are necessary, i.e. electrostatic or magnetic analyzers which have an energy resolution in the order of δE_d = 1 keV (see Sect.2.4.1.1). But these analyzers are more complicated and require long measuring times.

3.1.4. Energy loss and stopping cross section

If the analyzing ions penetrate the target and undergo a wide angle scattering process at a certain depth z from the surface, the energy of the outgoing ions is additionally influenced by the energy loss due to inelastic scattering processes with the electrons of the target atoms and due to small-angle elastic collisions with the target nuclei. These processes are considered in Sect.1.1.2. The energy loss dE per unit length ds, also called specific energy loss or stopping power, depends on the atomic number of the ion, on the composition of the target, and on the ion energy itself.

Relating the specific energy loss to the atomic density N of the target the stopping cross section ε is defind by

$$\varepsilon(E) = \frac{1}{N}\left(\frac{dE}{ds}\right) \quad . \tag{3.6}$$

The unit electron volt times square centimeter is commonly used for ε. Stopping cross section data for H and He ions are listed in the Appendices 5 and 6 for all elements. The situation for a scattering process of the incident ion with a target atom at the depth z is demonstrated in Fig.3.5. During its path from the surface to the depth z (angle to the target normal φ_1) due to inelastic scattering process the ion loses the energy

$$E_o - E_1 = \int_o^{z/\cos\varphi_1} ds \left(\frac{dE}{ds}\right)_{in} \quad . \tag{3.7}$$

E_1 is the ion energy immediately before the elastic scattering process. After the scattering process the ion has the energy KE_1. On the outgoing path (angle to the target normal φ_2) again there is an energy loss given by

$$KE_1 - E_2 = \int_o^{z/\cos\varphi_2} ds \left(\frac{dE}{ds}\right)_{out} \quad . \tag{3.8}$$

The subscripts "in" and "out" denote the values dE/ds along the inward and outward paths.

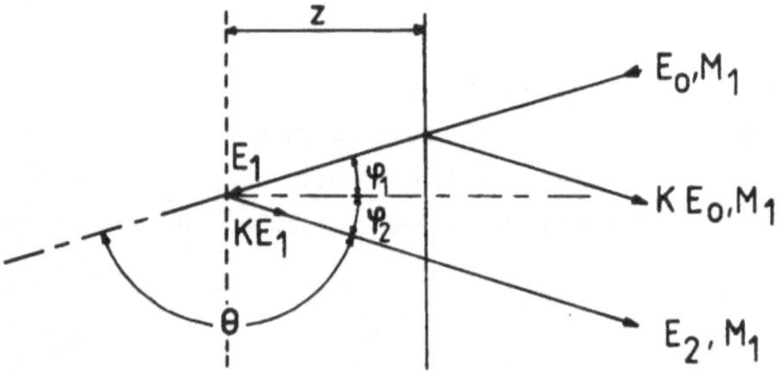

Fig.3.5. Geometry for backscattering processes within a single-elemental target

The ion leaves the target with the energy E_2. Note that the specific energy losses are always positive. With Eqs.(3.7) and (3.8) the ion energy which can be measured by the detector is

$$E_2(z) = K\left[E_o - \int_o^{z/\cos\varphi_1} ds \left(\frac{dE}{ds}\right)_{in}\right] - \int_o^{z/\cos\varphi_2} ds \left(\frac{dE}{ds}\right)_{out} \quad . \tag{3.9}$$

The energy of backscattered ions can be calculated by Eq.(3.9) if we know the energy and mass of the ions used, the masses and depths of the scattering atoms, the scattering angle, the energy losses, the entrance angle, and the exit angle for the ions.

But usually, in an RBS experiment the energy of the backscattered ions will be measured and the depth and/or the kind of the scattering target atoms is unknown. To get information about the depth of the scattering target atom, the energy scale has to be transformed into a depth scale. There are two possibilities to solve this problem.

(i) Tabulated values of $\varepsilon(E)$ are used and the integrations are carried out numerically. This is described in Sect.4.2.1.

(ii) A constant specific energy loss is assumed for the inward and outward paths.
For constant specific energy losses, Eq.(3.9) reduces to

$$E_2(z) = KE_o - \left[\frac{K}{\cos\varphi_1}\left(\frac{dE}{ds}\right)_{in} + \frac{1}{\cos\varphi_2}\left(\frac{dE}{ds}\right)_{out}\right] z \quad . \tag{3.10}$$

The energy KE_o corresponds to the energy of ions scattered from surface atoms of mass M_2. E_2 is the energy of ions scattered from atoms M_2 at the depth z.

Replacing the specific energy loss by the stopping cross section and introducing the difference $\Delta E_2(z) = KE_o - E_2(z)$, we obtain

$$\Delta E_2(z) = \left[\frac{K}{\cos\varphi_1}\,\varepsilon_{in} + \frac{1}{\cos\varphi_2}\,\varepsilon_{out}\right] Nz = [\varepsilon]\, Nz \quad , \tag{3.11}$$

where $[\varepsilon]$ is the stopping cross section factor.

For constant values of $[\varepsilon]$ the energy difference ΔE_2 between the edge of the spectrum (KE_o) and the energy E_2 corresponding to scattering events at depth z is directly proportional to the depth z. But note that in Eq.(3.11) ΔE_2 yields the number of atoms per unit area Nz. For the determination of z, the atomic density must be known. The application of Eq.(3.9) or Eq.(3.11) allows the transformation of the energy scale into a depth scale. There are two possibilities for the approximation of the energy losses by constant values.

(i) Surface energy approximation

For near surface layers the stopping cross sections ε_{in} and ε_{out} can be taken at the energies E_o and KE_o, respectively. In this approximation, which holds only for thin layers, the changes of ε with energy along the inward and outward paths are so small that they can be neglected. The stopping cross section factor in the surface energy approximation is then given by

$$[\varepsilon_o] = \frac{K}{\cos\varphi_1}\,\varepsilon(E_o) + \frac{1}{\cos\varphi_2}\,\varepsilon(KE_o) \quad . \tag{3.12}$$

$[\varepsilon_o]$ values for common conditions are tabulated (e.g. in Ref.[2]).

(ii) Mean energy approximation

For this approximation, the energy losses in the middle of the energy intervals for ingoing and outgoing ions are used

$$\bar{E}_{in} = \tfrac{1}{2}(E_o + E_1) \quad , \tag{3.13}$$

$$\bar{E}_{out} = \tfrac{1}{2}(KE_1 + E_2) \quad . \tag{3.14}$$

Then the stopping cross section factor in the mean energy approximation is given by

$$[\bar{\varepsilon}] = \frac{K}{\cos\varphi_1}\,\varepsilon(\bar{E}_{in}) + \frac{1}{\cos\varphi_2}\,\varepsilon(\bar{E}_{out}) \quad . \tag{3.15}$$

This approximation can be used for thicker layers for which the surface energy approximation fails.

But for the mean energy approximation the ion energy E_1 immediately before the scattering process must be known. A very useful procedure for the determination of E_1 has been described by R.F. Lever [81]. This will be discussed in Sect.4.2.1. For quick estimations it can be assumed that the energy $\Delta E_2 = KE_o - E_2$ is divided between the incident path and the outward path. Then, according to their lengths from Eqs.(3.13) and (3.14) we get

$$\bar{E}_{in} = E_o - \tfrac{1}{2}\Delta E_2/(1 + \cos\varphi_1/\cos\varphi_2) \quad , \tag{3.16}$$

$$\bar{E}_{out} = E_2 + \tfrac{1}{2}\Delta E_2/(1 + \cos\varphi_2/\cos\varphi_1) \quad . \tag{3.17}$$

This mean energy approximation is particularly good for $K \approx 1$.

RBS measurements are often performed using the channeling effect. It is well-known that the energy loss of particles travelling along channel directions may be considerably reduced (see Sect.1.3.1). This must be taken into consideration for the calculation of the energy-to-depth conversion in these cases. There is no general rule for taking this into account because the energy loss of channeled particles exhibits a complex dependence on different parameters (kind of crystal and channeling direction, crystal defects, properties of the ion beam, etc.).

3.1.5. Accessible depth

A very important quantity for RBS is the depth region under the surface of the sample which can be investigated. Generally, this accessible depth is in the order of 1 - 10 μm for backscattering analysis but the actual accessible depth depends on the ion energy, the ion mass, and the parameters of the target. The maximum measurable depth is determined by the lowest energy value of the backscattered ions to be recorded by the detector. Following the assumptions in Ref.[2], the limit for the accessible depth z_a can be presented by the somewhat arbitrary criterion $E_2 > \frac{1}{4} KE_o$ (see Fig.3.6). Thus, all ions with an energy in the region

$$\Delta E_2 = KE_o - \tfrac{1}{4} KE_o = \tfrac{3}{4} KE_o \tag{3.18}$$

will be detected. This energy interval corresponds to the accessible depth z_a (Eq.(3.11))

$$\Delta E_2 = \frac{3}{4} KE_o = [\varepsilon] Nz_a \tag{3.19}$$

The mean energy approximation (Eqs.(3.15 - 3.17)) for $\varphi_1 \approx \varphi_2$

$$\bar{E}_{in} = E_o - \frac{3}{16} KE_o \quad , \tag{3.20}$$

$$\bar{E}_{out} = E_2 + \frac{3}{16} KE_o = \frac{7}{16} KE_o \tag{3.21}$$

provides

$$z_a = \frac{3}{4} KE_o \frac{1}{N}\left[\frac{K}{\cos\varphi_1} \left.\frac{dE}{ds}\right|_{E_o-(3/16)KE_o} + \frac{1}{\cos\varphi_2} \left.\frac{dE}{ds}\right|_{(7/16)KE_o}\right]^{-1} . \tag{3.22}$$

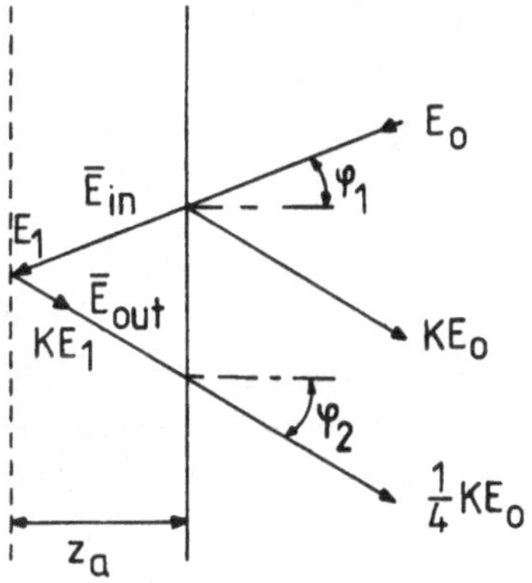

Fig.3.6. Scheme for the definition of the accessible depth

Values for the accessible depth for H and He ions backscattered on Si, Ni, Pd, and Au targets calculated by Eq.(3.22) are given in Table 3.1 for the conditions $\theta = 170^o$, $\varphi_1 = 0^o$, $\varphi_2 = 10^o$, $E_o(H^+) = 0.2$ MeV and 1.4 MeV, $E_o(He^+) = 1.4$ MeV and 2.0 MeV. From Table 3.1 it can be seen that the accessible depth for 1.4 MeV H ions is in the region between 4...10 μm and for 1.4 MeV and 2.0 MeV He ions in the order of about 1 μm. In practice, the use of low energy H ions (e.g. 0.2 MeV) allows only the investigation of very shallow layers in a depth region of about 0.3 μm.

As can be seen from Eq.(3.22), the accessible depth z_a depends considerably on the incidence and exit angles between the ion beam and the normal on the target. Glancing angle techniques which increase the depth resolution (see the following Sect.3.1.6) yield a drastic decrease of the depth which can be investigated.

TABLE 3.1

Accessible depth z_a for H and He ions backscattered at $\theta = 170^\circ$, $\varphi_1 = 0^\circ$, $\varphi_2 = 10^\circ$ for Si, Ni, Pd, and Au. Stopping cross section taken from J.F. Ziegler [30,31].

Ion	E_o/MeV	z_a/µm in			
		Si	Ni	Pd	Au
H	0.2	0.60	0.32	0.27	0.33
H	1.4	9.56	3.90	4.00	3.80
He	1.4	1.13	0.70	0.70	0.70
He	2.0	1.30	0.95	1.00	1.00

3.1.6. Depth resolution

The depth resolution δz is given by the uncertainty of E_2. According to Eq.(3.11) the depth resolution δz is related to the energy resolution δE_2 by (δz and δE_2 are FWHM values)

$$\delta z = \frac{\delta E_2}{N[\varepsilon]} = \frac{\delta E_2}{N\left[\frac{K}{\cos\varphi_1}\varepsilon_{in} + \frac{1}{\cos\varphi_2}\varepsilon_{out}\right]} . \quad (3.23)$$

There are different contributions δE_2 to the energy uncertainty, which will be discussed in the following. The width of the particular contribution is given by its full width half maximum value (FWHM). The main parts are (see Refs.[82-86])

Detector energy resolution δE_d

This contribution can be assumed to be independent of the analysis conditions. For surface barrier detectors, which are preferentially used, the energy resolution for MeV He beams is in the order of 12-15 keV and for MeV H beams it is about 5-8 keV. Improved detector resolution can be obtained by using electrostatic or magnetic analyzers which have an energy resolution of about 1 keV.

Energy straggling in the target δE_s

According to the treatment in Sect.1.2.2 the energy straggling along the path z consists of two contributions, the nuclear part and the electronic part. The total energy straggling is given by (see Eq.(1.112), $w_n = 1$ for $\bar{E} \gg E_\varepsilon$)

$$\Omega^2 = 260.5\, Z_1^2\left[Z_2^2\left(\frac{M_1}{M_1+M_2}\right)^2 + w_e(\bar{E})\, Z_2\right] Nz*10^{-15}\ (\text{eV cm})^2 , \quad (3.24)$$

where N is the atomic density and $\bar{E}$ is the average of the ion energy before and after penetrating the layer of thickness z. The factor w_e shows a weak energy dependence and takes into consideration deviations from the Bohr formula of the electronic part of the energy straggling due to screening and quantum mechanical effects. It is tabulated in the Appendix 7 according to Ref.[1]. In the case of backscattering events at the depth z, the ions undergo an energy straggling along the inward and along the outward path. The straggling for the incoming path has to be modified because at the end there is an elastic scattering event by which the energy will be changed to KE_1 yielding an alteration of the energy straggling by a factor K^2. Further, the directions of the ion beam relative to the target normal can be inclined so that we have a prolongation of the ion paths given by factors $1/\cos\varphi_1$ and $1/\cos\varphi_2$. The energy spread δE_s (FWHM) is equal to 2.355 Ω . Finally, we get

$$\delta E_s = 12.02\, Z_1 \left\{ \left[Z_2^2 \left(\frac{M_1}{M_1+M_2} \right)^2 + w_e(\bar{E}) Z_2 \right] \left[\frac{K^2}{\cos\varphi_1} + \frac{1}{\cos\varphi_2} \right] Nz \text{ cm}^2 \right\}^{1/2} * 10^{-7} \text{eV}. \tag{3.25}$$

The correction factor $w_e(\bar{E})$ exhibits a weak dependence on the ion energy for He ions (and also for higher values of Z_1). For H ions $w_e = 1$ is a good approximation in the energy region applied to RBS. The energy straggling scales with $(z)^{1/2}$.

Multiple scattering in the target δE_{ms}

Multiple scattering events occur along the ingoing and along the outgoing path of the ions, providing angular and lateral spread. The resulting energy spread contributions are given by

$$\delta E_{ms}^{in} = \frac{\partial E_2}{\partial K} \left[\frac{\partial K}{\partial \theta} + K\varepsilon_{in} \frac{Nz}{\cos\varphi_1} \frac{tg\varphi_1}{\Gamma^{in}} \right] \Delta\varphi^{in} \tag{3.26}$$

for the inward path and by

$$\delta E_{ms}^{out} = \left[\frac{\partial E_2}{\partial K} \frac{\partial K}{\partial \theta} + \varepsilon_{out} \frac{Nz}{\cos\varphi_2} \frac{tg\varphi_2}{\Gamma^{out}} \right] \Delta\varphi^{out} \tag{3.27}$$

for the outward path, where $\Delta\varphi^{in}(z/\cos\varphi_1)$ and $\Delta\varphi^{out}(z/\cos\varphi_2)$ are the multiple scattering angles (FWHM) for ingoing and outgoing particles taken from the tabulations of Sigmund and Winterbon [39]. $\Gamma^{in}(z/\cos\varphi_1)$ and $\Gamma^{out}(z/\cos\varphi_2)$ are the corresponding scaling functions of Marwick and Sigmund [87] which correlate the angular and the lateral spread. In Eqs.(3.26) and (3.27) the first terms in brackets exhibit the angular effect and the second terms correspond to the lateral spread.

Geometrical resolution δE_g

The finite detector acceptance angle and the finite beam size cause a geometric contribution to the energy spread. This is given by

$$\delta E_g = \left[\frac{\partial E_2}{\partial K}\frac{\partial K}{\partial \theta} + (KE_1 - E_2)\ \mathrm{tg}\varphi_2\right]\Delta\gamma_{det} \quad , \qquad (3.28)$$

where

$$\Delta\gamma_{det} = \frac{1}{D}\,(s^2 + d^2\cos^2\varphi_2/\cos^2\varphi_1)^{1/2} \qquad (3.29)$$

is the effective detector acceptance angle for the detector-target distance D, the incident beam width d in the scattering plane and the detector aperture width s.

The first term in the square brackets represents the energy spread caused by different scattering angles and the second term is due to the path length differences in the outgoing beam. There is also an additional contribution for the incident path resulting from the beam divergence. But for well-collimated beams as normally used this contribution can be neglected.

Energy spread δE_{sr} due to surface roughness

The surface roughness of the sample may have considerable influence on the total energy resolution especially for glancing angle experiments. In these cases the surface roughness must be determined very thoroughly, which may be difficult, but in most cases the surfaces of the samples are smooth enough so that this contribution can be neglected.

Assuming noncorrelated processes and Gaussian distributions for the different energy spreads, the total energy resolution is given by

$$(\delta E_2)^2 = (\delta E_d)^2 + (\delta E_s)^2 + (\delta E_{ms}^{in})^2 + (\delta E_{ms}^{out})^2 + (\delta E_g)^2 + (\delta E_{sr})^2 . \qquad (3.30)$$

As an example Fig.3.7 shows the magnitude of the main contributions to the energy resolution as a function of the angle of incidence φ_1 for a typical experimental arrangement (shown in the insert). Note that for small depths (46 nm Al) the detector resolution is the dominating term for incident angles $\varphi_1 < 80^\circ$. For angles $\varphi_1 > 85^\circ$ the lateral spread of the ingoing beam dominates. Those terms not shown in Fig.3.7 (e.g. angular spread) can be neglected.

Thus, the depth resolution for near-surface measurements in standard geometry ($\varphi_1 = 0^\circ$, $\varphi_2 = 10^\circ$) is only determined by the energy resolution δE_d of the detector. In Fig.3.8 the depth resolution at the surface

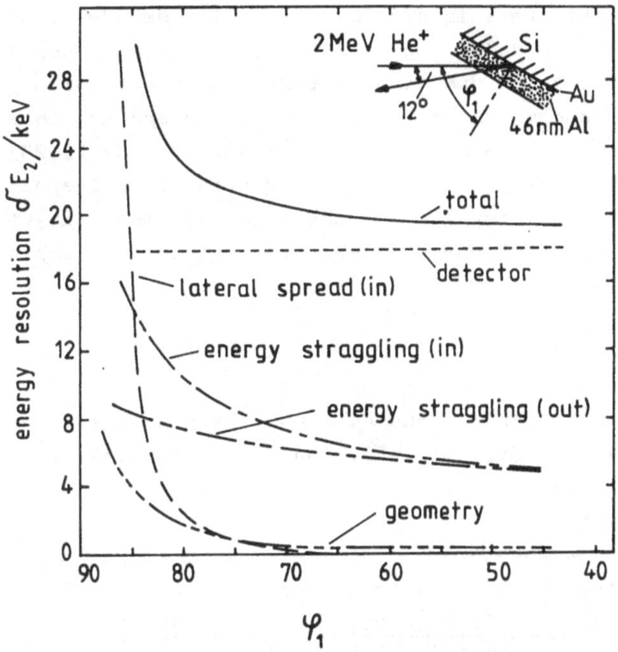

Fig.3.7. Individual contributions to the energy resolution for a depth of z=46nm, obtained from theoretical calculations for the experimental conditions shown in the insert (Ref.[85])

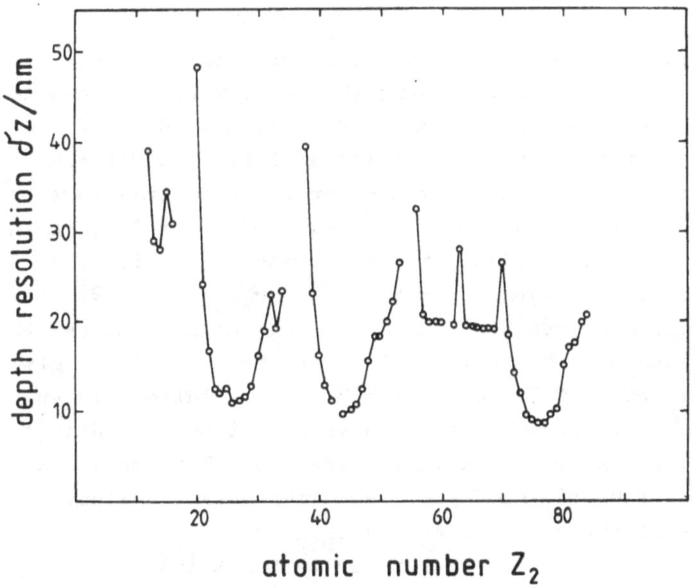

Fig.3.8. Depth resolution δz at the target surface as a function of the atomic number Z_2 for 1.4 MeV He ions (energy resolution δE_2 = 14 keV, scattering angle Θ = 170°, φ_1 = 0°, φ_2 = 10°)

versus atomic number of target atoms is displayed for 1.4 MeV He ions and a detector resolution of 14 keV. The large differences in depth resolution for the various atoms are due to different atomic densities.

In Table 3.2 the depth resolution at the surface for H and He ions for some elements and typical ion energies is summarized. From Eq.(3.23) it must be concluded that the depth resolution is essentially influenced by the value of the specific energy loss of the ions in the target material. The higher the specific energy loss, the better the depth resolution.

TABLE 3.2
Near-surface depth resolution for H and He ions backscattered at $\theta = 170^{o}$ ($\varphi_1 = 0^{o}$, $\varphi_2 = 10^{o}$) for Si, Ni, Pd, and Au. Stopping cross sections taken from J.F. Ziegler [30,31].

Ion	E_o/MeV	δz/nm in			
		Si	Ni	Pd	Au
H	0.2	42.7	19.1	16.3	17.2
(δE_d=8 keV)	1.4	115.5	40.9	41.0	36.8
He	1.4	27.0	11.1	10.1	9.7
(δE_d=14keV)	2.0	30.2	11.8	11.6	10.5

For various kinds of ions the energy regions for the greatest specific energy loss are different. For H and He ions the maximum energy loss is reached for energies between 100 keV and 200 keV, and 0.4 MeV and 1.5 MeV, respectively. Typical values are listed in Table 3.2. It can be seen that the depth resolution at the surface for standard measurements ($E_o(H^+)$ = 0.2 MeV, $E_o(He^+)$ = 1.4 MeV; $\theta = 170^{o}$, $\varphi_1 = 0^{o}$, $\varphi_2 = 10^{o}$) is in the order of 10 - 20 nm. For silicon the depth resolution is poorer (30 - 40 nm) because of its low atomic density and low mass M_2 = 28 amu. In this case the energy loss reaches maximum values at an energy of about 75 keV for H ions and of about 0.4 MeV for He ions (see also Sect.1.2.1.4). In the case of He ions, the depth resolution is only weakly influenced by an increase of the ion energy from 1.4 MeV to 2.0 MeV. For H ions the change of the energy from 0.2 MeV to 1.4 MeV yields a remarkable decrease of the depth resolution (higher values of δz) due to the decrease of the specific energy loss.

The surface approximation of the energy resolution holds only for small depths, usually up to about z = 50 nm. For larger thicknesses the energy straggling δE_s in the target has to be taken into consideration. In Table 3.3 the energy spread δE_s according to Eq.(3.25) is represented for some typical RBS conditions (the same as used in Table 3.2). For standard backscattering geometry ($\theta = 170^\circ$, $\varphi_1 = 0^\circ$, $\varphi_2 = 10^\circ$) the terms δE_{ms} (multiple scattering), δE_g (geometrical resolution) and δE_{sr} (surface roughness) can be neglected, so that the total energy spread given by Eq.(3.30) is reduced to

$$\delta E_2 = [(\delta E_d)^2 + (\delta E_s)^2]^{1/2} . \qquad (3.31)$$

With the energy straggling represented in Table 3.3 and the energy resolution of the detector (δE_d = 8 keV for H ions and δE_d = 14 keV for He ions) the total energy resolution δE_2 can be calculated. Then the depth resolution δz can be obtained by making use of Eq.(3.23). Because of larger thicknesses considered, the mean energy approximation (Eqs.(3.16) and (3.17)) is used instead of the surface energy approximation. Values for the depth resolution are given in Table 3.4.

TABLE 3.3
Calculated energy spread δE_s for H and He ions backscattered at z = 1 μm from Si, Ni, Pd, and Au ($\theta = 170^\circ$, $\varphi_1 = 0^\circ$, $\varphi_2 = 10^\circ$). Stopping cross sections taken from J.F. Ziegler [30,31].

Ion	E_0/MeV	δE_s/keV at z = 1 μm in			
		Si	Ni	Pd	Au
H	independent	13.5	26.6	29.7	36.5
He	1.4	21.9	40.2	42.0	44.0
He	2.0	22.2	37.4	38.8	40.4

TABLE 3.4
Depth resolution at z = 1.0 μm and depth z ($\delta E_s = \delta E_d$) for H and He ions backscattered at $\theta = 170^\circ$, $\varphi_1 = 0^\circ$, $\varphi_2 = 10^\circ$ for Si, Ni, Pd, and Au. Stopping cross sections taken from J. F. Ziegler [30,31].

Ion	E_0/MeV	δz/nm at z = 1 μm in				z($\delta E_s = \delta E_d$)/nm in			
		Si	Ni	Pd	Au	Si	Ni	Pd	Au
H	1.4	165.3	110.2	121.4	136.9	351	90	73	48
He	1.4	49.6	35.6	33.3	32.5	409	121	111	101
He	2.0	51.8	32.9	32.9	30.0	398	140	130	120

In Table 3.4 the depth values z ($\delta E_s = \delta E_d$) for which the energy straggling is equal to the energy resolution of the detector are also represented. This may be a rough estimation, up to which depth the depth resolution is dominated by detector resolution. It must be mentioned that the data in Tables 3.3 and 3.4 are intended only as an indication of the approximate energy straggling and depth resolution, because there are only very few experimental results on energy straggling. Note that a high depth resolution is always connected with a small value of the accessible depth (compare Table 3.1 and Table 3.4).

From Eq.(3.23) we find that the depth resolution can be improved considerably using glancing angles for the incidence and/or for the outgoing direction of the analyzing ions, that means for angles φ_1 and/or φ_2 near 90°. There are three ways to increase the path lengths of the ions and therefore the depth resolution as shown in Fig.3.9

(i) Both the inward and the outward path lengths are increased by the employment of glancing angles φ_1 and φ_2 (Fig.3.9a).

(ii) The normal incidence is preserved ($\varphi_1 = 0^\circ$) and the outward path is increased by the glancing exit angle φ_2 (Fig.3.9b).

(iii) The reverse of case 2, i.e. normal exit ($\varphi_2 = 0^\circ$) and glancing incidence angle φ_1 (Fig.3.9c).

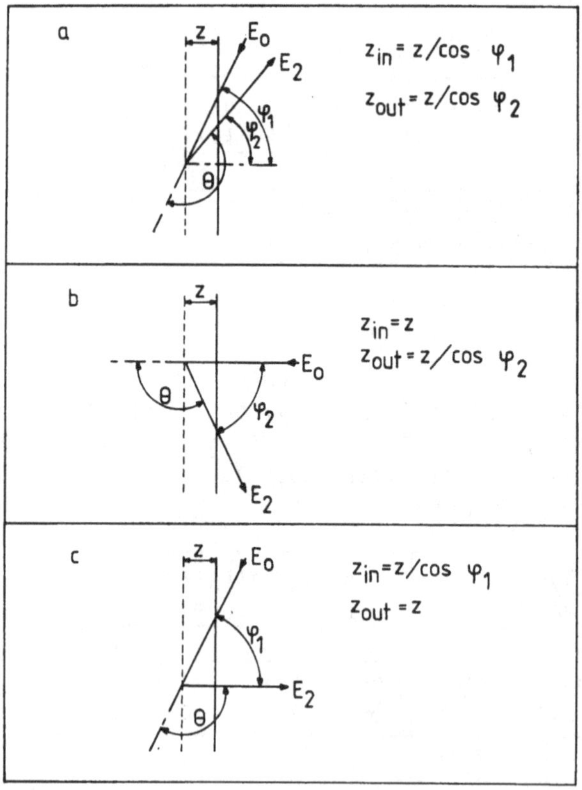

Fig.3.9. Glancing angle geometries for backscattering experiments: glancing angle incidence, glancing angle exit (a), normal incidence, glancing angle exit (b), glancing angle incidence, normal exit (c)

The highest depth resolution is obtained using glancing angle incidence and glancing angle exit (Fig.3.9a). Detailed investigations of optimized geometries and their applications were done by Williams [82-85]. At the surface of Si samples, where the energy resolution is only limited by the detector resolution δE_d ($\approx$14 keV for MeV He ions) a depth resolution of about δz = 2 nm can be obtained for $\varphi_1 \geq 85^o$, $\varphi_2 \geq 75^o$. But care must be taken using this optimized geometry because the sample must be large and flat enough, since the beam spot becomes much larger on the target than the actual beam size. For RBS investigations in combination with channeling techniques the geometry shown in Fig.3.9b is mostly used. For that, the normal orientation procedure to find the channel direction can be applied. Otherwise, using glancing angle incidence (Fig.3.9a and c), it is very difficult to orient the sample.

Generally, the depth resolution for investigations with glancing angles is influenced by all factors contributing to the total energy resolution (see Eq.(3.30)). As shown by Williams, the different energy resolution contributions depend on the thickness z of the analyzed layer. Investigations of Al layers with He ions (E_o = 2 MeV, $\varphi_1 = 85^o$, $\varphi_2 = 73^o$) provide the following results. Up to a depth of 20 nm, the depth resolution is solely determined by the detector resolution (and surface topography effects). The other terms are negligibly small. Between 20 - 40 nm the energy straggling begins to have an influence, whereby the depth resolution deteriorates about 20 - 25 %. For a depth of about 70 nm the detector resolution δE_d, the energy straggling δE_s and the multiple scattering δE_{ms} due to the lateral spreading have roughly the same effect. For z > 90 nm the lateral spreading becomes the dominant influence. For the usual configurations the beam divergence and the influence of the detector acceptance angle summarized by δE_g, and the angular spread due to the multiple scattering δE_{ms} can be neglected for all thicknesses. The depth resolution for larger depths is substantially influenced by the lateral spreading especially for large incident angles ($\varphi_1 > 85^o$).

For very thin layers (z < 10 nm) the combination of glancing angle scattering geometries with high energy-resolution detectors as electrostatic or magnetic analyzers, near-surface resolutions in the order of 1 nm in silicon can be obtained (see Ref.[88]).

3.1.7. Scattering cross section

For normal incidence of a uniform ion beam on a homogeneous elemental target, the total number of ions recorded by the detector is given by

$$dY = n\,(d\sigma/d\Omega)\,\Delta\Omega\,Hdz\;, \qquad (3.32)$$

with

n : total number of incident ions,

$(d\sigma/d\Omega)$: differential scattering cross section,

$\Delta\Omega$: solid angle of detection,

Ndz : number of target atoms per unit area within dz.

The scattering geometry is shown in Fig.3.1. Eq.(3.32) is valid under the following assumptions:

1. The solid detector angle $\Delta\Omega$ must be small enough to define a fixed scattering angle.
2. The thickness dz has to be so small that the energy loss of the ions in the layer dz can be neglected.

In most cases, the cross section for elastic backscattering of the ions from the target atoms can be calculated using the Coulomb potential. This was done first in 1911 by Rutherford [89] and therefore it is called the Rutherford cross section. From this the denotation Rutherford Backscattering Spectrometry (RBS) is derived.

The elastic scattering process is described in detail in Sect.1.1.1.5. For the laboratory system we get

$$\left(\frac{d\sigma}{d\Omega}\right)^R = \left(\frac{Z_1 Z_2 e^2}{8\pi\varepsilon_0 E_1}\right)^2 \frac{\left\{\cos\theta + [1-(M_1/M_2)^2\sin^2\theta]^{1/2}\right\}^2}{\sin^4\theta\,[1-(M_1/M_2)^2\sin^2\theta]^{1/2}} , \tag{3.33}$$

with

θ : scattering angle in the laboratory system,

E_1 : ion energy immediately before the scattering process (see Fig.3.5),

ε_0 : permittivity of vacuum,

e : elementary charge.

Now that in Eq.(3.32) the differential scattering cross section and the solid angle of detection are known, n can be obtained very easily by measuring the current charge of the beam and dY will be recorded by the detector. Thus, the number of target atoms per unit area Ndz can be determined. The great advantage of RBS is that values of Ndz are directly attainable so that a quantitative analysis is possible without use of etalons.

For a scattering angle $\theta = 180^\circ$ the Rutherford cross section (Eq.(3.33)) turns to its lowest value

$$\left(\frac{d\sigma}{d\Omega}\right)^R_{180^\circ} = \left(\frac{Z_1 Z_2 e^2}{16\pi\varepsilon_0 E_1}\right)^2 \left[1 - (M_1/M_2)^2\right]^2 . \tag{3.34}$$

For $M_1 << M_2$ Eq.(3.33) can be represented by a power series (Ref.[90])

$$\left(\frac{d\sigma}{d\Omega}\right)^R \approx \left(\frac{Z_1 Z_2 e^2}{16\pi\varepsilon_o E_1}\right)^2 \left[\sin^{-4}(\theta/2) - 2(M_1/M_2)^2 + \ldots\right] \quad , \tag{3.35}$$

where the first term omitted is in the order of $(M_1/M_2)^4$.

The scattering cross section in Eqs.(3.33 - 3.35) is essentially determined by the first term. The others are of order unity. To get an idea about the order of the cross section we choose the following example: A Si sample is investigated by 1.4 MeV He ions.
With $e^2/4\pi\varepsilon_o = 1.44*10^{-13}$ MeV cm and 10^{-24} cm^2 = 1 barn follows $(Z_1 Z_2 e^2/16\pi\varepsilon_o E_1)^2$ = 0.518 barn per steradian. In comparison to nuclear reaction cross sections this is a relatively large value.

The dependence of the Rutherford cross section on significant parameters (Eqs.(3.33 - 3.35)) are discussed in the following

1. $(d\sigma/d\Omega)^R$ is proportional to Z_1^2. This means that the application of ions with high atomic numbers increases the backscattering yield. But in practice, He ions are used most in the MeV region because the accessible depth for heavier ions would be too low and a remarkable amount of damage would be caused in the samples by such ions. This situation changes for higher ion energies, thus heavy ions ($Z_1 \leq 20$) may be very useful for energies in the region 10...50 MeV.
2. $(d\sigma/d\Omega)^R$ is proportional to Z_2^2. That means the sensitivity for detecting heavy target atoms is much higher than for light target atoms.
3. $(d\sigma/d\Omega)^R$ is proportional to E_1^{-2}. Thus, the backscattering yield increases with decreasing ion energy. As a consequence, the yield increases with increasing depth of the backscattering processes, due to the energy loss of the ions.
4. $(d\sigma/d\Omega)^R$ is approximately inversely proportional to $\sin^4(\theta/2)$. This provides a rapidly increasing scattering yield for reduced scattering angles.

   Near $\theta = 180^\circ$ the cross section exhibits only a weak dependence on the scattering angle. The difference between $\theta = 180^\circ$ and $\theta = 170^\circ$ is only about 1.5%.

Values of the Rutherford scattering cross section of various elements are listed in Appendix 10 for 1 MeV H ions and in Appendix 11 for 1 MeV He ions.

Deviations from the Rutherford scattering cross section arise from screening effects and from nuclear processes. As discussed in Sect.1.1.1.5, screening effects have to be taken into consideration for backscattering of low energy ions from heavy target atoms. They provide a correction of some percent. For sufficiently high energies the collision parameter can be very small, so that the nuclear forces may have an influence on the scattering events. Elastic and inelastic processes are possible. In all cases the scattering cross section will differ from the Rutherford cross section, exhibiting strong dependences on ion energy, on scattering angle, and on the kinds of collision partners.

At the beginning of this section we have determined the backscattering yield for normal incidence of the ion beam (Eq.(3.32)). For other angles of incidence, i.e. $\varphi_1 \neq 0$, the path length of an ion through the layer z is increased by a factor of $1/\cos\varphi_1$ (see e.g. Fig.3.9c). This causes an increase of the backscattering yield

$$dY = n \left(\frac{d\sigma}{d\Omega}\right)^R \Delta\Omega \frac{N dz}{\cos\varphi_1} \quad . \tag{3.36}$$

In the following, the abbreviation $\sigma^d = (d\sigma/d\Omega)^R$ will be used for the differential Rutherford scattering cross section in the laboratory system.

3.1.8. Spectrum height

3.1.8.1. Single-elemental layer in surface approximation

For backscattering events which take place in a thin surface layer, the ion energy before the scattering process can be assumed to be $E_1 = E_o$ and from Eq.(3.36) the height of the spectrum is given by

$$Y_o = n\, \sigma^d(E_o)\, \Delta\Omega\, \frac{N\Delta z_o}{\cos\varphi_1} \quad . \tag{3.37}$$

The thickness Δz_o of the layer corresponds to the width of an energy channel ϵ_{ch} of the analyzing system. According to Eqs.(3.11) and (3.12) $\Delta E_2 = \epsilon_{ch}$ is related to $z = \Delta z_o$ by

$$\epsilon_{ch} = [\epsilon_o]\, N\Delta z_o \quad . \tag{3.38}$$

Thus we can replace the unknown value $N\Delta z_o$ in Eq.(3.37) by the known value ϵ_{ch} yielding

$$Y_o = n\, \sigma^d(E_o)\, \Delta\Omega\, \frac{\epsilon_{ch}}{[\epsilon_o]\cos\varphi_1} \quad . \tag{3.39}$$

Eq.(3.39) shows that the height of a backscattering spectrum does not depend on the atomic density N of the target. On the other hand, the spectrum height is inversely proportional to $[\epsilon_o]\cos\varphi_1$ which decreases with increasing angle of incidence φ_1 as can be seen from Eq.(3.12).

3.1.8.2. Thick single-elemental layer

Now let us consider the spectrum height for scattering events from a thin layer Δz_i at a certain depth z_i. The circumstances are demonstrated in Fig.3.10. The corresponding spectrum height Y_i is given by

$$Y_i = n\,\sigma^d(E_{1,i})\,\Delta\Omega\,\frac{N\Delta z_i}{\cos\varphi_1} \quad . \tag{3.40}$$

The differential scattering cross section has to be evaluated for the ion energy $E_{1,i}$ immediately before the scattering process at the depth z_i. Again the thickness Δz_i is correlated to the channel width ϵ_{ch} of the detecting system. But note that the energy width after the scattering processes in Δz_i is no longer ϵ_{ch} but a quantity ϵ_i which cannot be measured (see Fig.3.10).

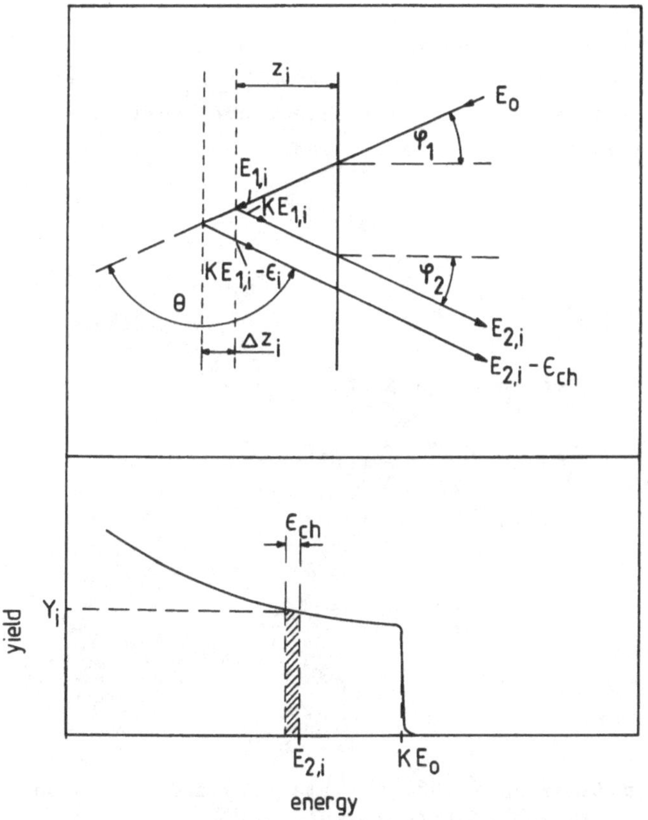

Fig.3.10. Scheme for backscattering processes in a thin layer Δz_i at depth z_i (single-elemental target); geometry and energy denotations (upper part), resulting energy spectrum (lower part)

To obtain a relation between ϵ_{ch} and Δz_i, as a first step the energy interval ϵ_i has to be related to the layer thickness Δz_i. The ions entering the layer Δz_i have the energy $E_{1,i}$. Thus, we can reduce the problem to that of near-surface scattering taking into account the energy loss in a cover layer of thickness z_i. Then from Eq.(3.38) it follows that

$$\epsilon_i = [\varepsilon(E_{1,i})]\, N\Delta z_i \quad . \tag{3.41}$$

The stopping cross section factor is given by Eq.(3.12)

$$[\varepsilon(E_{1,i})] = \frac{K}{\cos\varphi_1}\,\varepsilon(E_{1,i}) + \frac{1}{\cos\varphi_2}\,\varepsilon(KE_{1,i}) \quad . \tag{3.42}$$

With this the height of the spectrum becomes

$$Y_i = n\,\sigma^d(E_{1,i})\,\Delta\Omega\,\frac{\epsilon_i}{[\varepsilon(E_{1,i})]\cos\varphi_1} \quad . \tag{3.43}$$

The second step is to determine the energy width ϵ_i in such a way that the final energy width is equal to the measurable value ϵ_{ch}. As can be seen from Fig.3.10, the path lengths of the backscattered particles are the same in the energy region from $KE_{1,i}$ to $E_{2,i}$ and from $KE_{1,i} - \epsilon_i$ to $E_{2,i} - \epsilon_{ch}$. Thus we get

$$\int_{E_{2,i}-\epsilon_{ch}}^{KE_{1,i}-\epsilon_i} dE/\varepsilon = \int_{E_{2,i}}^{KE_{1,i}} dE/\varepsilon \quad . \tag{3.44}$$

The first integral can be divided in three parts

$$\int_{E_{2,i}-\epsilon_{ch}}^{KE_{1,i}-\epsilon_i} dE/\varepsilon = \int_{E_{2,i}-\epsilon_{ch}}^{E_{2,i}} dE/\varepsilon + \int_{E_{2,i}}^{KE_{1,i}} dE/\varepsilon + \int_{KE_{1,i}}^{KE_{1,i}-\epsilon_i} dE/\varepsilon \quad , \tag{3.45}$$

providing with Eq.(3.44)

$$\int_{E_{2,i}}^{E_{2,i}-\epsilon_{ch}} dE/\varepsilon = \int_{KE_{1,i}}^{KE_{1,i}-\epsilon_i} dE/\varepsilon \quad . \tag{3.46}$$

Within the integration limits in Eq.(3.46) the stopping cross section ε can be assumed to be energy independent (note that ϵ_i and ϵ_{ch} are very small compared with the energy $E_{1,i}$) yielding

$$\frac{\epsilon_i}{\epsilon_{ch}} = \frac{\varepsilon(KE_{1,i})}{\varepsilon(E_{2,i})} \quad . \tag{3.47}$$

Finally, the spectrum height in a given channel corresponding to Δz_i is achieved from Eq.(3.40) using Eqs.(3.41) and (3.47)

$$Y_i = n\,\sigma^d(E_{1,i})\,\Delta\Omega\,\frac{\epsilon_{ch}}{[\epsilon(E_{1,i})]\cos\varphi_1}\,\frac{\epsilon(KE_{1,i})}{\epsilon(E_{2,i})} \quad . \tag{3.48}$$

It must be mentioned that as in the case of the determination of the energy loss by the mean energy approximation (Sect.3.1.4) the energy E_1 immediately before the scattering process must be known. The solution of this problem will be discussed in detail in Sect.4.2.1.

The total yield Y_{tot} of backscattered particles due to scattering processes in a layer of thickness z is given by integration of Eq.(3.36). For thick samples, the energy dependence of the Rutherford cross section (Eq.(3.33)) has to be taken into account

$$\sigma^d(E_1) = \sigma^d(E_o)\,(E_o/E_1)^2 \quad . \tag{3.49}$$

The integration yields

$$Y_{tot} = n\,\sigma^d(E_o)\,\Delta\Omega\,\frac{N}{\cos\varphi_1}\,E_o^2\int_o^z dz\,E_1^{-2}(z) \quad . \tag{3.50}$$

Within the mean energy approximation (Eq.(3.13)) and using Eqs.(3.6) and (3.7) we obtain

$$Y_{tot} = n\,\sigma^d(E_o)\,\Delta\Omega\,\frac{N}{\cos\varphi_1}\,E_o^2\int_o^z dz\left[E_o - \frac{Nz}{\cos\varphi_1}\epsilon(\bar{E}_{in})\right]^{-2} \quad . \tag{3.51}$$

Performing the integration, the total backscattering yield is given by

$$Y_{tot} = n\,\sigma^d(E_o)\,\Delta\Omega\,\frac{Nz}{\cos\varphi_1}\left[1 - \frac{Nz}{\cos\varphi_1}\,\frac{\epsilon(\bar{E}_{in})}{E_o}\right]^{-1} \quad . \tag{3.52}$$

3.1.8.3. Thick multielemental layer

Now let us consider a thick sample which contains m different kinds of atoms. The circumstances are represented in Fig.3.11. The composition of a layer of thickness Δz_i at a certain depth z_i is described by $A^{(1)}_{n_1}A^{(2)}_{n_2}\ldots A^{(j)}_{n_j}\ldots A^{(m)}_{n_m}$ (e.g. Si_3N_4) where $A^{(j)}$ denotes the specific element j and n_j is the ralative number of atoms of kind j in the layer Δz_i. The atomic density of the element $A^{(j)}$ is given by $N_{A^{(j)}}$.

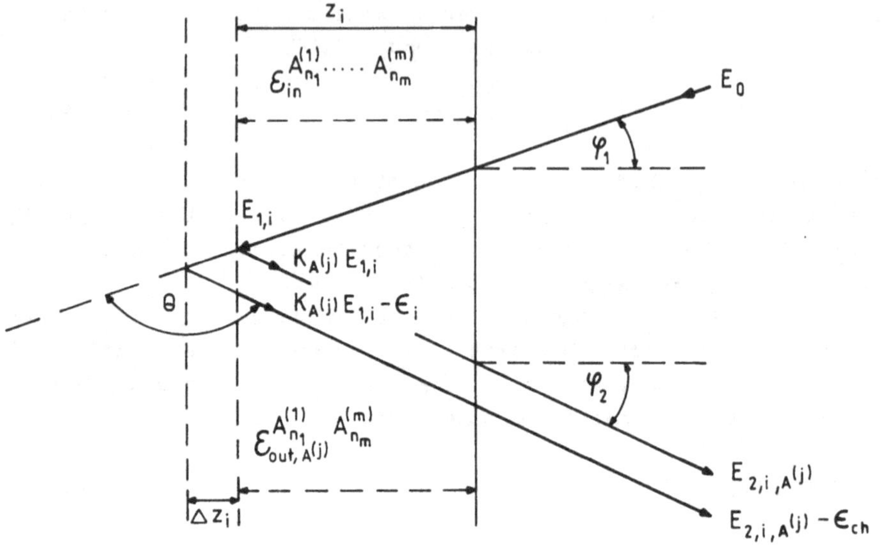

Fig.3.11. Scheme for backscattering processes in a thin layer Δz_i at a depth z_i (multielemental target)

According to Bragg's rule (see Eq.(1.118)) the mean stopping cross section per atom is given by

$$\epsilon^{A_{n_1}^{(1)} A_{n_2}^{(2)} \ldots A_{n_j}^{(j)} \ldots A_{n_m}^{(m)}} = \sum_{j=1}^{m} n_j \epsilon^{A^{(j)}} / \sum_{j=1}^{m} n_j \quad . \tag{3.53}$$

Analogous to the way described in Sect.3.1.8.2 for a single-elemental layer, the spectrum height for a multielemental layer can be attained taking into account that the Rutherford cross section $\sigma^d_{A(j)}(E_{1,i})$ has different values for the different elements. Thus, the spectrum height correlated to scattering processes at the atoms of kind $A^{(j)}$ in the layer Δz_i at depth z_i is given by

$$Y_{i,A^{(j)}} = n\, \sigma^d_{A^{(j)}}(E_{1,i})\, \Delta\Omega\, \frac{N_{A^{(j)}}\, \Delta z_i}{\cos\varphi_1} \quad . \tag{3.54}$$

Using the same procedure as in Sect.3.1.8.2 for determining the layer thickness Δz_i by the channel width ϵ_{ch} from Eq.(3.54), we get

$$Y_{i,A^{(j)}} = n\,\sigma^{d}_{A^{(j)}}(E_{1,i})\,\Delta\Omega\,\frac{n_j}{\sum_{j=1}^{m} n_j} * \frac{\epsilon_{ch}}{[\epsilon(E_{1,i})]_{A^{(j)}}^{A^{(1)}_{n_1}\ldots A^{(m)}_{n_m}}\cos\varphi_1}\,\frac{\epsilon^{A^{(1)}_{n_1}\ldots A^{(m)}_{n_m}}(K_{A^{(j)}}E_{1,i})}{\epsilon^{A^{(1)}_{n_1}\ldots A^{(m)}_{n_m}}(E_{2,i,A^{(j)}})}, \tag{3.55}$$

where

$$[\epsilon]_{A^{(j)}}^{A^{(1)}_{n_1}\ldots A^{(m)}_{n_m}} = \frac{K_{A^{(j)}}}{\cos\varphi_1}\,\epsilon_{in}^{A^{(1)}_{n_1}\ldots A^{(m)}_{n_m}} + \frac{1}{\cos\varphi_2}\,\epsilon_{out,A^{(j)}}^{A^{(1)}_{n_1}\ldots A^{(m)}_{n_m}} \tag{3.56}$$

is the stopping cross section factor for ions backscattered at an atom $A^{(j)}$ and with the stopping cross sections $\epsilon_{in}^{A^{(1)}_{n_1}\ldots A^{(m)}_{n_m}}$ and $\epsilon_{out,A^{(j)}}^{A^{(1)}_{n_1}\ldots A^{(m)}_{n_m}}$ for the inward path and for the outward path, respectively. Note that corresponding to the different kinematic factors $K_{A^{(j)}}$ there are also different energy values for particles scattered in the same depth layer Δz_i but on different kinds of atoms. Therefore the energy-to-depth conversion has to be determined separately for each kind of atom. On the other hand, the Rutherford cross section depends also on the identity of the ions. Thus, the backscattering spectra for the different elements may overlap and numerical treatments are necessary to get the concentration profiles for the different elements. For typical examples this is demonstrated in Chapter 4.

3.1.9. Rutherford backscattering spectra

By means of Eq.(3.55) general shapes of Rutherford backscattering spectra can be derived. An ion penetrating a target loses energy. As a consequence, the scattering cross section which scales with E_1^{-2} is increased. In the case where the specific energy loss of the ions is nearly constant, which is fulfilled near the maximum of the energy loss curves (specific energy loss versus energy), the spectra exhibit a clear E_1^{-2}-dependence. On the other hand, the specific energy loss can either increase or decrease with decreasing ion energy. If the specific energy loss increases with decreasing energy, the yield tends to decrease according to the inverse proportionality to $\epsilon(E_1)$. This compensates for the effect of the scattering cross section. When the specific energy loss decreases with decreasing energy, which occurs in energy regions below the maximum of the specific energy loss, the energy dependence becomes stronger than the E_1^{-2}-dependence given by the scattering cross section. The factor $\epsilon(KE_1)/\epsilon(E_2)$ only weakly depends on the energy.

As already mentioned in Sect.3.1.8.3, Rutherford backscattering measurements allow the determination of separate concentration profiles for each element in the sample if the difference of the corresponding masses is sufficiently large. The mass and depth resolution as well as the detection sensitivity depend on the beam parameters. This is illustrated by an example considering the shape of backscattering spectra obtained from a sample with a surface layer of a second element having a mass $M_2^{(2)}$ greater than the mass $M_2^{(1)}$ of the bulk material. A schematic representation of the spectra is given in Fig.3.12. As a starting point, let us consider the spectrum for normal incidence and conditions often used (1.4 MeV He ions, $\Theta = 170^{\circ}$; Fig.3.12a). The parts

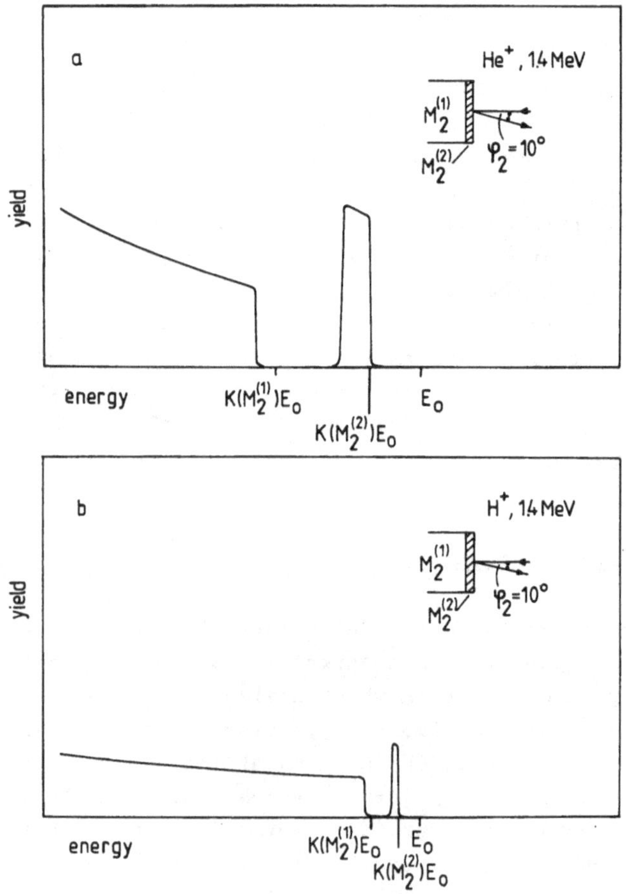

Fig.3.12. Diagrammatic representation of Rutherford backscattering spectra for various beam parameters (thin layer of mass $M_2^{(2)}$ on bulk material of mass $M_2^{(1)}$ with $M_2^{(2)} > M_2^{(1)}$)
(a) 1.4 MeV He ions, $\varphi_1 = 0^{\circ}$, $\varphi_2 = 10^{\circ}$,
(b) 1.4 MeV H ions, $\varphi_1 = 0^{\circ}$, $\varphi_2 = 10^{\circ}$

of the spectrum related to scattering from the surface layer and from the bulk material are well separated (e.g. the mass difference $M_2^{(2)} - M_2^{(1)}$ is large enough to prevent the overlapping of the spectrum parts). The thickness of the surface layer which can be investigated is given by the energy difference $E_0[K(M_2^{(2)}) - K(M_2^{(1)})]$.

Changing various beam parameters can have an influence on the backscattering spectra. If we use H ions instead of He ions without changing other parameters according to the decreased scattering cross section (Eq.(3.33)) and the lowered specific energy loss, the height of the spectrum (Eq.(3.55)) and the depth resolution (Eq.(3.23)) will be drastically decreased (Fig.3.12b). The only advantage of the use of H^+

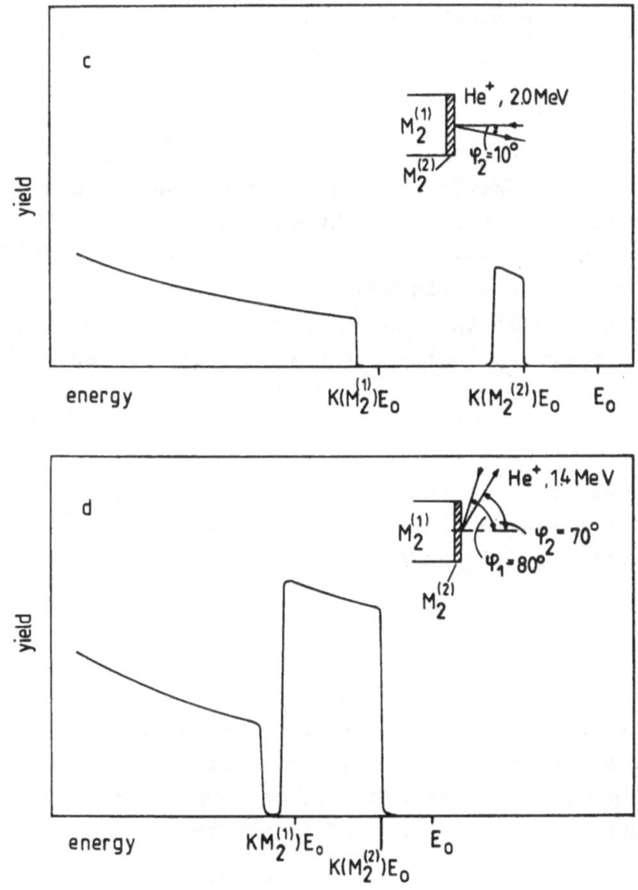

Fig.3.12. Diagrammatic representation of Rutherford backscattering spectra for various beam parameters (thin layer of mass $M_2^{(2)}$ on bulk material of mass $M_2^{(1)}$ with $M_2^{(2)} > M_2^{(1)}$)

(c) 2.0 MeV He ions, $\varphi_1 = 0°$, $\varphi_2 = 10°$,

(d) 1.4 MeV He ions, $\varphi_1 = 80°$, $\varphi_2 = 70°$

beams is the increased accessible depth (see Table 3.1).

With increased energy, the scattering cross section will be decreased due to its E_1^{-2}-depedence. The change of the energy spectrum, going from 1.4 MeV to 2.0 MeV He ions, can be seen by comparing Figs.3.12a and 3.12c. In this case as well, the depth resolution is somewhat decreased resulting from the lower specific energy loss of 2 MeV ions in comparison with 1.4 MeV ions. But there is the advantage of an increased mass resolution (see also Fig.3.4) and a higher value of the accessible depth (see Table 3.1).

The most distinct change of the energy spectrum can be obtained using glancing incidence and exit angles as shown in Fig.3.12d. With this method the depth resolution will be significantly increased (according to Eqs.(3.23) and (3.36)). On the other hand, the depth region which can be analyzed is drastically reduced.

3.2. ELASTIC RECOIL DETECTION (ERD)

As has been seen in Sect.3.1.2, Rutherford backscattering spectrometry shows some difficulties in the detection of light elements because the energy of backscattered ions is very low, due to the small value of the kinematic factor; where heavier elements are present in the sample, there is a considerable background in the energy spectra from scattering events on these atoms. But Rutherford scattering has the great advantage of a very high and well-known cross section, and so one is interested in applying this scattering technique to the investigation of light elements.

One way to do so is given by the investigation of the light atoms recoiled after being hit by a heavier ion. This analyzing method is called "Elastic Recoil Detection (ERD)" (Refs.[91-94]).

3.2.1. Principle

The scattering geometry is shown in Fig.3.13. An ion beam of energy E_0 is incident on a target at an angle φ_1 to the target normal. After an elastic scattering process between the ion and a light target atom, the latter recoils from the surface at a laboratory scattering angle θ_{rec} with an energy

$$E_2 = kE_0 \quad . \tag{3.57}$$

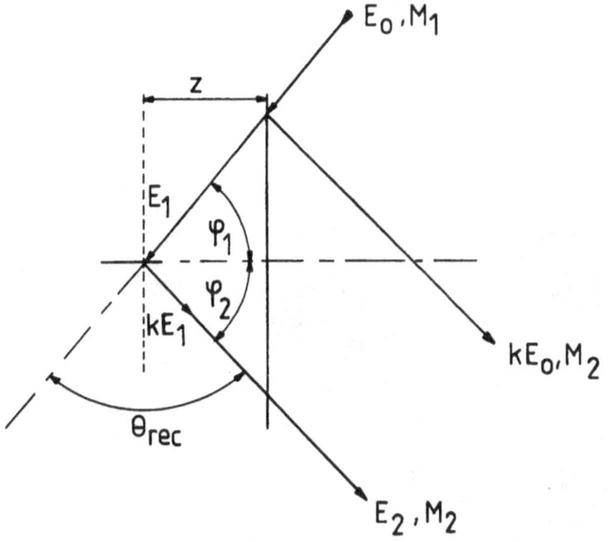

Fig.3.13. Scattering geometry for ERD

The kinematic factor is given by (see Sect.1.1.1.2)

$$k = \frac{4M_1M_2}{(M_1+M_2)^2}\cos^2\theta_{rec} \quad . \tag{3.58}$$

Due to conservation of energy and momentum, the recoiled atoms are only scattered in the forward direction (i.e. $k = 0$ for $\theta_{rec} \geq 90^o$).

In the following, only He ions as incident particles are considered. Fig.3.14 shows the kinematic factor k as a function of the mass of the recoiled atoms M_2 for different scattering angles θ_{rec}. The kinematic factor k is small for $M_1 < M_2$, passes through a maximum for $M_1 = M_2$ and decreases for $M_1 > M_2$. The highest values of k are obtained for $\theta_{rec} \approx 0^o$.

If the collision takes place at depth z, the energy of recoils coming out of the target will be

$$E_2 = kE_o - \left[\frac{k}{\cos\varphi_1}\left(\frac{dE}{ds}\right)_{in,1} + \frac{1}{\cos\varphi_2}\left(\frac{dE}{ds}\right)_{out,2}\right]z \quad , \tag{3.59}$$

where $(dE/ds)_{in,1}$ is the mean specific energy loss of the incident ion (mass M_1) and $(dE/ds)_{out,2}$ the mean specific energy loss of the outgoing particle (mass M_2). Introducing the stopping cross section factor $[\varepsilon']$ (considering that there are different particles in inward and outward paths) E_2 can be expressed as

$$E_2 = kE_o - [\varepsilon'] Nz \quad , \tag{3.60}$$

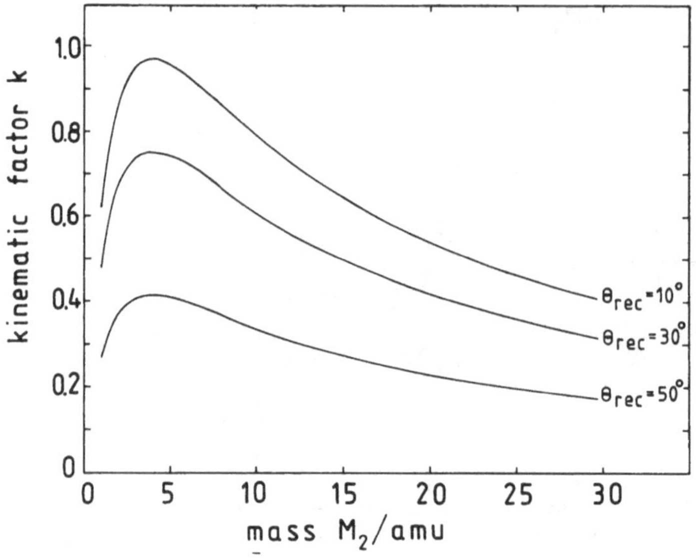

Fig.3.14. Kinematic factor k as a function of the mass of recoiled atoms M_2 for ^{4}He incident ions and recoil scattering angles $\theta_{rec} = 10^\circ, 30^\circ$, and 50°

where N is the atomic density of the target. In order to stop the large flux of the elastically scattered incident ions, an absorber foil must be mounted in front of the detector. The foil must be thick enough to stop all the scattered ions but the recoiled atoms of mass M_2 have to pass through the stopper foil. Thus, it is clear that ERD can only be used for $M_2 < M_1$. The application of He ions means that only H isotopes (^{1}H, ^{2}H, ^{3}H) can be investigated by such beams. The investigation of higher masses M_2 requires beams with larger ion masses M_1 (e.g. Fl^+, Cr^+, Ar^+, or Br^+) but also high ion energies in the order of some tens of MeV.

The particles M_2 undergo an energy loss ΔE_{sf} in the stopper foil, so the energy of the detected particles is given by

$$E_2' = kE_o - [\varepsilon'] Nz - \Delta E_{sf} \; . \qquad (3.61)$$

Eq.(3.61) gives a possible way to transform the energy scale into a depth scale. But for this the stopping power data for the particles M_1 and M_2 in the target and for the particle M_2 in the stopper foil must be known. The accuracy of the measurement of the stopper foil thickness is also important. In order to improve the depth scaling of ERD, spectra from the same samples are often used as reference standards.

The yield of recoiled particles recorded by the detector in the energy window dE_2 is given by

$$dY(E_2) = n \left(\frac{d\sigma}{d\Omega}\right)_{rec} \Delta\Omega \; N_{M_2}(z) \; dz \qquad (3.62)$$

with

n : total number of incident ions,
$(d\sigma/d\Omega)_{rec}$: differential recoil cross section,
$\Delta\Omega$: solid angle of detection,
$N_{M_2}(z)$: concentration of atoms of mass M_2 in depth z in the target,
dz : thickness corresponding to the energy window dE_2 of the detection system.

In the energy region of some MeV and for small scattering angles θ_{rec} the elastic scattering cross section for recoils is approximately that of Rutherford, which is given in the laboratory system by (Eq.(1.35))

$$\left(\frac{d\sigma}{d\Omega}\right)^R_{rec} = \begin{cases} \left[\dfrac{Z_1 Z_2 e^2 (M_1+M_2)}{8\pi\varepsilon_o E_1 M_2}\right]^2 \cos^{-3}\theta_{rec} & \text{for} \quad \theta_{rec} \le \frac{\pi}{2} \\ 0 & \text{for} \quad \theta_{rec} > \frac{\pi}{2} \end{cases} \qquad (3.63)$$

Generally, the depth profile $N_{M_2}(z)$ can be evaluated by measuring the yield of recoiled particles and the integrated ion beam current (which gives n), knowing the solid angle of detection $\Delta\Omega$, and calculating the differential scattering cross section with Eq.(3.63). There are some problems concerning the accuracy of the method for larger scattering angles θ_{rec}. Because of the influence of nuclear forces, the Rutherford recoil cross section (Eq.(3.63)) is no longer valid. So, deviations up to about one order of magnitude were found (see Ref.[94]). Therefore the most accurate way to obtain an absolute depth profile $N_{M_2}(z)$ is the use of a reference standard, e.g. an ion-implanted sample of nearly the same composition (see Ref.[93]).

3.2.2. Depth resolution, accessible depth, sensitivity

The depth resolution of this method is similar to that in Sect.3.1.6 (Eq.(3.23)), defined as

$$\delta z = \frac{\delta E_2'}{N[\varepsilon']} \quad , \qquad (3.64)$$

where $\delta E_2'$ is the total energy resolution under the given experimental conditions. The terms described already in Sect.3.1.6 (Eq.(3.30)) are again substantial contributions for the determination of the total energy resolution. But now we have to take into consideration that there are different particles on the incident path (M_1) and on the outgoing path (M_2). Further, there exists an energy spread δE_{sf} caused by the stopper foil. In addition to this energy straggling there is also an influence from inhomogenities of the absorber foil thickness, which cannot be calculated. For this reason, the whole spread has to be measured for each foil. Finally we get

$$(\delta E_2')^2 = (\delta E_d)^2+(\delta E_s)^2+(\delta E_{ms}^{in})^2+(\delta E_{ms}^{out})^2+(\delta E_g)^2+(\delta E_{sr})^2+(\delta E_{sf})^2. \quad (3.65)$$

A detailed description of the particular contributions and their dependence on the experimental conditions is given in Refs.[93,94]. The graphic representation of the magnitude of the individual contributions described in Ref.[93] is shown in Fig.3.15 for typical experimental conditions: ^1H recoils (hydrogen in silicon), 2.5 MeV ^4He incident ion beam, 9 µm thick Al stopper foil, analyzed depth z = 100 nm, $\varphi_1 = \varphi_2 = \varphi$. For angles between 85° and 63° the dominant term is the energy straggling in the stopper foil. At angles near 90° and 60° the

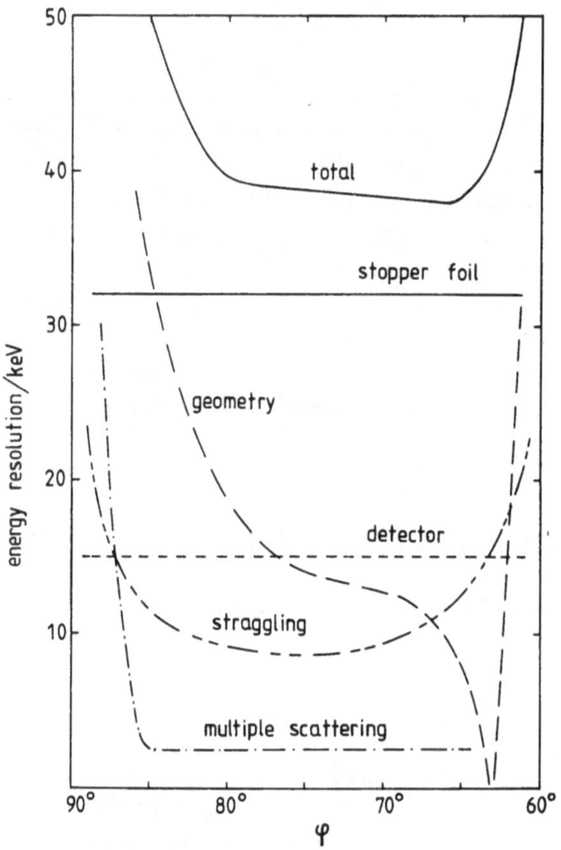

Fig.3.15. Magnitude of the individual contributions of energy resolution versus tilt angle (2.5 MeV ^4He ions, silicon target, 9 µm Al stopper foil, according to Ref.[93])

geometry has the most pronounced influence. For glancing angles ($\varphi_1 > 85^o$) the multiple scattering gives also a large contribution to the total energy spread. An increase of the analyzed depth increases the influence of the geometry, straggling and multiple scattering contributions.

According to Eq.(3.64) the best depth resolution will be achieved for the minimum value of $\delta E'/[\varepsilon']$. For the conditions given in Fig.3.15 the depth resolution as a function of the tilt angle is shown in Fig.3.16 (Ref.[93]). As can be seen from this figure, the best achievable depth resolution at the depth z = 100 nm is δz = 23 nm for a tilt angle $\varphi = 88^o$. An increase of the analyzing depth will decrease the depth resolution due to a poorer energy resolution.

Considering the angular dependence of the kinematic factor (Fig.3.14) one could expect that the accessible depth would be increased with decreasing scattering angle because the energy of the recoiled particles is increased. This is not the case, because a decreased scattering angle is always connected with an increase of the incident and/or exit angle (φ_1 and/or φ_2) by which the paths of the particles will be increased, yielding higher energy losses and therefore decreased accessible depths.

For the example so far considered, the accessible depth z_a as a function of the tilt angle for a fixed scattering angle $\theta_{rec} = 30^o$ is

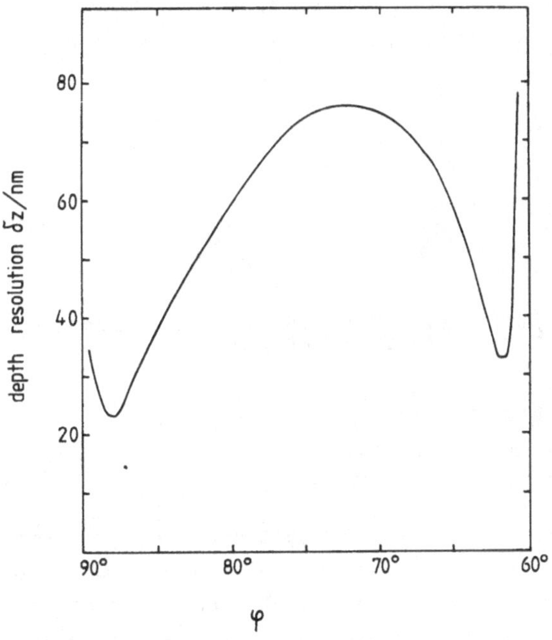

Fig.3.16. Depth resolution calculated for the depth z = 100 nm (for the same conditions as in Fig.3.15) versus tilt angle (according to Ref.[93])

represented in Fig.3.17 (Ref.[93]). The calculation was made assuming that the minimum proton energy which can be detected is 200 keV. The maximum values $z_a \approx 600$ nm are achieved for tilt angles in the region $\varphi = 13...22^o$. Comparing Fig.3.17 with Fig.3.16 we see again, as in the case of RBS, a large value of the accessible depth is always connected with a poor dépth resolution and vice versa.

The sensitivity of the ERD technique using a 2.5 MeV ^4He beam is about 0.1%. In practice the sensitivity is limited by the oontinuous background produced by protons which are recoiled from surface contamination atoms with sufficiently high energies to penetrate the stopper foil.

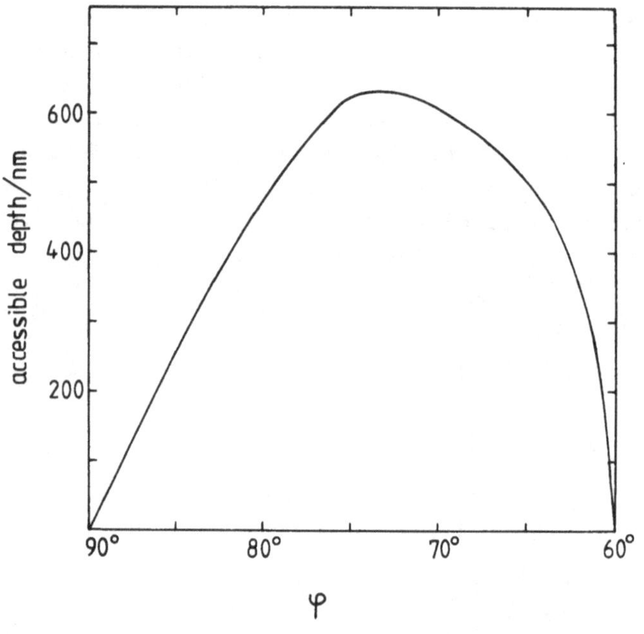

Fig.3.17. Accessible depth versus tilt angle for hydrogen in silicon at a scattering angle $\Theta_{rec} = 30^o$ (according to Ref.[93])

3.3. NUCLEAR REACTIONS (NR)

A further way to overcome the difficulties of RBS in investigating light elements (due to the fact that the high background of ions scattered from heavier atoms in the substrate covers the signal from light atoms) is the use of nuclear reactions or resonant nuclear scattering (Refs.[95-99]).

Here we will pay attention only to reactions which occur at ion energies not greater than 2 MeV. The most important ones are listed in Appendix 12 (for particle detection, Ref.[100]) and Appendix 13 (for proton induced γ-ray detection, Refs.[101-104]). Nuclear reactions exhibit a high selectivity because they are specific for one type of nucleus. In order to be of analytic interest, the following criteria must be fulfilled

(i) The spectrum must be easily recognizable, i.e. well-separated peaks in a relatively background-free region should exist.

(ii) Good depth resolution and reasonable cross section must be given.

As can be seen from Appendices 12 and 13, there are large differences for various nuclear reactions and the best conditions have to be selected very carefully.

3.3.1. Principle

The geometry is quite the same as shown for ERD in Fig.3.13, but now instead of the elastic scattering process we have a nuclear reaction. Thus the energy of the outgoing particle is essentially influenced by the Q-value of the reaction. Generally, those reactions which have high Q-values are advantageous, i.e., the outgoing particles have a higher energy than the elastically scattered ions. This has to be taken into consideration when calculating the energy of the particles leaving the target. The geometry for a nuclear reaction experiment is shown in Fig.3.18. A schematic representation of the collision process is drawn in the insert of Fig.3.18.

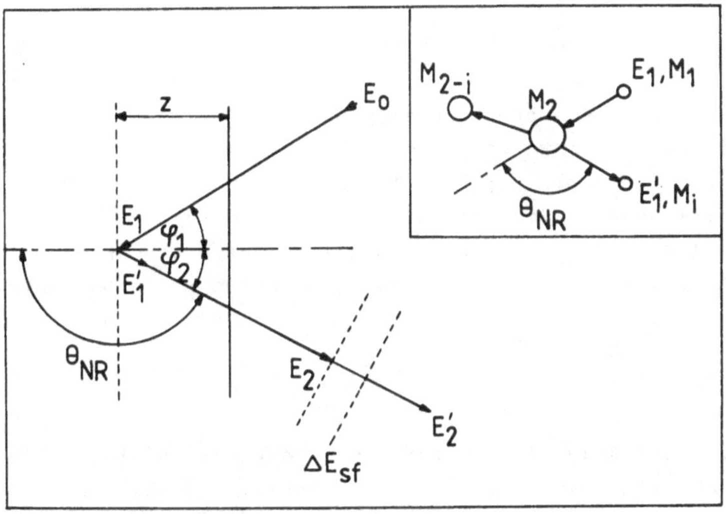

Fig.3.18. Geometry for nuclear reaction analysis

An ion with energy E_1 and mass M_1 undergoes a nuclear reaction with the target nucleus of mass M_2. After the reaction $M_2(M_1, M_i)M_{2-i}$, where M_{2-i} is the mass of the final nucleus and M_i the mass of the outgoing particle, the latter one will leave the target under the scattering angle θ_{NR}. If we assume mean energy losses in the target for the incident particle, the energy immediately before the reaction at depth z is given by

$$E_1 = E_o - \frac{z}{\cos\varphi_1} \left.\frac{dE}{ds}\right|_{in} . \tag{3.66}$$

After the nuclear reaction the outgoing particle M_i has the energy E_1' given by the well-known kinematic expression

$$E_1' = \frac{M_1 M_i E_1}{M_2^2} \left\{ 2\cos^2\theta_{NR} + \frac{M_{2-i}M_2}{M_1 M_i}\left(\frac{Q}{E_1} - \frac{M_1}{M_{2-i}} + 1\right) \right.$$
$$\left. + 2\cos\theta_{NR}\left[\cos^2\theta_{NR} + \frac{M_{2-i}M_2}{M_1 M_i}\left(\frac{Q}{E_1} - \frac{M_1}{M_{2-i}} + 1\right)\right]^{1/2}\right\} . \tag{3.67}$$

The energy loss of the outgoing particle can be also approximated by a mean value. Then the energy of the particle M_i coming out of the target is given by

$$E_2 = E_1' - \frac{z}{\cos\varphi_2} \left.\frac{dE}{ds}\right|_{out} . \tag{3.68}$$

Taking into consideration additionally the energy loss ΔE_{sf} of the particle in the stopper foil, the detected energy is

$$E_2' = E_2 - \Delta E_{sf} . \tag{3.69}$$

With these formulae, the scale of the energy of the detected particle can be transformed into a depth scale and the evaluation of depth profiles is generally possible. But for this we need the energy dependence of the cross section for the nuclear reaction used.

The yield of the detected particles for a given energy window dE_2 of the detection system is described by a similar expression as in the case of ERD (Eq.(3.62))

$$dY(E_2) = n \left(\frac{d\sigma}{d\Omega}\right)_{NR} \Delta\Omega \, N_{M_2}(z) \, dz , \tag{3.70}$$

where $(d\sigma/d\Omega)_{NR}$ is the differential nuclear reaction cross section. The other parameters are the same as in the case of Eq.(3.62) in Sect.3.2.

The height of the particle spectrum in the energy interval dE_2 is then directly proportional to the product of the concentration at depth z of nuclei to be analyzed and the differential reaction cross section for the energy E_1 of the particle M_1 at this depth, assuming a fixed observation angle θ_{NR}. The concentration profile can therefore be computed if the energy dependence of the nuclear reaction cross section is known. There are two methods to evaluate depth profiles, the energy-analysis method and the resonance method. The energy-analysis method can be applied if the reaction cross section is constant over a wide energy region or varies smoothly with energy. Then the depth information is provided by the energy analysis of the recorded particles according to Eqs.(3.66–3.70).

The resonance method for depth profiling requires a sharp peak in the energy dependence of the nuclear reaction cross section. Some of these reactions are listed in Appendix 13. The measurements are performed in such a way that the energy of the incident beam is changed in small increments, thereby changing the depth at which the resonance occurs. The profile is derived from the nuclear reaction yield as a function of the energy of the incident ions.

3.3.2. Depth resolution, accessible depth, sensitivity

In the case of depth profiling by the energy-analysis method, an expression for the depth resolution can be achieved by the derivation of Eq.(3.68) again using a mean value for the energy loss of the outgoing particle

$$\frac{dE_2}{ds} = \frac{dE_1'}{dE_1}\frac{dE_1}{ds} - \frac{N}{\cos\varphi_2}\,\varepsilon(\bar{E}_{out}) \quad . \tag{3.71}$$

With Eq.(3.66) we get for a mean value of the energy loss of the incident ions

$$\frac{dE_1}{ds} = -\frac{N}{\cos\varphi_1}\,\varepsilon(\bar{E}_{in}) \quad , \tag{3.72}$$

and the depth resolution is given by

$$\delta z = \frac{\delta E_2}{N\left[\frac{dE_1'}{dE_1}\frac{\varepsilon(\bar{E}_{in})}{\cos\varphi_1} + \frac{\varepsilon(\bar{E}_{out})}{\cos\varphi_2}\right]} \quad . \tag{3.73}$$

As can be seen from Eq.(3.73), the depth resolution depends on both the reaction kinematics (first term in the bracket) and the rate of the energy loss. Thus, for a given energy resolution (which itself depends on the detector resolution, the energy straggling in the sample and in the stopper foil, the multiple scattering, the geometrical resolution, and the surface roughness as discussed in Sects.3.1.6 and 3.2.2) the depth resolution is determined by the stopping cross sections, the incident energy, the detector angle, and the solid angle subtended by the detector. The depth resolution varies with the depth in the target. Thus, the depth resolution has to be calculated for the given conditions of every special case. However, the values are in the same order as those discussed for RBS and ERD, i.e. about a few nm for optimized geometries at the target surface and about 100 nm for normal incidence ($\varphi_1 = 0^o$, $\varphi_2 \approx 10^o$).

In the case of exothermic nuclear reactions, the outgoing particles may have such a high energy that the accessible depth is not limited by the lowest value of the energy which can be recorded by the detector. Thus, the maximum probing depth will be limited by other factors as the maximum value of the combined energy straggling of incident and outgoing particles that is acceptable to guarantee the necessary depth resolution. Also, the decrease of the nuclear reaction cross section at sufficiently low energies, together with upper limits on incident ion energy may limit the accessible depth.

For the resonance method, the depth resolution near the surface is determined by the resonance width of the nuclear reaction and the energy spread of the incident beam. The energy straggling of the incident beam becomes the dominating factor in limiting the depth resolution at depths where straggling exceeds the resonance widths. With very narrow resonance widths, in the order of 1 keV, depth resolutions of a few tens of nm can be achieved.

The accessible depth is usually limited by resonances or increased non-resonant yield at higher energies, unless the yields due to these reactions can be subtracted from the data with sufficient accuracy. Otherwise, the accessible depth will again be given by the maximum allowable broadening in the depth resolution due to the energy straggling in the beam, or by background contributions.

The sensitivity of nuclear reaction analysis for an isotope in the near-surface region is determined by the ratio of the yield from the reaction and the background observed by the detector. An additional practical limitation is set by the total fluence of the analyzing beam acceptable for the target (e.g. causing damage, blistering, sputtering). Otherwise, the sensitivity will be limited by the time at hand for the experiment. In each specific case, one has carefully to consider the different reactions which can be used to obtain the highest sensitivity (Appendices 12 and 13). The signal from the element to be investigated can be maximized by proper choice of incident ion energy, detection system (e.g. stopper foils), detector angle, and solid angle to give maximum detection efficiency. However, consideration of the background often requires significant compromises.

3.4. PARTICLE INDUCED X-RAY EMISSION (PIXE)

Besides elastic scattering of the projectiles (RBS) and recoil atoms (ERD) as well as nuclear reactions (NR), the interaction of high energetic light ions (1-2 MeV H^+ or He^+) with the atoms of a sample leads to the emission of X-rays by inner shell vacancy generation. Additionally, different kinds of bremsstrahlung are produced.

The characteristic energy spectrum of the X-rays allows the identification of elements and the determination of their concentrations. The accessible depth roughly corresponds to the projected ion range (e.g. ≈ 1 μm for 1 MeV He^+ and ≈ 10 μm for 1 MeV H^+ in silicon). PIXE is mainly used for the identification of elements in multielemental samples and for the investigation of trace elements in solids (Ref.[105]).

3.4.1. Principle

To give an introduction into the method, the experimental parameters will be discussed at first. In Fig.3.19 a schematic diagram of the set-up for PIXE analysis is represented. A collimated beam of light ions (mass M_1, energy E_0) interacts with a target containing different elements i (atomic number $Z_2^{(i)}$, atomic density $N(Z_2^{(i)})$). The characteristic X-rays of energy E_X are recorded by a detector within a solid angle $\Delta\Omega$. Often absorber foils are used to obtain energy-dependent attenuation of the X-ray intensity.

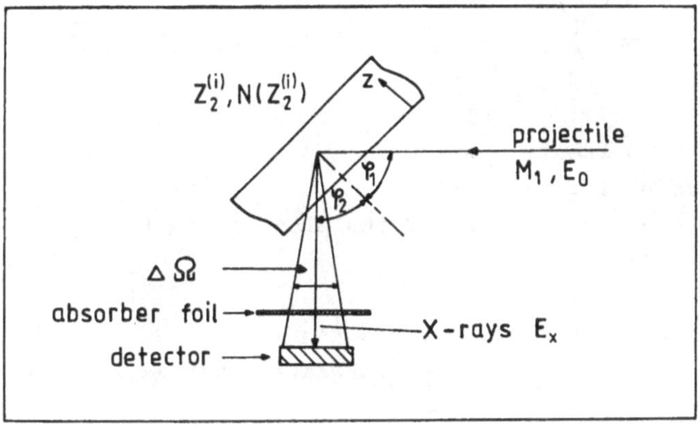

Fig.3.19. Schematic diagram of the experimental arrangement of a PIXE analysis system

Usually, the angle of incidence φ_1 and the exit angle φ_2 of the X-rays with respect to the surface of the target are set at 45°.

The collisions of the ions with the target atoms generate inner shell vacancies. The recombination of the excited atoms is accompanied by the emission of photons with characteristic energies according to the transitions of the electrons. The characteristic energies are given by Moseley's law

$$E_X = R' \, (Z_2^{(i)} - \zeta)^2 \left(\frac{1}{n_1^2} - \frac{1}{n_2^2}\right) \quad , \tag{3.74}$$

where R' is the modified Rydberg's constant, n_1, n_2 are the principal quantum numbers of electronic shells ($n_2 > n_1$) and ζ takes into account the screening effect ($\zeta = 1$ for $n_1 = 1$ (K-series)). It can be seen that the characteristic energies of the K-series scale with $(Z_2^{(i)}-1)^2$. Because of the fact that the native widths δE_X of the X-ray lines are much smaller than the energies E_X themselves ($\delta E_X/E_X \approx 10^{-4}$) the corresponding X-ray energies from elements with successive atomic numbers Z_2 and Z_2+1 are well separated. In practice, experimentally determined values of X-ray energies are used. In Appendix 14 the characteristic energies of the K-, L-,and M-series of elements Z_2 in the range from 11 to 100 are summarized.

With the knowledge of the characteristic energies, the elements of a sample can be identified. Furthermore, the yield Y of the recorded characteristic X-rays with the energy $E_X(Z_2^{(i)})$ is a measure of the amount of the element i in the target, where $E_X(Z_2^{(i)})$ denote the energies of the different electron transitions of the element i. The yield is given by

$$Y(E_X(Z_2^{(i)})) = \frac{n \, \eta(E_X) \, T_a(E_X)\Delta\Omega}{4\pi \cos\varphi_1} \int_0^{R_p+\Delta R_p} dz \; \sigma_X(E_1(z),Z_2^{(i)}) \; T_s(E_X,z) \; N(Z_2^{(i)},z) \tag{3.75}$$

with

- n : number of incident ions,
- $\eta(E_X)$: detection efficiency,
- $T_a(E_X)$: transmission coefficient of the absorber foil,
- R_p : projected ion range,
- ΔR_p : range straggling,
- $\sigma_X(E_1(z),Z_2^{(i)})$: X-ray cross section of the element i,
- $T_s(E_X,z)$: transmission coefficient of the layer excited ($\approx R_p$).

The X-ray cross section $\sigma_X(E_1(z),Z_2^{(i)})$ depends on the ion energy E_1 at depth z (see Sect.3.1.4). It consists of two factors, the cross section $\sigma_v(E_1,Z_2^{(i)})$ for the generation of inner shell vacancies and the fluorescence probability $\omega(Z_2^{(i)})$ for radiating recombination (see Eq.(1.62)). Calculations of the cross section for inner shell vacancy generation and the fluorescence probability are described in Ref.[14].

Experimental values of X-ray cross sections for various elements and ion energies are presented in Ref.[15].

The transmission coefficient $T_s(E_X,z)$ takes into consideration the self-absorption of X-rays with energy E_X in the sample material

$$T_s(E_X,z) = \exp\left(-\frac{\mu_s(E_X)\, z}{\cos\varphi_2}\right) , \qquad (3.76)$$

where $\mu_s(E_X)$ is the absorption coefficient of the sample. In the literature, usually the mass attenuation coefficient μ/ϱ (ϱ is the mass density) is tabulated (see Ref.[106]).

The transmission coefficient of the absorber foil is given by

$$T_a(E_X) = \exp(-\mu_a(E_X) d_a) , \qquad (3.77)$$

where μ_a and d_a are the absorption coefficient and the thickness of the foil, respectively. For absorber materials consisting of more than one element, the absorption has to be calculated by weighting according to the composition. The effect of energy-dependent attenuation of the X-rays by absorber foils is a very useful tool for suppressing the characteristic lines (K-edge absorption) of the matrix elements in order to allow the detection of trace elements (e.g. Cr-absorber for Fe, Al-absorber and Zn-absorber for Si and Ge, respectively). Also the application of Mylar absorber foils with different thicknesses for decreasing high background parts of a PIXE spectrum is commonly used (Ref.[107]).

For a quantitative analysis, i.e. the determination of $N(Z_2^{(i)})$ of an element, all parameters given in Eq.(3.75) have to be known. This is practically impossible in most cases. Therefore often standards (layer of similar composition) are used for reference. The analysis of <u>thin samples</u> (thickness $\Delta z \ll R_p$) has the advantage of neglecting the energy loss of the projectiles and therefore the X-ray cross section is constant over Δz. In addition, the self-absorption of the emitted X-rays in a thin sample can be neglected. To obtain information about the depth distribution of different elements, the change of the X-ray yield by varying the effective ion range has to be observed. This can be done by changing either the beam energy E_o or the incident angle φ_1 (see Ref.[108]), but the depth resolution is very poor ($\delta z \approx 1\,\mu m$).

Besides the characteristic X-rays from the elements in the sample, a continuous background arises from the slowing down of the projectiles and secondary electrons, from native background radiation, from Compton scattering of high-energy γ-rays in the detector and from carrier losses in the detector due to recombination processes. The main contributions to the background are the projectile bremsstrahlung and the secondary electron bremsstrahlung in thick target analysis.

(i) Projectile bremsstrahlung (PB)

According to Ref.[109] the yield is proportional to $(Z_1/M_1 - Z_2/M_2)^2$. The contribution of PB can be neglected for He ions because the ratio Z_2/M_2 for most elements is close to 1/(2amu), as is also the case for helium. But for protons there exists a considerable contribution.

(ii) Secondary electron bremsstrahlung (SEB)

Secondary electrons are generated by energy transfer from the projectile to free or bound electrons. The maximum energy E_{max} transferred is

$$E_{max} \approx \frac{4m}{M_1} E_o \quad , \tag{3.78}$$

where m is the electron mass. The slowing down of the secondary electrons due to Coulomb interaction with the matrix atoms generates the secondary electron bremsstrahlung. Therefore a continuous energy distribution up to the energy E_{max} exists. Because the secondary electrons are mainly scattered in a forward direction, this SEB background can be reduced by using very thin samples and detecting the X-rays at large angles with respect to the ion beam. Fig.3.20 shows an example of the distribution of projectile bremsstrahlung and secondary electron bremsstrahlung of a thick Al target irradiated with 2 MeV protons (Ref.[110]). It can be seen that the SEB-yield dominates at lower energies and decreases rapidly with increasing energies, whereas the PB-yield is nearly constant over a wide range of X-ray energies.

The other contributions to the background radiation are of minor importance, provided that the detector is shielded from the γ-radiation from any part of the experimental chamber.

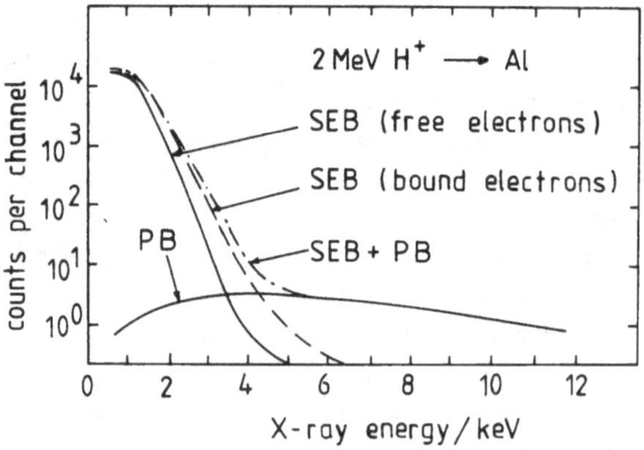

Fig.3.20. Distribution of projectile bremsstrahlung (PB) and secondary electron bremsstrahlung (SEB) due to the collision of 2 MeV protons on a thick Al target (Ref.[110])

A typical PIXE spectrum (1.8 MeV protons) of a thick Pb sample containing several impurity elements in the surface layer is shown in Fig.3.21. It can be seen that 16 peaks are well separated, corresponding

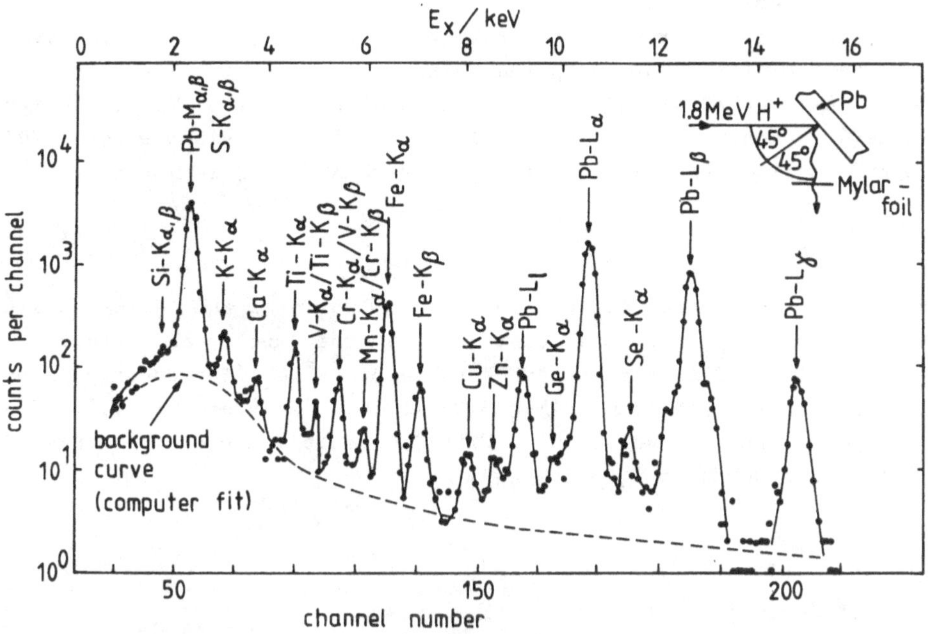

Fig.3.21. PIXE spectrum of a thick Pb target (1.8 MeV H^+, 10 μC collected charge, 20 μm Mylar absorber foil)

to 14 different elements. From most elements the K-lines are recorded, whereas for Pb the M- and L-lines are detected. The characteristic X-ray lines of the elements are superimposed on the continuous background (dashed curve). A 20 μm thick Mylar absorber was used to reduce the SEB-yield.

3.4.2. Mass resolution and energy resolution

As mentioned above, the characteristic energy values of the K-, L- or M-series of the elements imply on principle the unambiguous identification of every element, because the energy values are well separated and therefore a mass resolution M_2 (or better, atomic number resolution Z_2) below or equal to unity is available. But there are two restrictions: first, the limited energy resolution of the detection system and second, the limited range of measurable X-ray energies. In Table 3.5 characteristic X-ray energy values E_X of various elements and the corresponding energy differences ΔE_X between neighbouring lines are represented. The energy resolution of the detector depends on the X-ray energy and is in the order of about 110 eV at 1 keV and about 250 eV at 20 keV and consequently, interferences of different lines cannot be avoided (see Table 3.5).

In order to determine the energy resolution, a radioactive ^{55}Fe-source emitting the Mn-K_α and Mn-K_β lines free of background is commonly used (Ref.[111]). The FWHM value at the Mn-K_α energy (5.898 keV) is measured as shown in Fig.3.22. In this example the FWHM corresponds to 2.8 channels. The energy per channel ϵ_{ch} is given by

$$\epsilon_{ch} = \Delta E_X/\Delta n_{ch} \quad , \tag{3.79}$$

where ΔE_X is the energy difference between the Mn-K_α and Mn-K_β peaks (≙ 592 eV) and Δn_{ch} is the distance of the two peak positions (≙ 9.5 channels). This yields a FWHM value of 175 eV (The occurrence of escape peaks in the spectrum of Fig.3.22 will be discussed later.). The best values of energy resolution of Si(Li)-detectors are in the range of 150 eV to 180 eV FWHM at E_X = 5.9 keV.

TABLE 3.5
Examples of X-ray interferences

X-ray line	energy E_X/keV	energy difference ΔE_X/eV
Mo-$L_{\alpha1}$	2.293	
		15
S -K_α	2.308	
		38
Pb-$M_{\alpha1}$	2.346	
		97
Pb-M_β	2.443	
Cd-$L_{\alpha1}$	3.134	
		180
K -$K_{\alpha1}$	3.314	
		3
Cd-$L_{\beta1}$	3.317	
Ba-$L_{\alpha1}$	4.466	
		45
Ti-$K_{\alpha1}$	4.511	
Ba-$L_{\beta1}$	4.828	
		104
Ti-$K_{\beta1}$	4.932	
		20
V -K_α	4.952	
Pb-$L_{\alpha1}$	10.552	
		8
As-$K_{\alpha1}$	10.544	

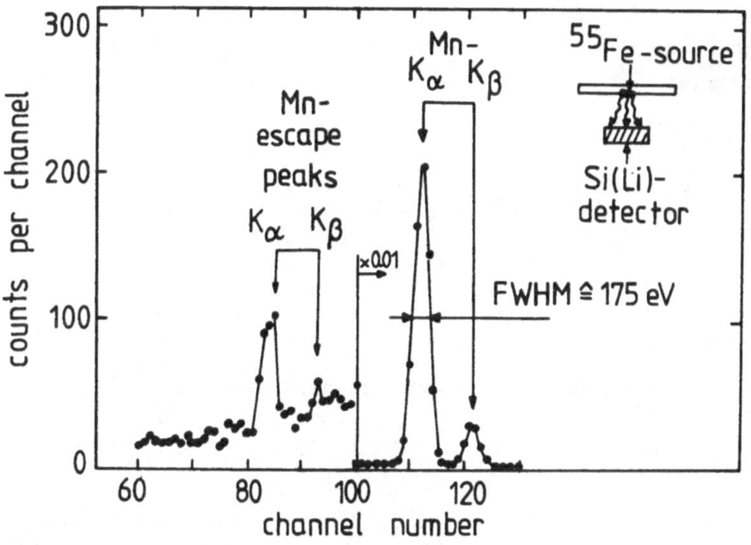

Fig.3.22. Energy spectrum of the X-rays of a ^{55}Fe-source and determination of the FWHM value at the Mn-K_{α} line

The principal way to reduce the problems of line interferences is to use a wavelength-dispersive detection system (based on diffraction by crystals) which guarantees an energy resolution in the order of a few tens of eV (Ref.[105]). However, the experimental expense is very high and the advantage of simultaneous detection of various elements, as is the case with energy-dispersive Si(Li)-detectors, is lost.

The detection system itself sets limitations for the lowest and highest energies measurable. This depends on the thickness of the window (mostly Be) in front of the detector (low energy edge) and the thickness of the active volume of the detector crystal (high energy edge). In Fig.2.18 (see Sect.2.4.1.2) the detection efficiency $\eta(E_X)$ as a function of X-ray energy was shown for different detector arrangements. Generally, the energy range of about 1 keV to 20 keV is suitable for recording the X-rays (Ref.[80]).

In Table 3.6 the Z_2-values of elements which correspond to the energy limits of the detector for K-, L- and M-series are presented. It can be seen that all elements heavier than sodium are detectable.

At this point it seems to be of interest that there are two additional effects which may affect the interpretation of characteristic X-ray lines. First, at high photon intensities an increasing pile-up effect of intense X-ray lines or high background yields gives additional lines by the superposition of two or more pulses and therefore the identification of the original peaks is more difficult (see Sect.2.4.3.3). Second, escape peaks also have to be taken into account if the X-ray beam is not well collimated. Escape peaks appear if photons

TABLE 3.6

Z_2-values of elements corresponding to the lower and upper limits of detectable X-ray energies

radiation	lower Z_2 limit (E_X = 1 keV)	upper Z_2 limit (E_X = 20 keV)
K-series	11[1)]	45
L-series	30	100
M-series	60	in practice unlimited

[1)] Windowless detectors are capable of recording energies down to that of the B-K_α line (Ref.[112]).

of fluorescence radiation generated in the detector material can leave the active volume. The energy E_{esc} of an escape peak is well determined by

$$E_{esc} = E_X - E_{Si-K_\alpha} \quad , \tag{3.80}$$

where E_X is the characteristic energy of an intense X-ray line and E_{Si-K_α} the energy of Si-K_α fluorescence radiation. The example in Fig.3.22 shows two escape peaks of Mn-K_α and Mn-K_β radiations.

3.4.3. Sensitivity

The main advantage of PIXE is its high sensitivity. The detectable concentration of trace elements is generally dependent on the experimental conditions. Commonly, the sensitivity is defined by the minimum detectable concentration of a trace element in a carbon matrix. The background radiation is in most cases the factor limiting the sensitivity. To record a significant peak of a trace element, it must exceed the statistic uncertainty of the background. Generally, it is assumed that this is fulfilled if

$$Y_p \gtrsim 3(Y_b)^{1/2} \quad , \tag{3.81}$$

where Y_b is the number of pulses due to background events in the energy region corresponding to the FWHM of the characteristic line and Y_p is the (separated) peak intensity (see Ref.[113, 114]). To separate the characteristic peaks, a computer fit of the background is performed by square well approximation of the spectrum as shown in Fig.3.21 (dashed curve).

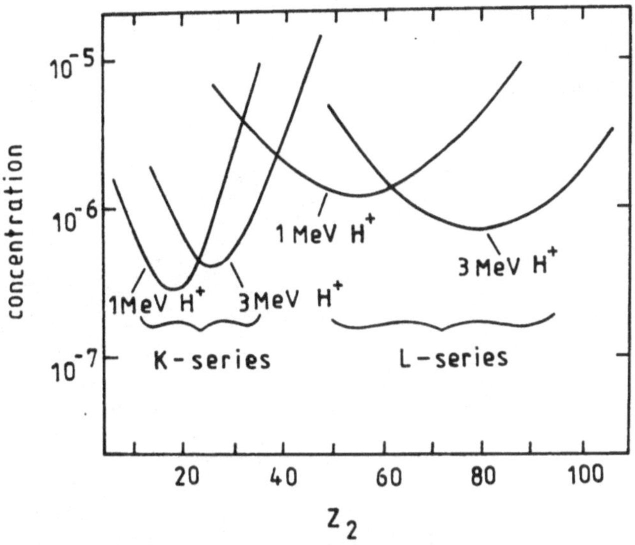

Fig.3.23. Minimum detectable concentration of trace elements as a function of atomic number and proton energy (experimental parameters: detector resolution 165 eV, solid angle 0.0377 sr, collected charge 10 μC, target thickness 0.1 mg/cm , Ref.[105])

The minimum detectable concentration as a function of atomic number Z_2 of trace elements is shown in Fig.3.23 for typical proton energy values E_o (Ref.[105]). An interesting fact of practical importance is that the highest sensitivity is obtained for $Z_2 \approx 20 - 30$, which happens to be the region of greatest interest in many applications. The most suitable proton energy is therefore in the range of 1 to 2 MeV. Using both the K- and L-series of the X-rays for analysis, a minimum detectable concentration of trace elements of about 10^{-6} can be achieved for nearly all elements.

Chapter 4

APPLICATION TO THE ANALYSIS OF COMPOSITION

B. Weber and A. Dittmar

In this chapter, examples are given of the application of RBS and PIXE to the investigation of amorphous and nonoriented crystalline materials, respectively. Mass and depth distribution of impurities, thickness and composition of thin layers or multilayered films, and the surface composition of thick multielemental targets are determined. Mathematical methods and approximations to calculate such quantities as the depth of backscattering, spectrum heights and the composition at various depths are discussed.

For most of the backscattering experiments a 1.4 MeV He ion beam is used. In most cases the scattering angle is $\theta = 170^{\circ}$ with normal incidence of the ion beam on the sample surface which leads to a depth resolution of about 30 nm in silicon (see Sect.3.1.6). For examples in which higher depth resolution is necessary, the angle of incidence is increased by tilting the sample. The backscattered He ions are collected by a silicon surface barrier detector with an energy resolution of 14 keV (FWHM).

PIXE is mostly used to identify the impurities (atomic number Z_2 and mass M_2, respectively), especially in such cases where the mass resolution of the RBS method is not sufficient for exact identification of elements with $M_2 > 30$ amu. The emission of characteristic X-rays of the elements is induced by application of H or He ion beams with energies of 1.4 to 1.8 MeV (see Sect.3.4). The energy resolution of the Si(Li) X-ray detector used is about 175 eV for the Mn-K_{α} line and therefore an exact identification of elements is possible.

4.1. SURFACE IMPURITIES

4.1.1. Mass identification

The first problem to be solved is the determination of the kind of impurity. Let us consider impurities with an areal density of about 10^{15} cm^{-2}, which corresponds to one monolayer. One way to identify the masses of surface impurities is RBS. In Sect.3.1.2 it is shown that the energetic positions of the signals in the backscattering spectrum depend only on the masses of the backscattering particles at the sample surface for a given scattering geometry. The heavier the target atom, the higher the energy E_2 of the detected He ions. This means that impurity atoms with masses lower than those from the substrate are hardly visible in the backscattering spectrum, because substrate signals are superimposed on the signals from the impurities. The mass identification of light elements on heavy substrates can be performed by PIXE as long as the elements are not lighter than sodium, as is discussed in Sect.3.4.

The principles of the RBS method are described in Sect.3.1. To begin an analysis, the energy width ϵ_{ch} of the channels of the multichannel analyzer and the zero point of the energy scale must be determined. For this calibration, special targets with light matrix elements, e.g. Be, C, Si, SiO_2, and with known impurity elements at the surface are used. Fig.4.1 shows a spectrum for 1.4 MeV He ions (normal incidence) backscattered from a silicon target with surface impurities of C, Co, and Au. The leading edge of the silicon signal is positioned at channel

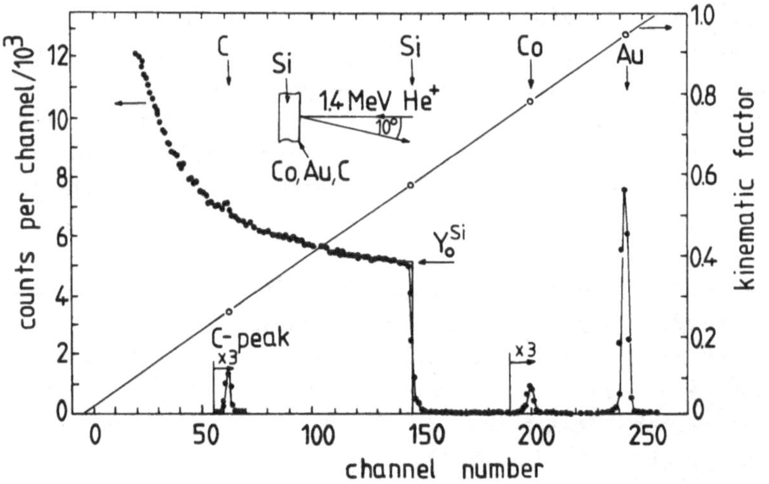

Fig.4.1. Backscattering spectrum of 1.4 MeV He ions incident on a Si target with surface impurities of C, Co, and Au ($\varphi_1 = 0°$, $\Theta = 170°$, $\Delta\Omega = 1.59$ msr, $\epsilon_{ch} = 5.18$ keV, ne = 20 μC). The right-hand scale gives the kinematic factors.

number 147; the peak positions of the impurities correspond to the channel numbers 63 for C, 200 for Co, and 244 for Au.

The right-hand scale of Fig.4.1 gives the kinematic factor K (Sect.3.1.2). The dependence of the energy KE_0 on the channel number n_{ch} is given by a straight line, the slope of which provides the energy width ϵ_{ch} of one channel

$$\epsilon_{ch} = \frac{K(M_2^{(1)}) - K(M_2^{(2)})}{n_{ch}(M_2^{(1)}) - n_{ch}(M_2^{(2)})} E_o \quad . \tag{4.1a}$$

By considering more than two surface impurities the value of ϵ_{ch} can be improved with linear regression. The point of intersection of the straight line with the scale of the channel number $n_{ch,o}$ provides the shift of the zero point of the energy scale. For the example in Fig.4.1 we get ϵ_{ch} = 5.18 keV and $n_{ch,o}$ = -5.5 which corresponds to 28.5 keV for the shift of the energy scale. Thus we obtain the detected energy $E_2 = (n_{ch} - n_{ch,o})5.18$ keV.

With the E_2-energy scale obtained from the results of the calibration measurement, the kinematic factors (and from these the impurity masses) can be determined. To examine whether the impurities are on the surface, the measurement is repeated after tilting the target (increasing the angle of incidence φ_1). By this, the positions of the signals in the energy spectrum from impurities at the surface will not be affected, but signals from atoms below the surface are shifted towards lower energies.

The identification of elements using the PIXE method is carried out as follows. First, one has to calibrate the energy scale of the recorded spectrum with the help of the characteristic X-ray energies of well-known elements and second, the unknown elements can then be identified. Note that in PIXE analysis usually more than one characteristic energy value of every detectable element is obtained, contrary to the situation in RBS analysis. The reason for this is the emission of all possible characteristic X-rays (see Sect.3.4).

Fig.4.2 shows a PIXE spectrum for calibrating the energy scale, where a sample consisting of the elements Al and Fe covered by an Au layer is considered. The energy E_o of the proton beam and the collected charge are 1.7 MeV and 1 μC, respectively. For Al we find only one peak at n_{ch} = 41 (Al-$K_{\alpha 1,2}$), for Fe two peaks are obtained at n_{ch} = 120 and n_{ch} = 130 (Fe-$K_{\alpha 1,2}$ and Fe-$K_{\beta 1}$). For Au we get five positions at n_{ch} = 51.5, 155, 173, 201, and 232 (Au-$M_{\alpha 1}$, Au-L_1, Au-$L_{\alpha 1,2}$, Au-$L_{\beta 1}$, and Au-$L_{\gamma 1}$). Because of the fact that beyond the X-ray energy range from 1 keV to about 60 keV the detection efficiency is close to zero (see Fig.2.18) the characteristic energies from Al-L and Al-M series, Fe-L and Fe-M series and Au-K series are not measureable. The Al-K_β line is close to the K_α position and due to the limited energy resolution of the detector they cannot be resolved (interference problem as discussed in Sect.3.4). With the energies E_X and the corresponding channel numbers mentioned above, the slope of the straight line shown in Fig.4.2 (right-hand scale) is given by

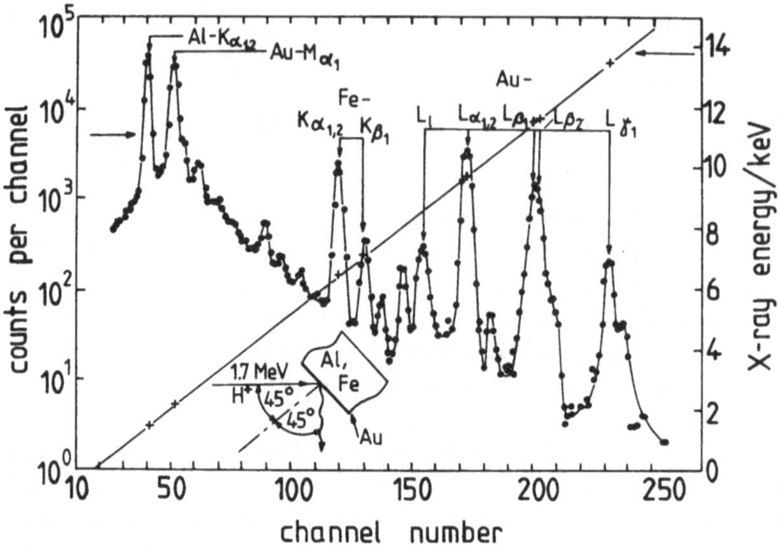

Fig.4.2. PIXE spectrum of 1.7 MeV protons incident on a Au/Al/Fe sample with $\varphi_1 = 45^o$, $\varphi_2 = 45^o$, ne = 1 μC, ϵ_{ch} = 62.6 eV, ΔΩ = 9 msr. The right-hand scale gives the X-ray energy.

$$\epsilon_{ch} = \frac{E_X^{(1)} - E_X^{(2)}}{n_{ch}(E_X^{(1)}) - n_{ch}(E_X^{(2)})} \tag{4.1b}$$

($E_X^{(1)}$ and $E_X^{(2)}$ are the energy values of the characteristic lines 1 and 2). The point of intersection of this line with the energy scale $n_{ch,o}$ gives the shift of the zero point of the energy scale. In the example considered we find ϵ_{ch} = 62.6 eV and $n_{ch,o}$ = 18, so the peak energy is $E_X = (n_{ch} - n_{ch,o})62.6$ eV.

An example of the application of PIXE for the identification of impurities with lower masses than those from the substrate elements is shown in Fig.3.21. There a thick Pb-target was analyzed and the elements Si, K, Ca, Ti, V, Cr, Mn, Fe, Cu, Zn, Ge, and Se are detected which is not possible by RBS.

Another advantage of PIXE in comparison to RBS is the higher mass resolution for heavy elements. In Fig.3.4, the mass resolution δM_2 as a function of the target mass M_2 is given for RBS using He ions. At an ion energy of 1.4 MeV the mass resolution for M_2 > 32 amu is greater than unity. It reaches values greater than 10 amu for heavy elements like Pt and Au. That means, elements with masses greater than 32 amu cannot be separated exactly by RBS. On the contrary, PIXE has a mass resolution $\delta M_2 \leq 1$ amu for M_2 > 22 amu independent of M_1 (see Sect.3.4). So the identification of elements with M_2 greater than ≈30 amu can be carried out by PIXE rather than by RBS.

4.1.2. Total amount

The second quantity of interest is the amount of impurities. Following Eq.(3.50) the number of impurities per unit area $(Nz)_{imp}$ can be determined directly from the total yield $Y_{tot,imp}$ of ions incident at an angle φ_1 to the surface normal and being backscattered from the impurity atoms. Approximating $\sigma^d_{imp}(E_1)$ by $\sigma^d_{imp}(E_o)$ we obtain

$$(Nz)_{imp} = \frac{Y_{tot,imp} \cos\varphi_1}{n\, \sigma^d_{imp}(E_o)\, \Delta\Omega} , \tag{4.2}$$

where $\Delta\Omega$ is the solid angle of detection, n is the number of incident ions collected during the measurement, and σ^d_{imp} is the differential scattering cross section for backscattering by the surface impurities discussed in Sect.3.1.7. The value of $Y_{tot,imp}$ is given by the sum of the counts per channel over the entire impurity signal. For the example in Fig.4.1 the results are listed in Table 4.1.

TABLE 4.1
Results of the impurity analysis corresponding to the example represented in Fig.4.1 (K, ε, and σ^d are given in the Appendices 3, 6, and 11)

Element	M_2/amu	K	$\sigma^d/10^{-24}cm^2$	$Y_{tot,imp}$	$(Nz)_{imp}/10^{15}$atoms cm^{-2} Eq.(4.2)	Eq.(4.5)
C	12	0.2526	0.0766	1226	80.52	80.01
Si	28	0.5657	0.5054	5050[1)]		
Co	59	0.7634	1.9393	1300	3.37	3.35
Au	197	0.9225	16.740	9602	2.89	2.87

[1)]Yield from the Si surface $Y_{o,Si}$.

To eliminate the parameters n and $\Delta\Omega$ of the experimental set-up, a second way to calculate $(Nz)_{imp}$ can be chosen using for reference the spectrum height $Y_{o,Si}$ from the substrate surface. Following Eq.(3.37) one obtains

$$Y_{o,Si} = n\, \sigma^d_{Si}(E_o)\, \Delta\Omega\, \frac{N_{Si}\, \Delta z_o}{\cos\varphi_1} , \tag{4.3}$$

where Δz_o is the thickness of the silicon surface layer corresponding to the channel width ϵ_{ch}. The target element considered is characterized by its chemical symbol as subscript of the symbols of the formulae. The depth interval Δz_o can be calculated following Eq.(3.38)

$$\Delta z_o = \frac{\epsilon_{ch}}{N_{Si}[\epsilon_o]_{Si}^{Si}} \quad , \tag{4.4}$$

where $[\epsilon_o]_{Si}^{Si}$ is the stopping cross section factor in surface energy approximation (Eq.(3.12)) for He ions penetrating a silicon target and being backscattered by silicon atoms according to the notation in Eq.(3.56). Inserting Eq.(4.4) into Eq.(4.3) and combining the latter one with Eq.(4.2) the areal density of the impurity atoms in surface energy approximation can be determined by

$$(Nz)_{imp} = \frac{Y_{tot,imp}}{Y_{o,Si}} \frac{\sigma_{Si}^{d}(E_o)}{\sigma_{imp}^{d}(E_o)} \frac{\epsilon_{ch}}{[\epsilon_o]_{Si}^{Si}} \; . \tag{4.5}$$

The experimental value of $Y_{o,Si}$ is obtained by extrapolating the silicon yield to the surface channel (in the example of Fig.4.1 $Y_{o,Si}$ = 5050 at channel number 147). The results for $(Nz)_{imp}$ are given in the last column of Table 4.1. The values obtained by Eqs.(4.2) and (4.5) differ by about 1% which is within the limit of experimental error. The accuracy of the RBS method with respect to the determination of the amount of impurities is in the order of 5 to 10%. In the case of Eq.(4.2) it depends on the error of the product $n\Delta\Omega$ and on the statistical error in the determination of Y_{tot}. Using Eq.(4.5), the accuracy is influenced by the precision of the ϵ-values and the statistical accuracy of $Y_{tot,imp}$ and $Y_{o,Si}$.

To increase the accuracy of the quantitative impurity analysis, samples with a known areal density of atoms determined by a more precise method are used to calibrate the measurement. As an example the application of thin Ni layers is discussed. The areal density $(Nz)_{Ni}$ of the Ni atoms is determined by the measurement of the total magnetic moment $\vec{M}$ of the layer, so that

$$(Nz)_{Ni} = \frac{\vec{M}}{\bar{\vec{\mu}}\, A} \quad , \tag{4.6}$$

where $\bar{\vec{\mu}}$ is the mean magnetic moment of one atom and A is the area of the target. The accuracy of the determination of $\vec{M}$ by a dynamic method is about 1%. Another method to produce such precise standards is ion implantation. Using these standards to compare the unknown amount of impurities which has to be determined with the known amount of the standard, an accuracy of about 1% can be achieved.

The problem of sensitivity of RBS in detection of impurities will now be discussed briefly. If there were no background counts in the spectrum, no limitation in the detectable amount of impurities would exist. Increasing the number n of the incident ions, the sensitivity could be increased without any other limit than the measuring time, which can be shortened using high ion beam currents. In reality, however, the background noise increases with increasing counting rates, which is mainly caused by pulse pile up effects in the detection system as it is discussed in Sect.2.4.3.3. The height of the background is actually small compared to the height of the substrate signal, but is sufficient to limit the detection of impurities. The pulse pile up effect can be reduced by lowering the primary beam current and using fast electronic devices.

An increased sensitivity can also be achieved by tilting the target up to produce glancing incidence of the ions. From all these facts it follows that the sensitivity depends on the experimental conditions. Under normal conditions, a simple empirical estimation for the minimum detectable amount of impurities for Z_2(substrate) < Z_2(impurity) can be made by

$$(Nz)_{imp} \approx \frac{Z_2^2(\text{substrate})}{Z_2^2(\text{impurity})} * 10^{14}\ \text{cm}^{-2} \quad , \tag{4.7}$$

derived from Eq.(4.5) taking into account that σ^d is proportional to Z_2^2 (Ref.[2]). For the example in Fig.4.1, the minimum detectable areal density is about $3*10^{13}$ cm^{-2} for cobalt and about $3*10^{12}$ cm^{-2} for gold. This shows that for heavy elements on light substrates less than one thousandth of a monolayer is detectable.

4.2. THIN LAYERS

4.2.1. Methods of calculation

The problems to be solved in the investigation of thin layers and layered systems are the identification of layer components, the calculation of layer thicknesses, and the determination of composition profiles.

The identification of the layer components can be performed similarly to the previous section. If we are concerned with one-elemental or homogeneously composed layers, the masses can be identified by RBS as long as the mass resolution is sufficient, otherwise by PIXE. In the case of impurities at a certain depth within the sample, the mass no longer can be determined by RBS.

To calculate layer thicknesses and composition profiles it is necessary to convert the energy scale of the backscattering spectrum into a depth scale.

4.2.1.1. Energy-to-depth conversion by numerical methods

The basic relation for energy-to-depth conversion is given by Eq.(3.9) with $dE/ds = N\varepsilon$

$$E_2(z) = K\left[E_o - \int_o^{z/\cos\varphi_1} ds\, N\varepsilon_{in}\right] - \int_o^{z/\cos\varphi_2} ds\, N\varepsilon_{out} \quad . \tag{4.8}$$

Using ε-values tabulated in Appendices 5 and 6 the integration of Eq.(4.8) can be performed numerically. There are two general ways.

(i) Taking into account the experimental conditions, the sample can be divided into thin slabs of different thicknesses where the ions scattered at the two boundaries of each slab have the same energy difference ε_{ch} (channel width)

$$\varepsilon_{ch} = E_{2,i-1,0} - E_{2,i,0} \ . \tag{4.9}$$

Fig.4.3 illustrates the procedure and the quantities necessary for the calculation with:

$E_{1,i}$: ion energy before backscattering at the frontside of slab i,

$KE_{1,i}$: energy immediately after backscattering at the frontside of slab i,

$E_{2,i,j}$: energy of ions backscattered at the frontside of slab i and after passage of slab j on the outward path.

Within this notation, the energy of outcoming particles is $E_{2,i,0}$, where the subscript 0 means the passage of the surface layer and Δz_i expresses the thickness of the i-th slab.

The calculation starts from the surface layer. The corresponding energy difference is given by

$$\varepsilon_{ch} = E_{2,0,0} - E_{2,1,0} \tag{4.10}$$

with $E_{2,0,0} = KE_o$ and (according to Eq(3.10))

$$E_{2,1,0} = K\left[E_o - \frac{N\Delta z_o}{\cos\varphi_1}\varepsilon(E_o)\right] - \frac{N\Delta z_o}{\cos\varphi_2}\varepsilon(KE_{1,1}) \quad , \tag{4.11}$$

where the stopping cross section within a slab can be assumed to be constant.

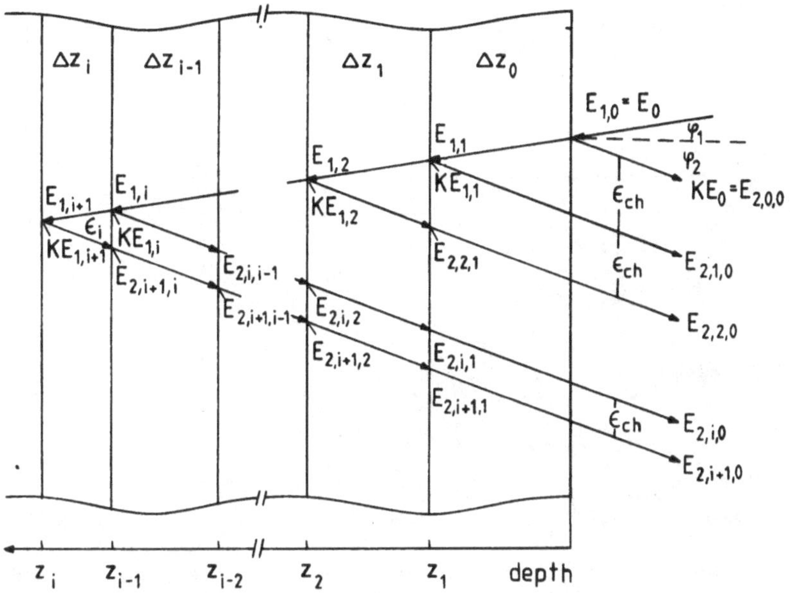

Fig.4.3. Schematic illustration of the energy-to-depth conversion in the case of a fixed value of ϵ_{ch}

Eqs.(4.10) and (4.11) immediately provide

$$N\Delta z_o = \frac{\epsilon_{ch}}{\frac{K}{\cos\varphi_1}\varepsilon(E_o) + \frac{1}{\cos\varphi_2}\varepsilon(KE_{1,1})} \quad , \tag{4.12}$$

with

$$E_{1,1} = E_o - \frac{N\Delta z_o}{\cos\varphi_1}\,\varepsilon(E_o) \quad . \tag{4.13}$$

Approximating $\varepsilon(KE_{1,1})$ by $\varepsilon(KE_o)$ in Eq.(4.12) we find $N\Delta z_o$ and $E_{1,1}$ in the first approximation. This value of $E_{1,1}$ can be used for a second approximation of $N\Delta z_o$ and $E_{1,1}$ and so on. The iteration has to be continued until the values of $N\Delta z_o$ and $E_{1,1}$ do not change within a given limit.

The same procedure for the layer Δz_1 yields

$$\epsilon_{ch} = E_{2,1,0} - E_{2,2,0} \tag{4.14}$$

with

$$E_{2,1,0} = KE_{1,1} - \frac{N\Delta z_o}{\cos\varphi_2}\,\varepsilon(KE_{1,1}) \tag{4.15}$$

and

$$E_{2,2,0} = K\left[E_{1,1} - \frac{N\Delta z_1}{\cos\varphi_1}\varepsilon(E_{1,1})\right] - \frac{N}{\cos\varphi_2}\left[\Delta z_1\,\varepsilon(KE_{1,2}) + \Delta z_o\,\varepsilon(E_{2,2,1})\right]. \quad (4.16)$$

This provides

$$N\Delta z_1 = \frac{\epsilon_{ch}\left[1 + \dfrac{N\Delta z_o}{\cos\varphi_2}\,\dfrac{\varepsilon(KE_{1,1}) - \varepsilon(E_{2,2,1})}{\epsilon_{ch}}\right]}{\dfrac{K}{\cos\varphi_1}\varepsilon(E_{1,1}) + \dfrac{1}{\cos\varphi_2}\varepsilon(KE_{1,2})} \quad (4.17)$$

with

$$E_{1,2} = E_{1,1} - \frac{N\Delta z_1}{\cos\varphi_1}\,\varepsilon(E_{1,1}) \quad (4.18)$$

and

$$E_{2,2,1} = KE_{1,2} - \frac{N\Delta z_1}{\cos\varphi_2}\,\varepsilon(KE_{1,2}) \quad . \quad (4.19)$$

This equation system for the three unknown quantities Δz_1, $E_{1,2}$, and $E_{2,2,1}$ can be solved iteratively. Approximating the stopping cross section values of the unknown energies in Eq.(4.17) by

$$\varepsilon(KE_{1,2}) = \varepsilon(KE_{1,1}) \text{ and } \varepsilon(E_{2,2,1}) = \varepsilon(KE_{1,1}) \quad (4.20)$$

the thickness of the first slab is given by

$$N\Delta z_1 = \frac{\epsilon_{ch}}{\dfrac{K}{\cos\varphi_1}\varepsilon(E_{1,1}) + \dfrac{1}{\cos\varphi_2}\varepsilon(KE_{1,1})} \quad , \quad (4.21)$$

and the energies $E_{1,2}$ and $E_{2,2,1}$ can be calculated in first approximation. By the same iterative procedure as described for the surface slab $N\Delta z_o$ the results for the energies and the layer thickness can be improved.

Generally, we get for the i-th layer according to Eqs.(4.17), (4.18), and (4.19)

$$N\Delta z_i = \frac{\epsilon_{ch}}{\dfrac{K}{\cos\varphi_1}\varepsilon(E_{1,i}) + \dfrac{1}{\cos\varphi_2}\varepsilon(KE_{1,i+1})}\; X \quad (4.22)$$

with

$$X = 1 + \frac{N}{\epsilon_{ch}\cos\varphi_2}\left[\Delta z_{i-1}\,\varepsilon(KE_{1,i}) + \sum_{j=1}^{i-1}\Delta z_{j-1}\,\varepsilon(E_{2,i,j}) - \sum_{j=1}^{i}\Delta z_{j-1}\,\varepsilon(E_{2,i+1,j})\right]$$

and

$$E_{1,i+1} = E_{1,i} - \frac{N\Delta z_i}{\cos\varphi_1}\,\varepsilon(E_{1,i}) \quad , \tag{4.23}$$

$$E_{2,i+1,i} = KE_{1,i+1} - \frac{N\Delta z_i}{\cos\varphi_2}\,\varepsilon(KE_{1,i+1}) \quad , \tag{4.24}$$

$$E_{2,i+1,j} = E_{2,i+1,j+1} - \frac{N\Delta z_j}{\cos\varphi_2}\,\varepsilon(E_{2,i+1,j+1}) \tag{4.25}$$

for $j = i...0$, where $E_{2,i+1,0}$ are the energy values of the detected particles. The values of all quantities with the subscript i are known from the previous steps. Approximating the unknown values of the stopping cross section in Eq.(4.22) by

$$\varepsilon(KE_{1,i+1}) = \varepsilon(KE_{1,i}),$$

$$\varepsilon(KE_{2,i+1,i}) = \varepsilon(KE_{1,i}), \tag{4.26}$$

and

$$\varepsilon(E_{2,i+1,j}) = \varepsilon(E_{2,i,j})$$

with $j = 1...i-1;\ i > 3$,

the thickness of the i-th slab in first order is given by

$$N\Delta z_i = \frac{\epsilon_{ch}}{\frac{K}{\cos\varphi_1}\varepsilon(E_{1,i}) + \frac{1}{\cos\varphi_2}\varepsilon(KE_{1,i})} \quad , \tag{4.27}$$

which means a surface energy approximation for the stopping cross section factor of the i-th slab

$$\Delta z_i = \frac{\epsilon_{ch}}{N[\varepsilon_i]} \tag{4.28}$$

with

$$[\varepsilon_i] = \frac{K}{\cos\varphi_1}\,\varepsilon(E_{1,i}) + \frac{1}{\cos\varphi_2}\,\varepsilon(KE_{1,i}) \quad , \tag{4.29}$$

(which corresponds to Eqs.(3.41) and (3.42) with $\epsilon_i = \epsilon_{ch}$). Now the unknown energies (Eqs.(4.23, 4.24, 4.25)), can be calculated in first approximation. Adding an iteration process, the results can be improved again. By this method the energy scale provided by the multichannel analyzer can be transferred into a depth scale.

(ii) A further way to calculate depth values corresponding to the E_2-values starts by subdividing the sample into thin slabs of the same thickness Δz. The slabs must be thin enough so that the energy loss over the whole width Δz is practically constant. The calculation starts at the surface layer. The energy difference between the two boundaries of slab i for the incoming particle is given by the relation

$$E_{1,i+1} - E_{1,i} = \frac{N\Delta z}{\cos\varphi_1}\,\varepsilon(E_{1,i}) \quad . \tag{4.30}$$

The procedure is illustrated in Fig.4.4.

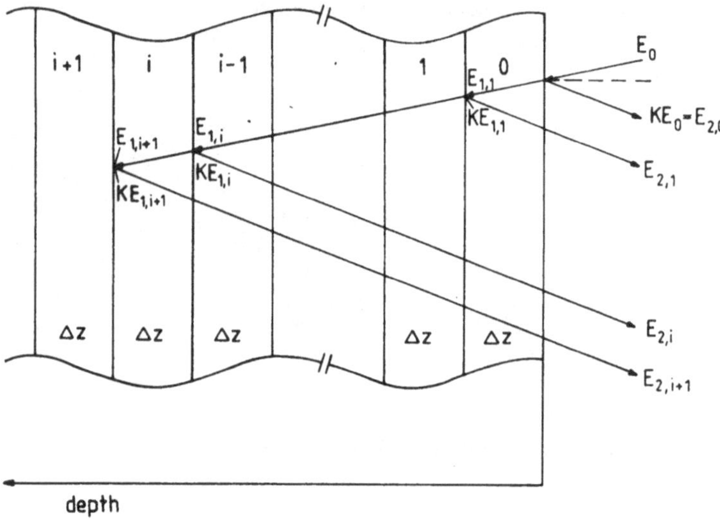

Fig.4.4. Schematic illustration of the energy-to-depth conversion in the case of a fixed layer thickness Δz

Particles backscattered after passing the first slab have the final energy

$$E_{2,1} = KE_{1,1} - \frac{N\Delta z}{\cos\varphi_2}\,\varepsilon(KE_{1,1}) \quad . \tag{4.31}$$

Ions backscattered from the backside of the second slab are detected with an energy

$$E_{2,2} = \{KE_{1,2} - \frac{N\Delta z}{\cos\varphi_2}\varepsilon(KE_{1,2})\} - \frac{N\Delta z}{\cos\varphi_2}\varepsilon(\{E\}) \quad , \tag{4.32}$$

where $\varepsilon(\{E\})$ is the stopping cross section at the energy $\{E\}$ calculated by the expression in braces of Eq.(4.32). Continuing this procedure, we obtain for the energy of ions backscattered from the i-th slab

$$E_{2,i} = \left\{\cdots\left\{\left\{\left\{KE_{1,i} - \frac{N\Delta z}{\cos\varphi_2}\varepsilon(KE_{1,i})\right\} - \frac{N\Delta z}{\cos\varphi_2}\varepsilon(\{E\})\right\} - \frac{N\Delta z}{\cos\varphi_2}\varepsilon(\{\{E\}\})\right\} - \cdots\right\}, \tag{4.33}$$

where the ε-value for a certain slab has to be evaluated at an energy given in the braces of the preceding term. Using this method, the energy differences $(E_{2,i} - E_{2,i-1})$ for different i-values are not equal. Therefore, to calculate spectrum heights Y_i (see Eq.(3.48)) the yield must be redistributed to an equal energy difference ε_{ch} corresponding to the measured spectrum.

4.2.1.2. Approximation methods

The assumption of constant values of the stopping cross section ε over the depth z leads to a linear relation

$$\Delta E_2 = KE_o - E_2 = Nz[\varepsilon] \tag{4.34}$$

with

$$[\varepsilon] = \frac{K}{\cos\varphi_1}\varepsilon_{in} + \frac{1}{\cos\varphi_2}\varepsilon_{out} \tag{4.35}$$

between the exit energy E_2 and the depth z of backscattering as discussed in Sect.3.1.4. The depth interval for which this relation provides good results depends on the choice of the value of $[\varepsilon]$. This is shown in Fig.4.5 (Ref.[2]) where the numerically calculated curve $\Delta E_2(z)$

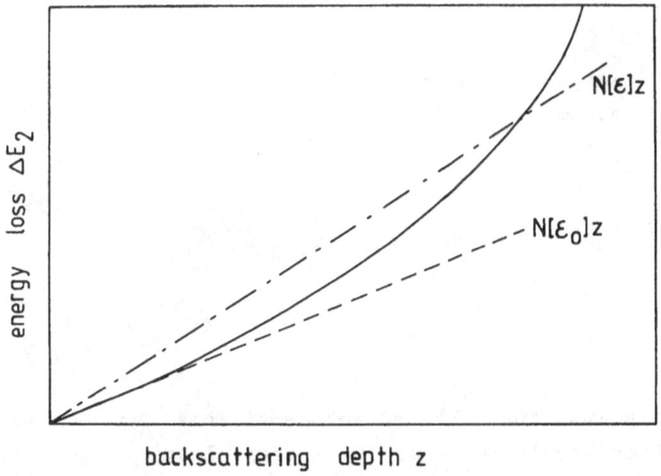

Fig.4.5. Energy width ΔE_2 versus z curves numerically calculated (full line), in the surface energy approximation (dashed line), and for an arbitrary fixed value (dash-dotted line)

is drawn, together with two curves for linear approximations. The linear relation gives an exact result for one pair of ε_{in} and ε_{out} only at a certain depth z, namely when the relation

$$Nz[\varepsilon] = \frac{K}{\cos\varphi_1} \int_{in} ds\ N\ \varepsilon(E) + \frac{1}{\cos\varphi_2} \int_{out} ds\ N\ \varepsilon(E) \tag{4.36}$$

holds.

Useful assumptions for the constant stopping cross sections are the surface energy approximation (Eq.(3.12)) and the mean energy approximation (Eq.(3.15)). The surface energy approximation provides good results in the case of near-surface regions where the thickness is small enough so that the relative changes of energy along the inward path as well as along the outward path are small.

Using the mean energy approximation, the energy E_1 before scattering must be determined in order to calculate the mean energies E_{in} (Eq.(3.13)) and E_{out} (Eq.(3.14)). One method for quick estimation of E_1 is discussed in Sect.3.1.4 (Eqs.(3.16) and (3.17)), where the energy difference ΔE_2 is divided between the inward path and the outward path according to their path length.

Other possibilities of the calculation of E_1 are the energy loss ratio method described by Lever [81], the iterative method, and analytical methods using expressions which approximate the energy dependence of the energy loss analytically. We want to discuss here the energy loss ratio method and the iterative method briefly.

(i) Energy loss ratio method

The basic assumption is the steadiness of the ratio α of the energy lost along the outward path to the energy lost along the inward path

$$\alpha = \frac{\Delta E_{out}}{\Delta E_{in}} \quad , \tag{4.37}$$

which is approximately independent of z. With $\Delta E_{out} = KE_1 - E_2$ and $\Delta E_{in} = E_o - E_1$ the ratio α becomes

$$\alpha = \frac{KE_1 - E_2}{E_o - E_1} \quad , \tag{4.38}$$

from what follows

$$E_1 = \frac{E_2 + \alpha E_o}{K + \alpha} \tag{4.39}$$

for the energy E_1 before scattering. The advantage of this method is that not the energy loss itself but only the energy loss ratio is necessary for the calculation of E_1. Because the value of α is largely independent of depth z, it can be calculated by the surface energy approximation

$$\alpha \approx \frac{\varepsilon(KE_o)}{\varepsilon(E_o)} \frac{\cos\varphi_1}{\cos\varphi_2} \quad , \tag{4.40}$$

where the stopping cross section values can be taken from the Appendices 5 and 6 for H and He ions, respectively. Using constant values of α, this method is most accurate for the analysis of thin layers where the surface energy approximation is valid and also for thick layers if the composition does not change very greatly, because the change of α with energy is much smaller than the change of ε. The method provides incorrect results if the composition changes considerably within the layer. But the method can be improved by the calculation of different values of α for regions of different compositions.

(ii) Iterative method

This method starts from the surface energy approximation to calculate E_1 in first order

$$E_1 = E_o - \frac{Nz}{\cos\varphi_1} \varepsilon(E_o) \quad , \tag{4.41}$$

and with this an improved energy can be calculated using the mean energy approximation

$$E_1 = E_o - \frac{Nz}{\cos\varphi_1} \varepsilon(\tfrac{1}{2}(E_1+E_o)) \quad . \tag{4.42}$$

This procedure can be iterated to find still better estimates of E_1. The method converges rapidly.

4.2.1.3. Determination of composition profiles

(i) Energy loss ratio method using a constant value of α

One of the problems in the application of RBS is the determination of the depth dependence of the composition of thin layers consisting of more than one element. A simple method to determine the concentration as a function of the depth is based on the energy loss ratio method (see Eq.(4.39)) for the calculation of the energy E_1 before scattering, using a constant value of α. This method was developed by R.F. Lever [81]. The procedure is illustrated schematically in Fig.4.6. A layer of substance B is assumed to overlie a substrate of a lighter substance A with a compound layer A_xB_{1-x} between them. B is assumed to have double the stopping cross section of A and four times the scattering cross section. Under typical experimental conditions, the factor $(K + \alpha)$ is taken as 2. For simplicity, the effects of the increase in scattering cross section with decreasing energy are not considered. The method presupposes that backscattering signals from different elements are separated. The starting point for the calculation is the backscattering spectrum (counts per channel versus the measured exit energy E_2) shown in Fig.4.6a. First, the energy scale is converted from E_2 into the energy E_1 (by Eq.(4.39)) before scattering. After the transformation, the new channel sizes are different for element A and element B and an interpolation is required to redraw the transformed spectra with a common channel width. In Fig.4.6b, the same channel width as in Fig.4.6a is assumed. The number of counts per channel is doubled because a value of 2 for $(K + \alpha)$ is assumed. Now, using Eq.(3.54) it is possible to convert the ordinate of Fig.4.6b from counts per channel to atoms cm^{-2} per channel for each element.

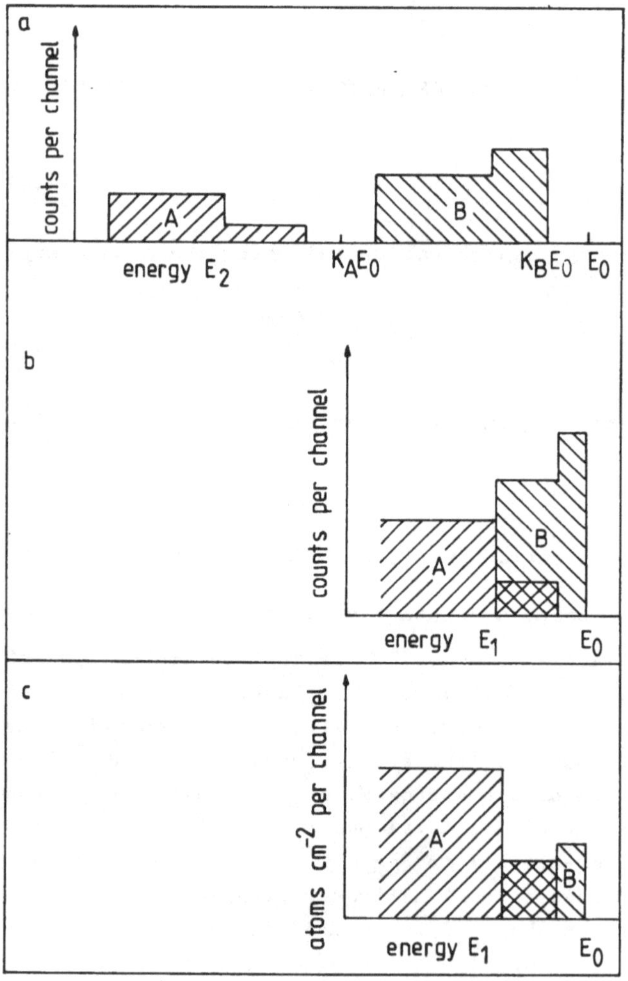

Fig.4.6. Schematic representation of the determination of composition profiles by Lever [81]

For the i-th channel and the element $A^{(j)}$ one obtains ($\varphi_1 = 0$)

$$(N_{A^{(j)}}\Delta z)_i = \frac{Y'_{i,A^{(j)}}}{n\,\sigma^d_{A^{(j)}}(E_{1,i})\,\Delta\Omega} \quad , \tag{4.43}$$

where $Y'_{i,A(j)}$ is the transformed spectrum height and $N_{A(j)}$ is the atomic density of the element $A^{(j)}$. This result is shown in Fig.4.6c. Then the concentration $c_{i,A(j)}$ of element $A^{(j)}$ at depth z_i is given by

$$c_{i,A^{(j)}} = \frac{(N_{A^{(j)}}\Delta z)_i}{\sum_j (N_{A^{(j)}}\Delta z)_i} \quad , \tag{4.44}$$

and the corresponding depth is obtained by integrating the spectra in Fig.4.6c for all elements from the surface to the i-th slab

$$(Nz)_i = \sum_{k=0}^{i} \sum_j (N_{A^{(j)}}\Delta z)_k \quad . \tag{4.45}$$

Because of the assumption of a constant value of α (Eq.(4.37)) the method described works very quickly. Since the outgoing particles traverse the same range of composition as before recoil, the value of α in fact will not change widely with depth even though the composition may change considerably. However, in the case of strong variations of composition with depth, the accuracy of this assumption must be proved.

(ii) Energy loss ratio method with different α-values

In the case of strong variations of composition, the value of α can change up to 30%. The method described above can be improved by calculating the energy $E_{1,i+1}$ before scattering at the backside of slab i, not from the energies E_o and $E_{2,i+1,0}$ as in Eq.(4.39) but from the energies $E_{1,i}$ and $E_{2,i+1,i}$ (see Fig.4.3). Then $E_{1,i+1}$ reads

$$E^{A^{(j)}}_{1,i+1} = \frac{E^{A^{(j)}}_{2,i+1,i} + \alpha_{i,A(j)}\, E^{A^{(j)}}_{1,i}}{\alpha_{i,A^{(j)}} + K_{A^{(j)}}} \tag{4.46}$$

for backscattering at different target elements $A^{(j)}$ (for simplicity, in the following, the index $A^{(j)}$ is omitted in the description of this procedure; it is only written if concentration ratios are considered). The value of α_i can be determined for each channel using Bragg's rule (see Eqs.(1.118) and (3.53)) to calculate the stopping cross section of the compound layer. But in practice it is not necessary to calculate α_i for each channel because α is approximately constant over a wide range of depth. Only considerable changes of the concentration versus depth give rise to a noticeable variation of α .

For i > 0 the energies $E_{2,i+1,i}$ in Eq.(4.46) are unknown. They can be eliminated from Eq.(4.46) by subtraction of two succesive E_1-values

$$\Delta E_{1,i} = E_{1,i} - E_{1,i+1} = \frac{\epsilon_i}{K + \alpha_i} \tag{4.47}$$

with $\epsilon_i = KE_{1,i} - E_{2,i+1,i}$. The energy difference ϵ_i can be related to the energy per channel ϵ_{ch} (see Eq.(3.47)) utilizing the equality of the path lengths between $KE_{1,i}$ and $E_{2,i,0}$ and between $E_{2,i+1,i}$ and $E_{2,i+1,0}$. This procedure is outlined in Sect.3.1.8.2 (Eqs.(3.44 - 3.47)). Inserting Eq.(3.47) in Eq.(4.47) we obtain

$$\Delta E_{1,i} = \frac{\epsilon_{ch}}{K + \alpha_i} \frac{\varepsilon(KE_{1,i})}{\varepsilon(E_{2,i,0})} = \frac{\epsilon_{ch}}{[K + \alpha_i]} . \tag{4.48}$$

with the abbreviation

$$[K + \alpha_i] = (K + \alpha_i)\frac{\varepsilon(E_{2,i,0})}{\varepsilon(KE_{1,i})} .$$

The number of particles backscattered from element $A^{(j)}$ within a depth interval related to the energy interval $\Delta E_{1,i}^{A(j)}$ is given by

$$Y_{i,A^{(j)}} = n\, \sigma^d_{A^{(j)}}(E_{1,i})\, \Delta\Omega\, c_{i,A^{(j)}} \frac{\Delta E_{1,i}^{A^{(j)}}}{\varepsilon^{A^{(1)}_{n_1}\ldots A^{(m)}_{n_m}}(E_{1,i})} , \tag{4.49}$$

which follows from Eq.(3.54) with

$$\Delta z_{i,A^{(j)}} = \frac{\Delta E_{1,i}^{A^{(j)}} \cos\varphi_1}{N_{A^{(j)}}\, \varepsilon^{A^{(1)}_{n_1}\ldots A^{(m)}_{n_m}}(E_{1,i})} . \tag{4.50}$$

Inserting Eq.(4.48) in Eq.(4.49) we get

$$Y_{i,A^{(j)}} = n\, \sigma^d_{A^{(j)}}(E_{1,i})\, \Delta\Omega\, c_{i,A^{(j)}} \frac{\epsilon_{ch}}{[K_{A^{(j)}}+\alpha_{i,A^{(j)}}]\, \varepsilon^{A^{(1)}_{n_1}\ldots A^{(m)}_{n_m}}(E_{1,i})} . \tag{4.51}$$

From Eqs.(4.46) and (4.48) it can be seen that the transformation of the measureable energy E_2, given by equidistant values (distance ϵ_{ch}), to the energy before scattering leads to element and depth dependent values $E_{1,i}^{A(j)}$ and $\Delta E_{1,i}^{A(j)}$. To calculate the concentration ratios within a certain depth interval, the spectra of the different elements must be redistributed in such a way that unified intervals $\Delta E_{1,i}$ are obtained. Practically, one element is chosen as a reference element and all other spectrum heights have to be corrected. For now, element $A^{(1)}$ will be used for reference. The correction factor with respect to the energy difference $\Delta E_{1,i}$ can be calculated from Eq.(4.48)

$$f_{i,j} = \frac{\Delta E_{1,i}^{A^{(1)}}}{\Delta E_{1,i}^{A^{(j)}}} = \frac{[K_{A^{(j)}} + \alpha_{i,A^{(j)}}]}{[K_{A^{(1)}} + \alpha_{i,A^{(1)}}]} \quad . \tag{4.52}$$

Since the $E_{1,i}^{A(j)}$-values are different for different elements (see Eq.(4.48)) the parts of the measured $Y_{i,A(j)}$ (for each channel width ϵ_{ch}) which correspond to the corrected energy difference are not equal for the different elements. Consequently, the counts per channel measured must be rearranged, starting from the surface channel. The transformation is demonstrated in the following example ($Y'_{i,A(j)}$ is the transformed spectrum height) for one element j

$$f_{0,j} = 0.5 \longrightarrow Y'_{0,A^{(j)}} = 0.5\ Y_{0,A^{(j)}}$$

$$f_{1,j} = 0.8 \longrightarrow Y'_{1,A^{(j)}} = 0.5\ Y_{0,A^{(j)}} + 0.3\ Y_{1,A^{(j)}}$$

$$f_{2,j} = 0.6 \longrightarrow Y'_{2,A^{(j)}} = 0.6\ Y_{1,A^{(j)}}$$

$$f_{3,j} = 1.2 \longrightarrow Y'_{3,A^{(j)}} = 0.1\ Y_{1,A^{(j)}} + Y_{2,A^{(j)}} + 0.1\ Y_{3,A^{(j)}}$$

$$f_{4,j} = 1.2 \longrightarrow Y'_{4,A^{(j)}} = 0.9\ Y_{3,A^{(j)}} + 0.3\ Y_{4,A^{(j)}}$$

Following Eq.(4.49) the concentration profiles of the different elements can be calculated using the corrected spectrum heights. Because of equal intervals $\Delta E_{1,i}^{A(j)}$ for all elements (which means equal depth intervals) a simple relation for the calculation of the concentration of one element in the i-th depth interval is obtained

$$c_{i,A^{(j)}} = \frac{Y'_{i,A^{(j)}} / \sigma^{d}_{A^{(j)}}(E_{1,i})}{\sum_{j}(Y'_{i,A^{(j)}} / \sigma^{d}_{A^{(j)}}(E_{1,i}))} \quad . \tag{4.53}$$

4.2.2. Distribution of low impurity concentrations

4.2.2.1. Implantation profiles

The RBS method offers the possibility to characterize ion implanted samples with regard to the implantation dose, the range of the implanted ions, the location of impurity atoms in the lattice, and the radiation damage induced by the passage of the ions through the crystal. In this section, we want to concern ourselves with the determination of the areal density of implanted ions and their depth distribution. Fig.4.7 shows an energy spectrum of 1.4 MeV He ions ($\varphi_1 = 4^o$, $\varphi_2 = 76^o$, $\theta = 100^o$) backscattered from a silicon target implanted with 10^{16} Sbcm^{-2} of an energy E_o = 65 keV. The Sb signal has a Gaussian distribution with the peak position at 1218 keV and a width of 66 keV (FWHM). The peak is shifted by $\Delta E_{2,Sb}$ = 78 keV from the surface position of Sb. The leading edge of the silicon signal lies at 1000 keV. From the spectrum of Fig.4.7 we want to calculate the amount, the projected range, and the range straggling of the Sb atoms using both the surface energy approximation and the computer program described in Sect.4.2.1.3 for comparison.

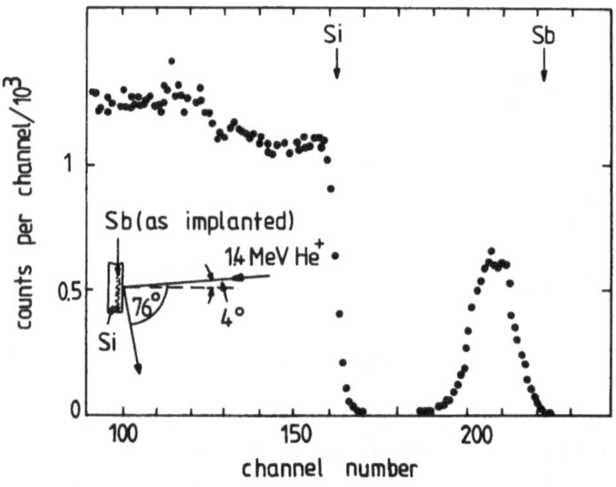

Fig.4.7. Backscattering spectrum of 1.4 MeV He ions incident on a Sb implanted Si sample with $\varphi_1 = 4^o$, $\varphi_2 = 76^o$, $\theta = 100^o$, $\Delta\Omega$ = 1.7 msr, ne = 5 μC, and ϵ_{ch} = 4.921 keV

Using the <u>surface energy approximation</u>, the areal density can be calculated from Eqs.(4.2) or (4.5)

$$(Nz)_{Sb} = \frac{Y_{tot,Sb} \cos\varphi_1}{n\, \sigma^d_{Sb}(E_o)\, \Delta\Omega} = 9.6*10^{15} \text{ cm}^{-2} \tag{4.54}$$

or

$$(Nz)_{Sb} = \frac{Y_{tot,Sb}}{Y_{o,Si}} \frac{\sigma^d_{Si}(E_o)}{\sigma^d_{Sb}(E_o)} \frac{\epsilon_{ch}}{[\epsilon_o]^{Si}_{Si}} = 9.9*10^{15} \text{ cm}^{-2} \quad . \tag{4.55}$$

The two results agree within 3%. The maximum concentration of antimony atoms can be estimated from the peak height of the antimony signal if the FWHM of the antimony distribution is greater than the FWHM of the detector resolution

$$\frac{N_{Sb}}{N_{Si}} = \frac{Y_{max,Sb}}{Y_{o,Si}} \frac{\sigma^d_{Si}(E_o)}{\sigma^d_{Sb}(E_o)} \frac{[\varepsilon_o]^{Si}_{Sb}}{[\varepsilon_o]^{Si}_{Si}} \tag{4.56}$$

with

$$[\varepsilon_o]^{Si}_{Sb} = \frac{K_{Sb}}{\cos\varphi_1} \varepsilon_{Si}(E_o) + \frac{1}{\cos\varphi_2} \varepsilon_{Si}(K_{Sb}E_o) \tag{4.57}$$

and

$$[\varepsilon_o]^{Si}_{Si} = \frac{K_{Si}}{\cos\varphi_1} \varepsilon_{Si}(E_o) + \frac{1}{\cos\varphi_2} \varepsilon_{Si}(K_{Si}E_o) \quad . \tag{4.58}$$

The influence of antimony on the stopping cross section factor $[\varepsilon_o]$ results only from the kinematic factor K_{Sb} if the ions are backscattered from antimony. The type of scattering center is denoted by the subscript of the stopping cross section factor. The contribution of Sb to the energy loss of the He ions is negligible. Only the energy loss of the He ions in silicon has to be taken into account. This fact is expressed by the superscript Si on the stopping cross section factor. For the example shown in Fig.4.7, a maximum concentration of $2.1*10^{21}$ Sb cm^{-3} (corresponding to 4.2 at %) is obtained.

The projected range R_p of implanted antimony can be determined from the energy shift $\Delta E_{2,Sb}$ = 78 keV of the peak position below the surface edge

$$N_{Si}R_p = \frac{\Delta E_{2,Sb}}{[\varepsilon_o]^{Si}_{Sb}} = 2.7*10^{17} \text{ cm}^{-2} \quad . \tag{4.59}$$

With the atomic density of silicon $N_{Si} = 4.98*10^{22}$ cm^{-3} a projected range of R_p = 54 nm is obtained. In the case of a Gaussian distribution of the implanted ions, the depth profile can be described by the projected range R_p and by the range straggling ΔR_p. The standard deviation ΔR_p is related to the FWHM-value by

$$\text{FWHM} = 2\,(2\ln 2)^{1/2}\,\Delta R_p = 2.355\,\Delta R_p \quad . \tag{4.60}$$

The FWHM-value of the antimony peak in Fig.4.7 corresponds to ΔE_{FWHM} = 66 keV which gives a range straggling of

$$\Delta R_p = \frac{\Delta E_{FWHM}}{2.355\, N_{Si}[\varepsilon_o]^{Si}_{Sb}} = 19.3 \text{ nm} \quad . \tag{4.61}$$

The computer program provides the depth profile of antimony shown in Fig.4.8 with $N_{Si}R_p = 2.65*10^{17}$ cm^{-2} which corresponds to $R_p = 53.2$ nm and $\Delta R_p = 19$ nm.

The results of ΔR_p derived by the two methods are somewhat too large because the influence of the energy resolution of the detector and the energy straggling of the He ions (see also Sect.3.1.6) are not taken into account. The energy resolution of the detecting system is quite independent of the energy and it can be determined from the slope of the silicon edge in the spectrum by measuring the energy spread of the step from 12% to 88% of the step height. From Fig.4.7 we obtain 18.3 keV.

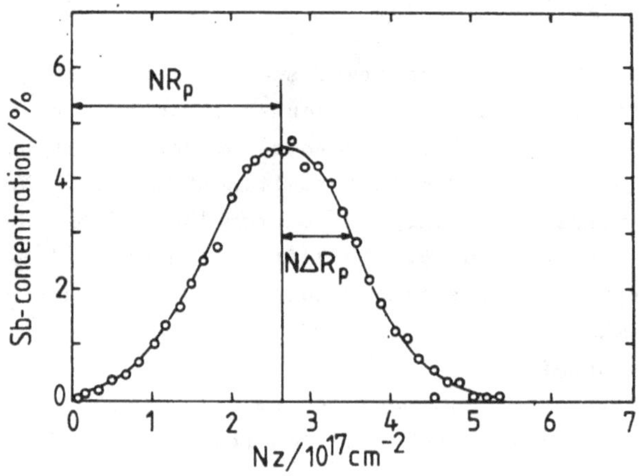

Fig.4.8. Concentration profile of the implanted Sb atoms of the sample of Fig.4.7

The energy straggling can be estimated from Bohr's theory (see Sect.1.2.2). For the energy spread due to the straggling, we obtain 7.82 keV following Eq.(3.25). For the corrected FWHM we achieve

$$\Delta E_{FWHM_c} = [(66 \text{ keV})^2 - (18.3 \text{ keV})^2 - (7.82 \text{ keV})^2]^{1/2} \tag{4.62}$$

$$= 62.9 \text{ keV}$$

with all three distributions assumed to be Gaussian. The range straggling calculated from ΔE_{FWHM_c} provides $\Delta R_p = 18.4$ nm.

The results obtained by the surface energy approximation and by the computer program show no significant differences as can be seen in Table 4.2. The results for $(Nz)_{Sb}$ agree within the limits of error. Therefore we can conclude that for shallow depths, where the energy shift ΔE_2 is small compared to E_2 and the energy loss changes little over ΔE_2, the surface energy approximation is adequate for the analysis of implanted distributions.

TABLE 4.2

Comparison of R_p, ΔR_p and $(Nz)_{Sb}$ of the spectrum of Fig.4.7 determined by means of the surface energy approximation and by means of the computer program

Method	R_p/nm	ΔR_p/nm		$(Nz)_{Sb}/10^{16}$ cm^{-2}
		noncorrected	corrected	
Surface energy approximation	54	19.3	18.4	0.96 (Eq.(4.2)) 0.99 (Eq.(4.5))
Computer program	53.2	19	18.1	1.0

4.2.2.2. Diffusion profiles

Fig.4.9 shows the backscattering spectrum of 1.4 MeV He ions for a Sb implanted Si wafer after a laser annealing process.

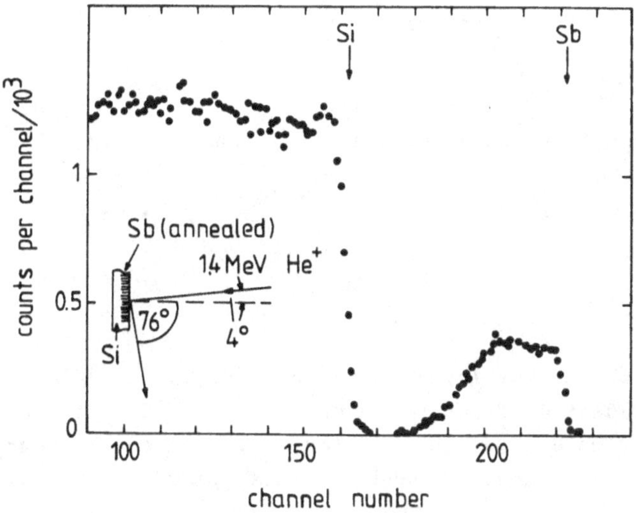

Fig.4.9. Backscattering spectrum of 1.4 MeV He ions incident on a silicon sample implanted with Sb atoms and laser irradiated (same experimental parameters as given in Fig.4.7)

Now the Sb ions are distributed from the sample surface to a depth two to three times greater than in the example represented in Fig.4.7. Since the surface energy approximation is no longer valid the mean energy approximation (see Sect.3.1.4) or the computer program must be used for the calculation of the composition profile. Fig.4.10 shows the Sb concentration as a function of the depth computed by the program described in Sect.4.2.1.3.

The number N_{Sb} of Sb atoms per cm^3 at a depth z has to be calculated from the spectrum height at energy E_1 which corresponds to the depth z, in analogy to Eq.(3.48)

$$Y_{Sb}(E_1) = n\,\sigma^{d}_{Sb}(E_1)\,\Delta\Omega\;\frac{N_{Sb}(z)}{N_{Si}}\;\frac{\epsilon_{ch}}{[\epsilon(E_1)]^{Si}_{Sb}}\;\frac{\epsilon(K_{Sb}E_1)}{\epsilon(E_2)}\quad . \tag{4.63}$$

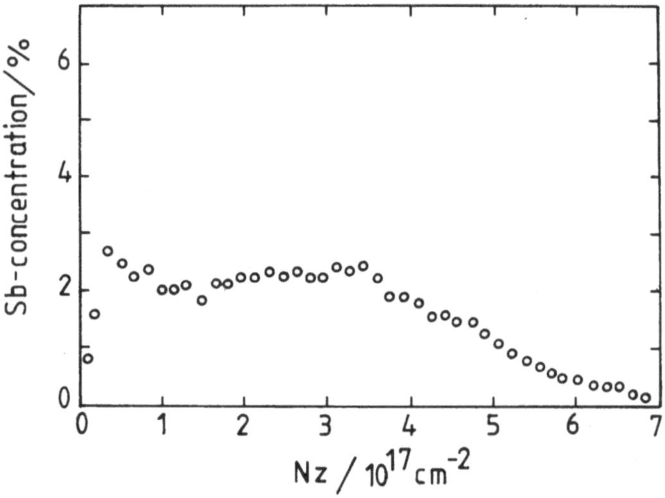

Fig.4.10. Diffusion profile of the Sb atoms for the example of Fig.4.9

4.2.3. Elemental layers

4.2.3.1. Single layer

There are two ways to calculate the layer thickness. First, the layer thickness z can be calculated from the energy width ΔE_2 of the backscattering spectrum, which is the energy difference between particles backscattered at the frontside and particles backscattered at the backside of the layer

$$\Delta E_2 = [\epsilon]\,Nz \quad , \tag{4.64}$$

where $[\varepsilon]$ is the stopping cross section factor already discussed in Sect.3.1.4. This method is only applicable if ΔE_2 really represents the layer thickness. This depends on the depth resolution δz as discussed in Sect.3.1.6, which is mainly determined by the detector resolution in the case of thin layers. Fig.4.11 shows the influence of the detector resolution on the shape of the ideal ($\delta z \rightarrow 0$) backscattering spectra of Ni layers calculated for different thicknesses, ranging from 3 nm up to 160 nm. The parameters used for the calculation are $E_o = 1.4$ MeV, $\varphi_1 = 0^o$, $\Theta = 170^o$, $\epsilon_{ch} = 5$ keV, $\delta E_d = 14$ keV, $n = 1.25*10^{14}$, and $\Delta\Omega = 4$ msr. In the case of thicker layers ($z \geq 40$ nm) the half-widths of the spectra are independent of the detector resolution.

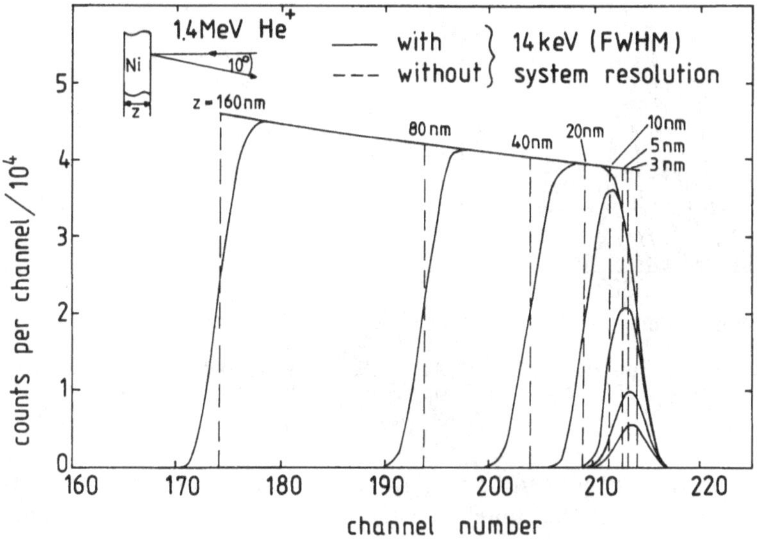

Fig.4.11. Calculated backscattering spectra of 1.4 MeV He ions incident on Ni layers of different thicknesses with a system resolution of 14 keV (solid lines) and without system resolution (broken lines) with the following parameters: $\varphi_1 = 0^o$, $\Theta = 170^o$, $\epsilon_{ch} = 5$ keV, ne = 20 μC, and $\Delta\Omega = 4$ msr

With decreasing film thickness, however, the half-widths and the heights of the spectra are more and more influenced by the detector resolution. If the width of the ideal spectrum is smaller than the detector resolution, the height is lowered. Then the calculation of the layer thickness from ΔE_2 is not possible.

In this case, the second way to determine layer thickness must be chosen using the total number of counts Y_{tot} originating from backscattering events throughout the whole layer in a similar manner as is discussed in Sect.4.2.2 for the calculation of the impurity concentration. Fig.4.12 shows the measured spectra of 1.4 MeV He ions backscattered from very thin Ni layers. Only one spectrum reaches the height of the ideal spectrum and the layer thickness can be determined from ΔE_2 by Eq.(4.64). Using the bulk value of the atomic density $N_{Ni} = 9.133*10^{22}$ cm^{-3} and the surface energy approximation for the stopping cross section factor $[\epsilon] = [\epsilon_o]$ we obtain $z = 40$ nm for the thickest layer. With the surface energy approximation $[\epsilon_o]$ a linear relation between ΔE_2 and z is obtained.

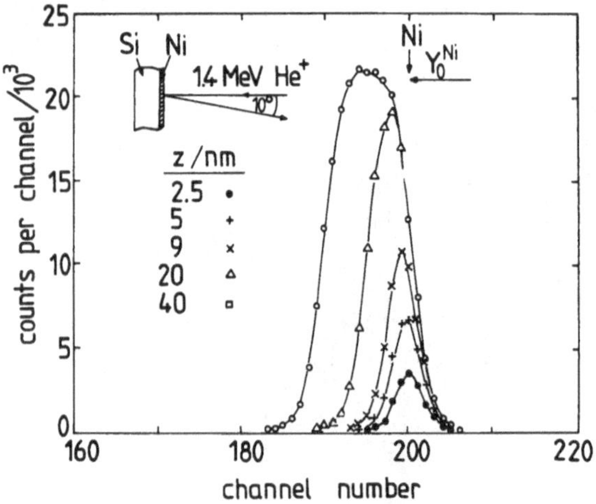

Fig.4.12. Backscattering spectra of 1.4 MeV He ions incident on thin Ni layers with $\varphi_1 = 0^o$, $\Theta = 170^o$, $\Delta\Omega = 2.3$ msr, $\epsilon_{ch} = 5.089$ keV, ne = 20 µC

This approximation yields good results up to layer thicknesses of about 200 nm in the case of 1.4 MeV He ions incident on Ni layers and under the conditions of normal incidence and $\Theta = 170^o$. This is illustrated in Fig.4.13 where the solid line shows the relation between ΔE_2 and z in the surface energy approximation and the dashed line represents the results of a computer calculation using energy dependent ϵ-values. For increasing layer thickness, the uncertainty of the surface energy approximation increases. The results obtained by the two methods differ by about 1% for each 100 nm of film thickness. The thicknesses of the other layers given in Fig.4.12 must be determined from $Y_{tot,Ni}$ in the case of surface energy approximation, which is a good approximation for such thin layers. We obtain from Eq.(3.51)

$$Y_{tot,Ni,o} = n\, \sigma^d_{Ni}(E_o)\, \Delta\Omega\, \frac{Nz}{\cos\varphi_1} \quad . \tag{4.65}$$

The calculated layer thicknesses are depicted in Fig.4.12. Using Y_{tot} for the calculation of larger thicknesses, the energy dependence of the scattering cross section has to be taken into account as is discussed in

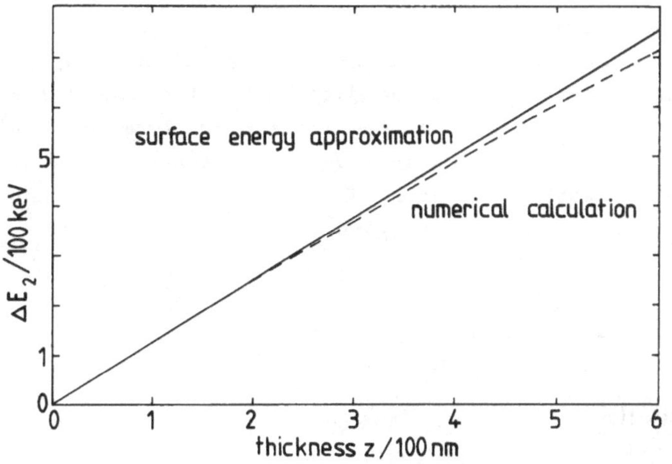

Fig.4.13. $\Delta E_2(z)$ calculated numerically and for the surface energy approximation

Sect.3.1.8.2. The total amount of backscattered particles up to the depth z is given by Eq.(3.52) from which follows the number of particles per unit area

$$Nz = \frac{Y_{tot} \cos\varphi_1}{n\,\sigma^d(E_o)\,\Delta\Omega} \left(1 + \frac{Y_{tot}\ \varepsilon(\bar{E}_{in})}{n\,\sigma^d(E_o)\ \Delta\Omega\ E_o}\right)^{-1} \quad . \tag{4.66}$$

With the abbreviation

$$(Nz)_o = \frac{Y_{tot}\ \cos\varphi_1}{n\,\sigma^d(E_o)\,\Delta\Omega} \tag{4.67}$$

we can rewrite the expression for Nz in the form

$$Nz = (Nz)_o \left[1 + \frac{(Nz)_o}{\cos\varphi_1} \frac{\varepsilon(\bar{E}_{in})}{E_o}\right]^{-1} \quad . \tag{4.68}$$

Usually, the second term in brackets is small compared to unity. Therefore the determination of the number of atoms per unit area in the layer based on the total number of counts Y_{tot} is widely independent of the exact knowledge of the stopping cross section of the element in the layer. However, the solid angle $\Delta\Omega$ and the total number of incident particles n must be known, i.e. an absolutely calibrated system is required. The correction term of Eq.(4.68) is proportional to $(Nz)_o$ and ε for a fixed incident energy E_o. Since ε increases with M_2, the value of the correction term is greater for heavier target elements, i.e. the error of the Nz-values calculated by Eq.(4.65) increases with increasing mass of the elements of the layer.

There is another way to determine the thickness of such thin layers as shown in Fig.4.12. An increase of the ion path length throughout the layer can be achieved by changing the backscattering geometry as discussed in Sect.3.1.9. Fig.4.14 shows the backscattering spectrum of the 20 nm thick Ni layer with increased angles of incidence and emergence, $\varphi_1 = 65^o$ and $\varphi_2 = 55^o$, respectively.

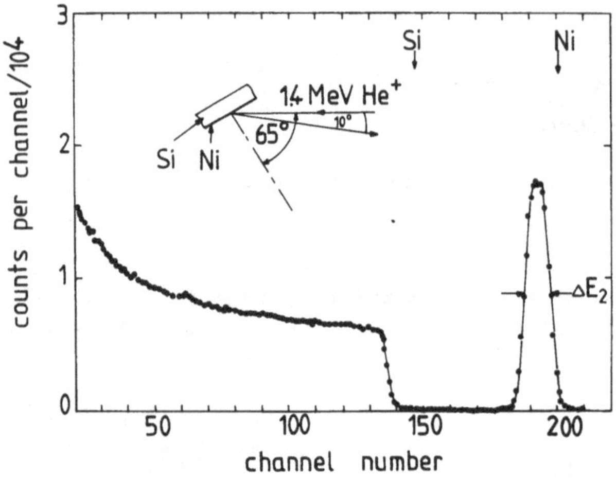

Fig.4.14. Backscattering spectrum of 1.4 MeV He ions incident on a 20 nm thick Ni layer on silicon with increased angles of incidence and emergence ($\varphi_1 = 65^o$, $\varphi_2 = 55^o$, $\theta = 170^o$, $\Delta\Omega = 1.59$ msr, $\epsilon_{ch} = 5.18$ keV, ne = 20 μC)

The backscattering spectrum becomes broadened and reaches the height of the ideal spectrum. Now the energy width ΔE_2 can be used again for the calculation of the layer thickness. With Eq.(4.64) and

$$[\varepsilon_o]^{Ni}_{Ni} = \frac{K_{Ni}}{\cos 65^o}\,\varepsilon_{Ni}(E_o) + \frac{1}{\cos 55^o}\,\varepsilon_{Ni}(K_{Ni}E_o) \qquad (4.69)$$

one obtaines $Nz = 188.2*10^{15}$ cm^{-2}. Using the bulk density of Ni, $N_{Ni} = 9.133*10^{22}$ cm^{-3}, we get for the layer thickness z = 20.6 nm.

The method of glancing incidence ($\varphi_1 \geq 85^o$) is useful for layer thicknesses smaller than about 10 nm. For thicker layers, the path lengths of the ions through the layer become so long that energy straggling must be considered. With the methods described the accessible depth for 1.4 MeV He ions depends on the energy loss of He ions in the target as already discussed in Sect.3.1.5. For example, ΔE_2 = 500 keV will give z_a = 970 nm for silicon, 400 nm for nickel, and only 310 nm for platinum. Since protons lose much less energy than He ions in the same material, layers up to ten times the thickness can be analyzed by proton backscattering.

4.2.3.2. Elemental multilayers

Now we turn to an investigation of a sequence of elemental layers. Figs.4.15(a-c) show as an example the backscattering spectra of 1.4 MeV He ions incident on a Si/Ni-sandwich consisting of 6 layers deposited on a silicon substrate for different angles of incidence $\varphi_1 = 0^\circ$ (a), $\varphi_1 = 45^\circ$ (b), $\varphi_1 = 60^\circ$ (c).

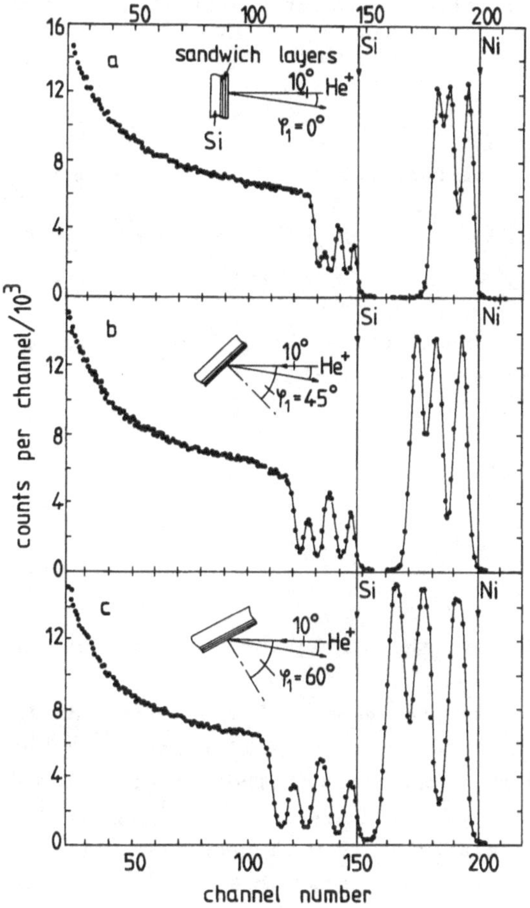

Fig.4.15. Backscattering spectra of 1.4 MeV He ions incident on a Si/Ni-sandwich system with different angles of incidence: $\varphi_1 = 0^\circ$ (a), $\varphi_1 = 45^\circ$ (b), and $\varphi_1 = 60^\circ$ (c) ($\theta = 170^\circ$, $\Delta\Omega$ = 1.59 msr, ne^1 = 20 µC, ϵ_{ch} = 5.18 keV)

By comparing these spectra, we can find out whether the surface layer consists of silicon or nickel. The energy of the ions backscattered at the surface must not change, whereas the signals originating from backscattering at the front-sides of the deeper layers are shifted towards lower energies because of the increased path length. In the example, the high energy edge of the silicon signal is not shifted, i.e. the surface layer is a silicon layer. The energy shift of the deeper layers increases with increasing depth of the layer. So, from the amount of energy shift, the sequence of the layers can be determined. In the case of well-separated backscattering signals from the single layers, the energy E_2 of ions backscattered at the frontside of the second layer can be calculated in the following manner.

1. Calculation of the thickness z_1 of the first layer by one of the methods described above.
2. Calculation of the energy E_1 of the ions after passing the first layer

$$E_1 = E_o - \int_0^{z_1/\cos\varphi_1} ds\, N_{Si}\, \varepsilon_{Si}(E) \quad ; \qquad (4.70)$$

in the case of thin layers, the surface energy approximation can be used

$$E_1 = E_o - \frac{N_{Si} z_1}{\cos\varphi_1}\, \varepsilon_{Si}(E_o) \quad . \qquad (4.71)$$

3. Calculation of the energy E_2 of ions backscattered from the surface of the second layer (in the example it is a nickel layer)

$$E_2 = K_{Ni}E_1 - \int_0^{z_1/\cos\varphi_2} ds\, N_{Si}\, \varepsilon_{Si}(E) \quad , \qquad (4.72)$$

$$E_2 = K_{Ni}E_1 - \frac{N_{Si} z_1}{\cos\varphi_2}\, \varepsilon_{Si}(K_{Ni}E_1) \quad . \qquad (4.73)$$

In the same way the thicknesses of the following layers can be determined.

In Fig.4.15, however, the backscattering spectra of the different Si layers and also of the different Ni layers are not well separated. The layer thicknesses are derived from the concentration profile calculated from the computer program, which can be applied as long as the backscattering signals of different elements are well separated. This is fulfilled in all three cases of Fig.4.15. From Fig.4.15c it can be concluded that a further increase of φ_2 leads to an overlap of the nickel and the silicon signals. The result of the computer calculation is shown in Fig.4.16. The concentration of 100% is not reached for the single layers because of the restricted depth resolution. Using bulk

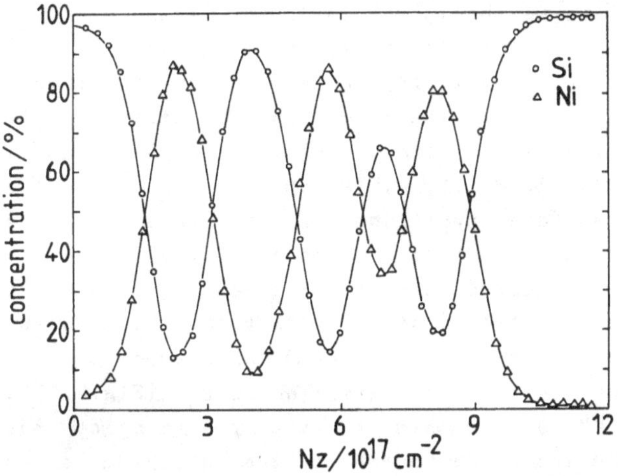

Fig.4.16. Concentration profile calculated for the sample of Fig.4.15

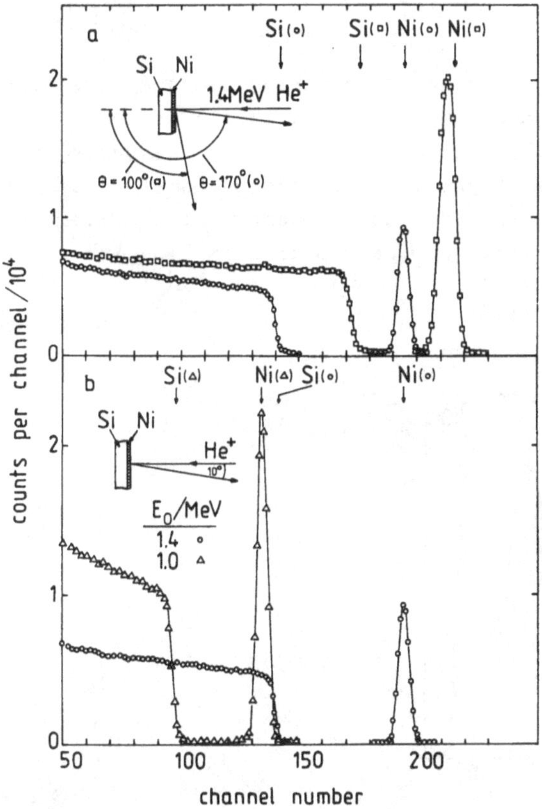

Fig.4.17. Backscattering spectra of He ions incident on a 20 nm thick Ni layer on silicon by changing the backscattering angle (a) and by changing the He ion energy (b)

values for the atomic densities we obtain the following layer sequence 31 nm Si, 16 nm Ni, 40 nm Si, 16 nm Ni, 18 nm Si, and 16 nm Ni.

Now, the influence of the backscattering angle θ and the ion energy E_0 on the shape of the backscattering spectrum as discussed in Sect.3.1.9 is shown in Figs.4.17a and b with the example of the 20 nm Ni layer of Fig.4.12. A variation of θ (Fig.4.17a) changes the kinematic factor and with this the surface positions of the backscattering signals. Since the angle of incidence is not changed, the angle of emergence changes with θ. A decreased backscattering angle causes an increase of φ_2 and thereby an increase of depth resolution. The spectrum height is increased on the strength of the θ-dependence of the differential scattering cross section. A variation of E_0 (Fig.4.17b) shifts the high energy edges KE_0 and changes the energy widths of the layer signals because of the energy dependence of the stopping cross section. The spectrum height is mainly influenced by the E^{-2}-dependence of the scattering cross section.

4.2.4. Compound layers

The problems to be solved in the case of compound layers are the determination of layer thickness and composition of the layers.

In the case of a homogeneously composed layer, the concentration ratio can be determined from the height ratio of the surface edges of the components' backscattering signals. As an example of a layer of known composition, the backscattering spectrum of a $NiSi_2$ layer on silicon is shown in Fig.4.18.

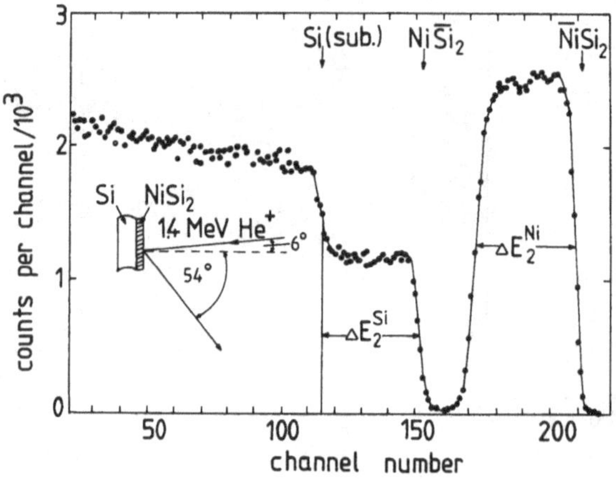

Fig.4.18. Backscattering spectrum of 1.4 MeV He ions incident on a $NiSi_2$ layer on silicon with $\varphi_1 = 6^\circ$, $\varphi_2 = 54^\circ$, $\theta = 120^\circ$, $\Delta\Omega = 1.59$ msr, $ne = 10\,\mu C$, $\epsilon_{ch} = 3.94$ keV

The Ni signals are well separated from the high energy silicon edge and the analysis can be easily carried out.

For the nickel spectrum height of normally incident ions backscattered at the surface, Eq.(3.55) yields

$$Y_{o,Ni}^{NiSi_2} = n\,\sigma_{Ni}^{d}(E_o)\,\Delta\Omega\;c_{Ni}\,\frac{\epsilon_{ch}}{[\epsilon_o]_{Ni}^{NiSi_2}}\quad, \tag{4.74}$$

where c_{Ni} is the relative concentration of Ni in $NiSi_2$ ($c_{Ni} = 1/3$). The stopping cross section in $NiSi_2$ is calculated using Bragg's rule of linear additivity (Eq.(3.53))

$$\epsilon^{NiSi_2} = \frac{1}{3}\,\epsilon^{Ni} + \frac{2}{3}\,\epsilon^{Si}\quad. \tag{4.75}$$

$[\epsilon_o]_{Ni}^{NiSi_2}$ means the stopping cross section factor of $NiSi_2$ for backscattering at nickel atoms in surface energy approximation. The distinct step in the silicon spectrum represents the interface between the compound layer and the substrate. For the height of the silicon spectrum at the surface, Eq.(3.55) gives

$$Y_{o,Si}^{NiSi_2} = n\,\sigma_{Si}^{d}(E_o)\,\Delta\Omega\;c_{Si}\,\frac{\epsilon_{ch}}{[\epsilon_o]_{Si}^{NiSi_2}}\quad, \tag{4.76}$$

where c_{Si} is the relative concentration of Si in $NiSi_2$ ($c_{Si} = 2/3$).

For an unknown composition Ni_xSi_{1-x} ($x = c_{Ni}$) the ratio of the heights provided by Eqs.(4.74) and (4.76) yields

$$\frac{c_{Ni}}{c_{Si}} = \frac{Y_{o,Ni}^{Ni_xSi_{1-x}}}{Y_{o,Si}^{Ni_xSi_{1-x}}}\;\frac{\sigma_{Si}^{d}(E_o)}{\sigma_{Ni}^{d}(E_o)}\;\frac{[\epsilon_o]_{Ni}^{Ni_xSi_{1-x}}}{[\epsilon_o]_{Si}^{Ni_xSi_{1-x}}}\quad. \tag{4.77}$$

For a first-order approximation, the ratio of the stopping cross section factors can be assumed to be unity. With the concentration ratio obtained in this way, the $[\epsilon_o]$ ratio can be calculated in first order and so on. The iteration works very quickly because the $[\epsilon_o]$ ratio for backscattering from different elements in the same matrix hardly differs from unity.

The thickness of a compound film can also be calculated from the energy width ΔE_2 of the backscattering spectrum of the layer if the film is thick enough so that the energy shift is well defined (discussed in connection with Figs.4.11 and 4.12). Here it is possible to calculate the layer thickness from the two energy widths $\Delta E_{2,Ni}$ and $\Delta E_{2,Si}$ of the components of the layer

$$Nz = \frac{\Delta E_{2,Ni}}{[\varepsilon_o]_{Ni}^{NiSi_2}} \tag{4.78}$$

and

$$Nz = \frac{\Delta E_{2,Si}}{[\varepsilon_o]_{Si}^{NiSi_2}} , \tag{4.79}$$

where the surface energy approximation for the stopping cross section factor has been used and N is the total atomic density. From Fig.4.18 we get $\Delta E_{2,Ni}$ = 145.8 keV and $\Delta E_{2,Si}$ = 138 keV. Following Eqs.(4.78) and (4.79) this yields Nz = $8.61*10^{17}$ cm^{-2} and Nz = $8.64*10^{17}$ cm^{-2}, respectively. The difference between the two results is well beyond the limit of experimental error. The mean value $8.625*10^{17}$ cm^{-2} is equivalent to an 113 nm thick $NiSi_2$ layer when a bulk density of $7.62*10^{22}$ cm^{-3} is used for the $NiSi_2$ film.

Using the mean energy approximation to calculate the layer thickness

$$Nz = \frac{\Delta E_{2,Ni}}{[\bar{\varepsilon}]_{Ni}^{NiSi_2}} , \tag{4.80}$$

with

$$[\bar{\varepsilon}]_{Ni}^{NiSi_2} = \frac{K_{Ni}}{\cos\varphi_1} \varepsilon^{NiSi_2}(\bar{E}_{in}) + \frac{1}{\cos\varphi_2} \varepsilon^{NiSi_2}(\bar{E}_{out,Ni}) \tag{4.81}$$

and the adequate equations for $\Delta E_{2,Si}$, the mean energies $\bar{E}_{in}$ and $\bar{E}_{out}$ have to be determined. Therefore the energy E_1 before backscattering from the backside of the film must be calculated, for instance by means of the energy loss ratio method (Eq.(4.39)). For the example shown in Fig.4.18 one obtains E_1 = 1320 keV for E_2 taken at the low energy edges of both the Ni spectrum and the Si spectrum of the layer. This provides for the mean energies $\bar{E}_{in}$ = 1360 keV, $\bar{E}_{out,Ni}$ = 1034 keV, and $\bar{E}_{out,Si}$ = 812 keV and finally Nz = $8.37*10^{17}$ cm^{-2}. The resulting layer thickness of 110 nm differs by only 2.6% from that derived by the surface energy approximation.

The energy loss ratio method (Eq.(4.39)) can also be used with multilayer films to calculate the energies before scattering at the various interfaces. Since the particles traverse exactly the same range of compositions on the outward path as on the inward path, the value of α will not change greatly even though the composition changes in the different layers. So the surface energy approximation is a rather good approximation in calculating the value of α. In Table 4.3, values of K and α are listed for different elements of the target.

TABLE 4.3

Values of K and α for 1.4 MeV He ions backscattered from a thin film at normal incidence with a scattering angle at 170°. The values of α are determined from the surface energy approximation using Eq.(4.38).

Target element	K	α
C	0.253	1.15
O	0.362	1.09
Si	0.566	1.17
Ni	0.763	1.04
Mo	0.847	1.07
Pt	0.92	1.02
Au	0.92	1.03

Fig.4.19 shows a spectrum for a sample with a Ni_2Si layer formed between a thin nickel layer and the silicon substrate. The high energy edge of the Ni signal at E_2 = 1066 keV is related to the Ni surface. The height $Y_{o,Ni}$ = 2610 counts/channel corresponds to the height of a pure nickel layer. The high energy edge of the silicon signal is shifted by ΔE_2 = 98 keV with respect to the surface edge (792 keV). This corresponds to the total energy loss of the ions in the nickel layer.

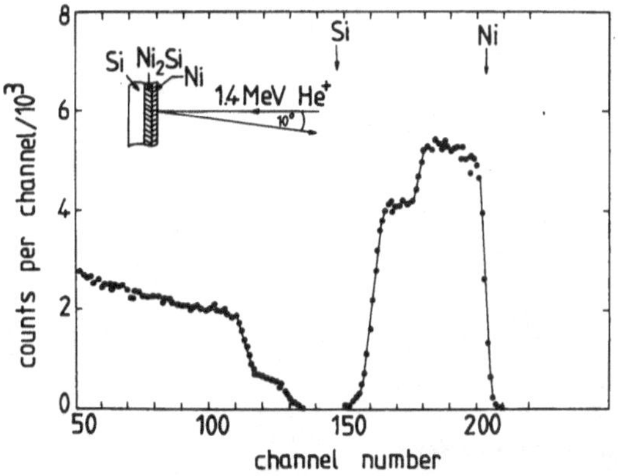

Fig.4.19. Backscattering spectrum of 1.4 MeV He ions incident on a layer system of Ni/Ni_2Si on silicon with φ_1 = 0°, Θ = 170°, $\Delta\Omega$ = 2.1 msr, ne = 5 μC, ϵ_{ch} = 5.3 keV

The energies before scattering at the various interfaces listed in Table 4.4 are obtained using the energy loss ratio method with α_{Ni} = 1.04 and α_{Si} = 1.17.

TABLE 4.4

Energies E_1 and E_2 at the different interfaces corresponding to the spectrum in Fig.4.19

Interface	Scattering element	E_2/keV	E_1/keV
Ni/Ni_2Si	Ni	954	1337
	Si	694	1343
Ni_2Si/Si	Ni	864	1287
	Si	615	1298

Good agreement is found for the energies E_1 at a given interface as computed from the measured energies E_2 for backscattering from Ni and Si atoms which proves the accuracy of the energy loss ratio method. The composition profile shown in Fig.4.20 is calculated using the computer program described in Sect.4.2.1.3. A layer sequence of $Ni/Ni_2Si/Si$-substrate is obtained with the thicknesses $(Nz)_{Ni} = 8.8*10^{17}\ cm^{-2}$, $(Nz)_{Ni_2Si} = 23.4*10^{17}\ cm^{-2}$. With the assumption of bulk densities ($N_{Ni} = 9.133*10^{22}\ cm^{-3}$, $N_{Ni_2Si} = 8.97*10^{22}\ cm^{-3}$) for the films we obtain $z_{Ni} = 96$ nm and $z_{Ni_2Si} = 87$ nm.

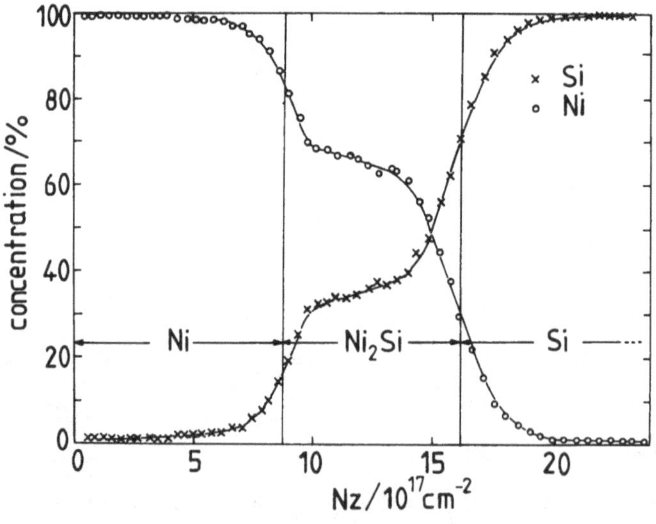

Fig.4.20. Concentration profile of Ni and Si calculated from the backscattering spectrum in Fig.4.19

In the case of inhomogeneously composed layers, the analysis of the backscattering spectra can only be carried out with the computer program.

4.3. THICK TARGETS

4.3.1. Homogeneously distributed impurities of low concentration

As an example Fig.4.21 shows the energy spectrum of 1.8 MeV He ions backscattered from a silicon wafer homogeneously doped with As (Ref.[115]). The impurity concentration can be determined from the ratio of the spectrum heights Y_o at the surface because of the homogeneous distribution of the impurities over the depth. The signal of the impurity at the surface is given by Eq.(3.55) in surface energy approximation

$$Y_{o,As} = n\,\sigma^{d}_{As}(E_o)\,\Delta\Omega\;c_{As}\,\frac{\epsilon_{ch}}{[\epsilon_o]^{Si}_{As}} \quad , \qquad (4.82)$$

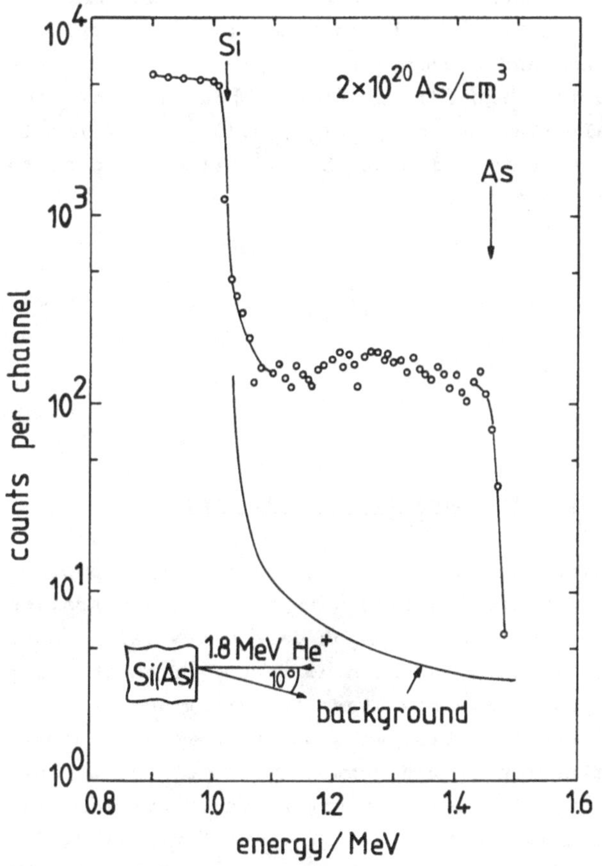

Fig.4.21. Backscattering spectrum of 1.8 MeV He ions incident on a silicon sample containing $2*10^{20}$ As cm^{-3}. The solid line shows the background at energies greater than $K_{Si}E_o$ obtained on samples without As (Ref.[115]).

where the change of the Si density by the impurities is neglected, and in $[\varepsilon_0]^{Si}_{As}$ the influence of the impurities is considered only by the kinematic factor. For the stopping cross section, the values of silicon can be used because of the low concentration of impurities as was discussed in Sect.4.2.2. The relative concentration of As is given by

$$c_{As} = \frac{N_{As}}{N_{Si}} = \frac{Y_{o,As}}{Y_{o,Si}} \frac{\sigma^d_{Si}(E_o)}{\sigma^d_{As}(E_o)} \frac{[\varepsilon_o]^{Si}_{As}}{[\varepsilon_o]^{Si}_{Si}} \quad . \tag{4.83}$$

For the example of Fig.4.21, a concentration N_{As} of $2*10^{20}$ cm^{-3} is obtained ($Y_{o,As}$ = 113 counts per channel, $Y_{o,Si}$ = 5200 counts per channel). The $[\varepsilon_o]$ ratio amounts to 1.07. In general, for MeV He ions this ratio is within 10% of unity for a wide variety of impurities and substrates (Ref.[2]).

The sensitivity for such measurements depends on the problem and on the experimental conditions. In Ref.[2] it was found that for typical conditions of normal incidence and a scattering angle of 170°, the detectable height of an impurity signal is about one thousandth of the height of the substrate signal, that means $Y_{imp}/Y_{sub} \approx 10^{-3}$. If it is assumed that the $[\varepsilon_o]$ ratio is unity and that the σ^d ratio is equal to the Z^2 ratio, the sensitivity limit is given by

$$\frac{N_{imp}}{N_{sub}} \geq \frac{Z^2(\text{substrate})}{Z^2(\text{impurity})} * 10^{-3} \quad , \tag{4.84}$$

which yields $2*10^{-4}$ (10^{19} As cm^{-3}).

4.3.2. Homogeneous samples containing more than one element

The aim of Rutherford backscattering analysis of homogeneously composed targets containing various components (samples with a high concentration of homogeneously distributed impurities, compounds, mixtures) is also the determination of the composition. The main difference from the case of the preceding section is that now the contribution of all constituents to the stopping cross section must be taken into account. As an example, the determination of the composition of a ceramic material is discussed. First, the constituents of the material are identified by PIXE. The PIXE spectrum of Fig.4.22 shows X-ray lines for the elements Si, P, and Ca. The backscattering spectrum of 1.4 MeV He ions incident on the ceramic material is shown in Fig.4.23.

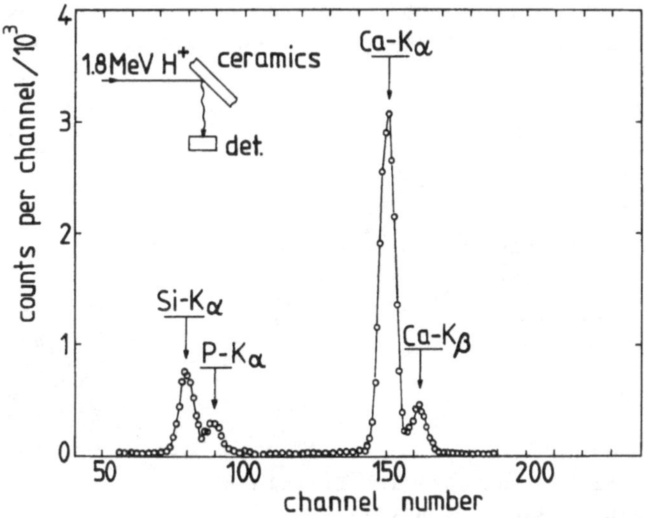

Fig.4.22. PIXE spectrum of 1.8 MeV protons incident on a ceramics

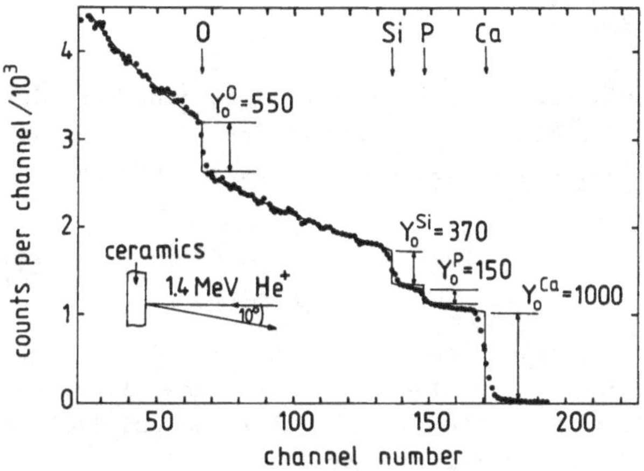

Fig.4.23. Backscattering spectrum of 1.4 MeV He ions incident on a ceramic material with $\varphi_1 = 0^\circ$, $\theta = 170^\circ$, $\Delta\Omega = 2.48$ msr, ne = 10 μC, $\epsilon_{ch} = 4.14$ keV

The positions of the surface channels of the elements identified by PIXE are marked by arrows. Clear steps in the backscattering spectrum are found for Si, Ca, and P. The additional step at channel number 66 corresponds to the surface position of oxygen. From the surface heights of the backscattering signals due to calcium, phosphorus, silicon, and oxygen, the relative concentration of each constituent can be determined. In Fig.4.23 lines are drawn over the points of the spectrum to form a ladder where the vertical distances give the surface spectrum heights $Y_{o,A(j)}$ for the elements characterized by the subscript $A^{(j)}$ which can be calculated using Eq.(3.55) in surface energy approximation

$$Y_{o,A^{(j)}} = n\,\sigma^{d}_{A^{(j)}}(E_o)\,\Delta\Omega\;c_{A^{(j)}}\,\frac{\epsilon_{ch}}{[\epsilon_o]^{cer}_{A^{(j)}}}\;. \tag{4.85}$$

The spectrum height ratio of two elements $A^{(1)}$ and $A^{(2)}$ is given by

$$\frac{Y_{o,A^{(1)}}}{Y_{o,A^{(2)}}} = \frac{c_{A^{(1)}}}{c_{A^{(2)}}}\,\frac{\sigma^{d}_{A^{(1)}}}{\sigma^{d}_{A^{(2)}}}\,\frac{[\epsilon_o]^{cer}_{A^{(2)}}}{[\epsilon_o]^{cer}_{A^{(1)}}}\;, \tag{4.86}$$

from which the concentration ratio $c_{A(1)}/c_{A(2)}$ can be calculated. On the assumption that the $[\epsilon_o]$ ratio is unity, which is within the error limit of 10% (see Sect.4.3.1), the concentration ratio is

$$\frac{c_{A^{(1)}}}{c_{A^{(2)}}} \approx \frac{Y_{o,A^{(1)}}}{Y_{o,A^{(2)}}}\,\frac{\sigma^{d}_{A^{(2)}}}{\sigma^{d}_{A^{(1)}}} \approx \frac{Y_{o,A^{(1)}}}{Y_{o,A^{(2)}}}\left(\frac{Z_2^{(2)}}{Z_2^{(1)}}\right)^2\;. \tag{4.87}$$

With the step heights taken from Fig.4.23 (1000 counts for Ca, 150 counts for P, 370 counts for Si, and 550 counts for O), we obtain $c_{Ca}/c_{Si} = 1.324$, $c_P/c_{Si} = 0.353$, and $c_O/c_{Si} = 4.552$. With these results and using

$$\sum_j c_{A^{(j)}} = 1 \tag{4.88}$$

the composition of the ceramic material is found to be $Ca_{0.18}P_{0.05}Si_{0.14}O_{0.63}$. To improve the results, an iteration procedure can follow using Eq.(4.85) with $[\epsilon_o]^{cer}_A$ values calculated on the basis of Bragg's rule with the composition obtained.

Commonly, only the combination of RBS and PIXE analysis enables the determination of the complete composition of complex materials.

Chapter 5

APPLICATION TO THE ANALYSIS OF CRYSTALSTRUCTURE

E. Glaser and W. Wesch

5.1. PERFECT CRYSTALS

Basic problems in the analysis of crystals by means of RBS are

- the identification of the crystalline structure and the pre-orientation of the sample to be investigated and
- the alignment of the crystal with respect to the analyzing ion beam as a first step in experiments to determine lattice location of crystal atoms or impurities as well as the amount of lattice disorder.

The basis for carrying out such experiments is the occurrence of channeling and blocking of the ions incident on a crystalline target. The physical foundations of channeling and blocking phenomena are described in Sect.1.3.

5.1.1. Crystal structure

Information about the crystalline structure and about the orientation of unknown crystalline samples can be obtained by carrying out RBS measurements in the blocking regime. In this regime, the analyzing ions are incident on the crystal in a random direction, i.e. they penetrate the material by ungoverned motion. As a consequence, the ions can interact with the target atoms with impact parameters small compared to the lattice constant, thus suffering wide angle single collisions. After this interaction, some of the ions return to the crystal surface by governed motion due to channeling and leave the crystal at a certain angle with respect to the low-index axial (or planar) direction (see Sect.1.3.2.4.2). In this way, the atoms in a row acting as scattering centers prevent the ions from leaving the crystal in a cone around the atomic row (this behaviour corresponds to ions emitted from lattice sites in the case of radioactive ions incorporated in the crystal). For the structural analysis and an exact determination of the orientation of crystals using this effect, a photographic technique is suitable, representing the crystal axes and planes as points and lines on the film. This method enables the simultaneous registration of ions backscattered over a wide angular range and allows the determination of the crystallographic indices.

Such ionograms are obtained using the principal experimental arrangement shown in Fig.5.1.

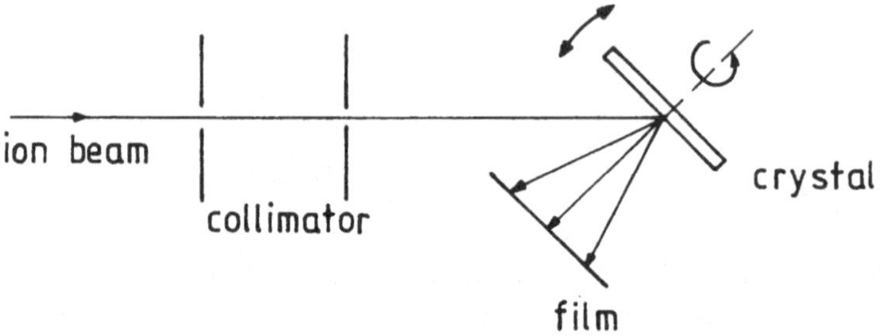

Fig.5.1. Schematic experimental arrangement for recording of ionograms by means of high-energetic ion backscattering

As an example, in Fig.5.2a a protonogram of a <001> oriented Mo crystal is shown (Ref.[116]). The dots in the figure correspond to the projections of crystal axes onto the film, the black lines are connected with the crystal planes. In Fig.5.2b the distribution of crystalline axes and planes for a <001> oriented cubic crystal calculated according to Ref.[116] is illustrated. A comparison with Fig.5.2a allows the identification of crystal axes and planes in the protonogram.

This method is also useful for the investigation of special structural growth processes. This is illustrated in Fig.5.3 for the example of an epitaxial <001> Au layer grown on an NaCl crystal. In the protonogram several lines are split. This is a consequence of the existence of crystalline regions with slightly different orientations. A quantitative analysis of the splitting of the (100) line gives a tilt angle of 1.5° between the two crystalline regions.

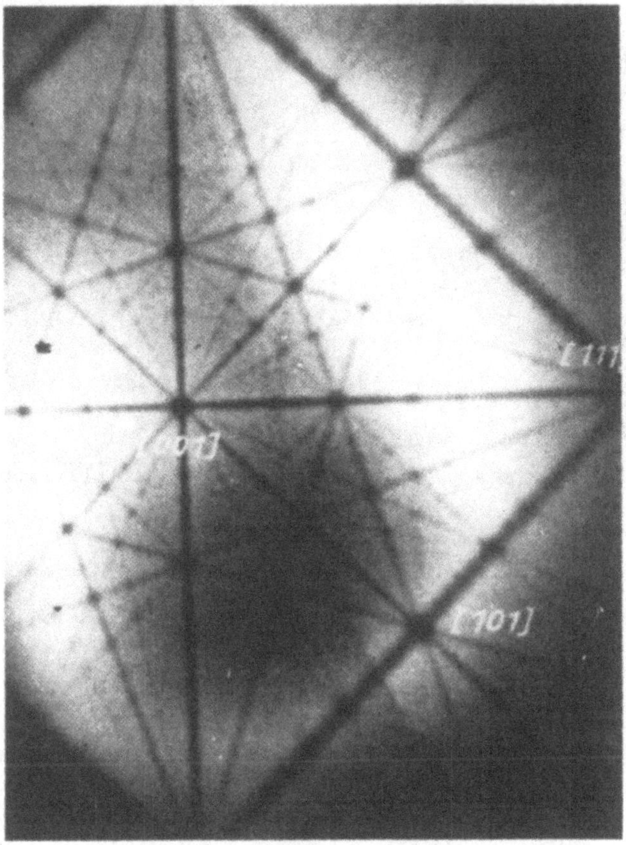

a)

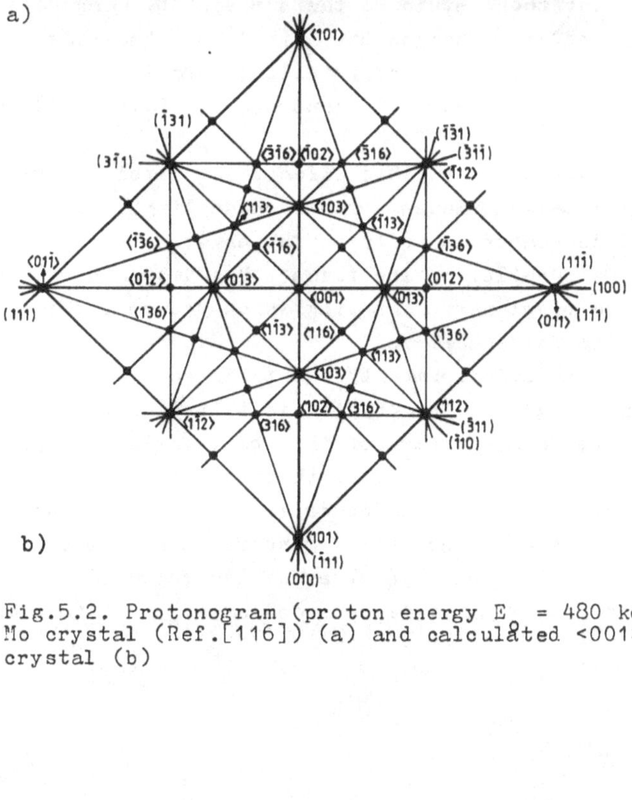

b)

Fig.5.2. Protonogram (proton energy E_p = 480 keV) of a <001> oriented Mo crystal (Ref.[116]) (a) and calculated <001> orientation of a cubic crystal (b)

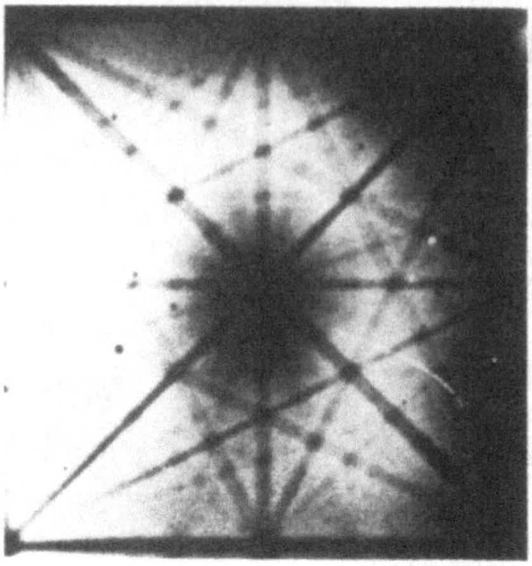

Fig.5.3. Protonogram (E_0 = 1.0 MeV) of a thin Au layer epitaxially grown on NaCl monocrystal

5.1.2. Orientation

For the orientation of crystal axes and planes with respect to the analyzing ion beam, the channeling regime is used (see Sect.1.3.2.4.1). In this regime the ions enter the crystal under a small angle with respect to a low-index direction and penetrate the crystal by governed motion due to correlated small-angle collisions with the rows or planes of atoms until they reach a certain depth, where they suffer a wide-angle single collision and then return to the surface by ungoverned motion. This channeling effect changes the yield of backscattered particles as the orientation of the monocrystalline target is varied with respect to the incident beam of ions (see Eq.(1.171) in Sect.1.3.2.4.1).

For a more detailed illustration of this effect, we first consider channeling between closely packed planes (planar channeling). In this case the scattering yield decreases markedly as the crystal is tilted so that the beam direction is parallel to the plane. The decrease of the scattering yield is caused by planar channeling which prevents the ions from closely approaching the lattice atoms.

Second, channeling along an axis formed by the intersection of planes is considered (axial channeling). If the crystal is tilted with respect to this axis, this results in a decrease of the backscattering yield along the planes. When the beam is aligned with this axis, a further decrease of the backscattering yield is observed. In Fig.5.4 as an example, the normalized backscattering yields measured across the <110> axis and the (110) plane of a silicon crystal are illustrated. As can be seen, the normalized backscattering yield recorded at $\psi_{in} = 0^o$ (minimum yield) is remarkably lower in the case of axial channeling compared to planar channeling. Typical values of the minimum yield of perfect crystals are χ_{min} = (2-5)% for axes and χ_{min} = (5-30)% for planes depending on the crystal direction.

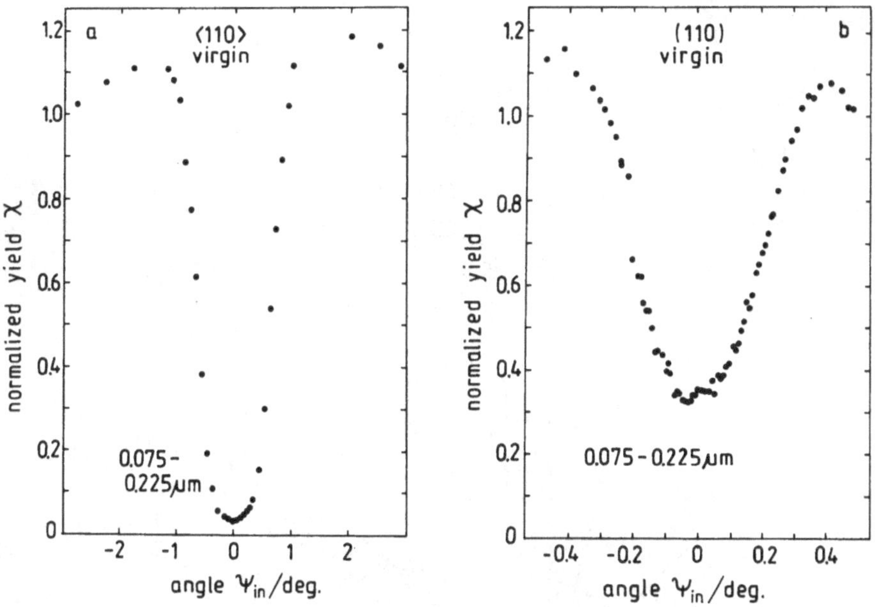

Fig.5.4. Normalized backscattering yields χ as a function of the angle of incidence ψ_{in} measured across the <110> axis (a) and the (110) plane (b) of a perfect silicon crystal.

A supposition for the alignment of a crystallographic axial direction with respect to the ion beam is the existence of well-defined planar minima in the backscattering yield. Thus, in the alignment procedure the crystal is tilted with respect to the beam direction. If the crystal is tilted by an angle θ_t off the direction of the incident beam and rotated about its surface normal, minima in the backscattering yield occur at rotation angles where the beam is aligned with planar directions in the crystal. By plotting the angular positions of the planar minima on polar coordinate graph paper, the direction of the crystal axis with respect to the ion beam is found at the intersection of lines connecting the minima. With the coordinates θ_t and Φ determined in this way, the crystal axis can be aligned with respect to the beam.

As an example, in the following the alignment of a wafer of monocrystalline silicon will be described, which is cut in such a way that the <111> axis is nearly normal to the sample surface (Fig.5.5).

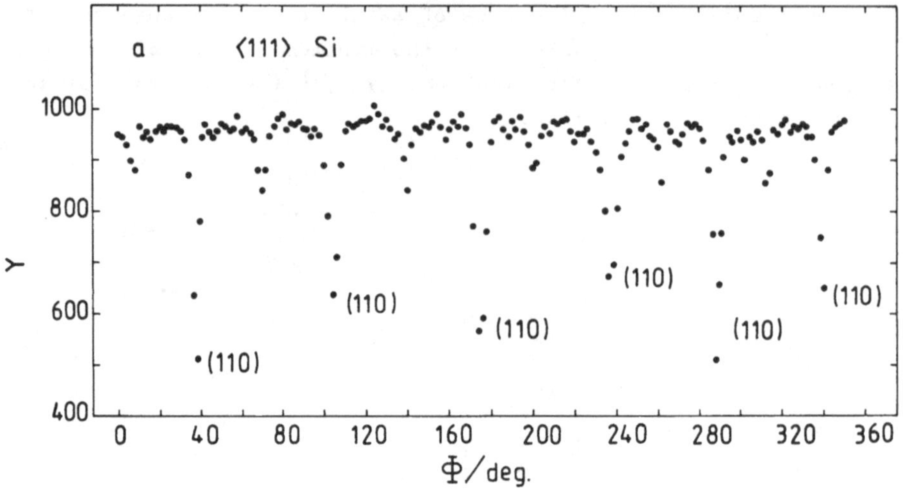

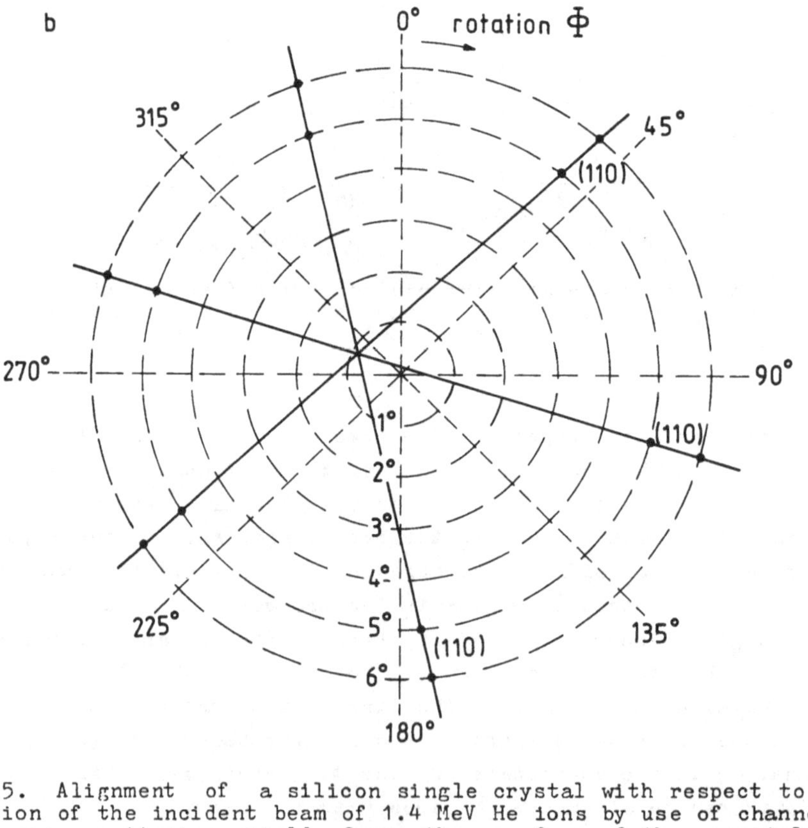

Fig.5.5. Alignment of a silicon single crystal with respect to the direction of the incident beam of 1.4 MeV He ions by use of channeling technique: scattering yield from the surface of the crystal as a function of rotation angle for 360° rotation (a) and location of the planar minima on polar coordinates for tilt angles $\theta_t = 5^\circ$ and 6°

The rotation axis of the goniometer used is tilted by 6° from the incident beam direction. Now the sample is rotated and the backscattering yield is recorded as a function of the angle of rotation. In Fig.5.5a the measured yield is plotted versus the rotational angle. As can be seen, there are six pronounced minima in the angular plot at about 36°, 104°, 175°, 236° and 339° corresponding to the three (110) planes around the <111> axial direction. The measured values are recorded in polar coordinates in Fig.5.5b. For a greater accuracy, the procedure can be repeated for a different value of Θ_t (in the example done for $\Theta_t = 5^{\circ}$). By connecting the minima corresponding to the planes (solid lines in Fig.5.5b) the <111> axial direction is found at the intersection of these lines. In our case the coordinates of the <111> axis with respect to the surface normal of the silicon sample are $\Theta_t = 0.9^{\circ}$ and $\Phi = 298^{\circ}$. To align the crystal with respect to the ion beam, the goniometer must be adjusted to this position.

In another relatively simple and widely used alignment procedure, the crystal to be aligned is tilted to a fixed angle φ with respect to the surface normal. Then the crystal is tilted by an angle ϑ in a direction perpendicular to the direction of φ and the backscattering yield is measured as a function of the angle ϑ. This procedure is done for two angles (normally $\varphi = +5^{\circ}$ and $\varphi = -5^{\circ}$). In a second step the backscattering yield is recorded for two fixed angles of $\vartheta = \pm 5^{\circ}$ as a function of the angle φ. Fig.5.6 a shows the result of such an alignment procedure for a <111> silicon single crystal. For $\varphi = -5^{\circ}$ pronounced planar dips are found at $\vartheta = -2.6^{\circ}$ and $\vartheta = 1.4^{\circ}$, for $\varphi = +5^{\circ}$ a minimum at $\vartheta = -2^{\circ}$ is recorded. If ϑ is kept constant, planar dips occur at $\varphi = -2.8^{\circ}$ and $\varphi = +3.6^{\circ}$ ($\vartheta = +5^{\circ}$) and $\varphi = -0.7^{\circ}$ ($\vartheta = -5^{\circ}$). As in the method described above, the connection of the measured minima (solid lines in Fig.5.6b) yields the <111> axial direction at the intersection of the planes. In the case described, the coordinates of the <111> axis are $\varphi = -1.8^{\circ}$, $\vartheta = +0.3^{\circ}$.

Generally, the number and position of the planes around the axis and the measured backscattering yield in the minima of the planes and axes are different for different types of planes and axes.

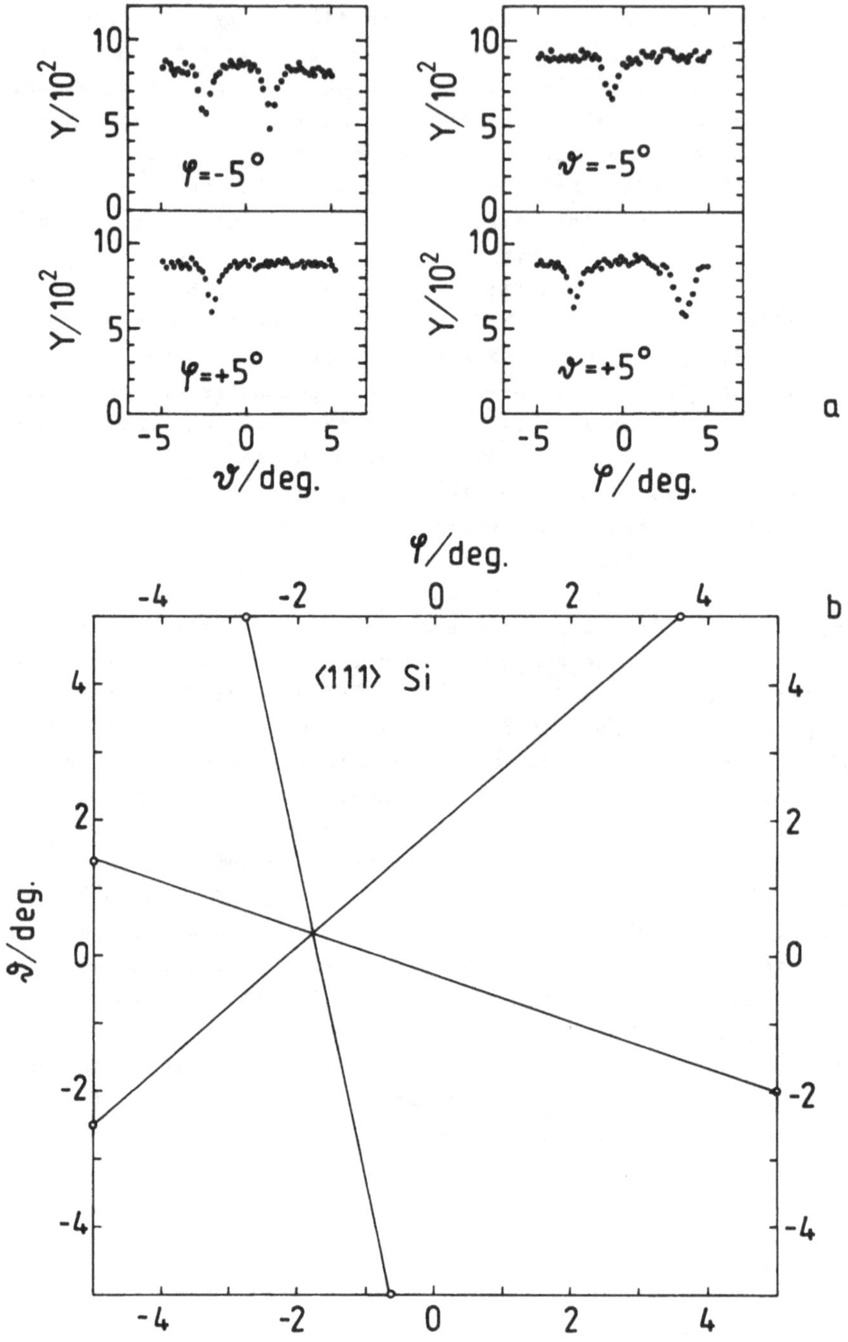

Fig.5.6. Alignment of a <111> silicon single crystal: backscattering yield from the surface of the crystal for fixed angles $\varphi = \pm 5°$ as a function of the tilt angle ϑ and vice versa (a) and location of the planar minima (b)

In Table 5.1 the major planar and axial orientations in cubic structures are listed in order of decreasing magnitude of the amount of reduction in yield for the given structure as the crystal is tilted from "random" to "aligned" orientations.

TABLE 5.1

Planar and axial indices for cubic structures listed in order of decreasing magnitude of the amount of reduction in backscattering yield (data from blocking patterns according to Barrett et al. [117])

	diamond	fcc	bcc		diamond	fcc	bcc
planes (hkl)	110	111	110	axes <hkl>	110	110	111
	111	100	100		111	100	100
	100	110	112		100	111	110

Fig.5.7 shows the orientation of the major planes around the <111>, <100>, and <110> axes in cubic crystals. The angles between planes are also drawn in the figure. Around the <111> axis the (110) planar minima are most pronounced, whereas around the <100> axis the (100) and (110) planes show the highest decrease in the magnitude of the backscattering

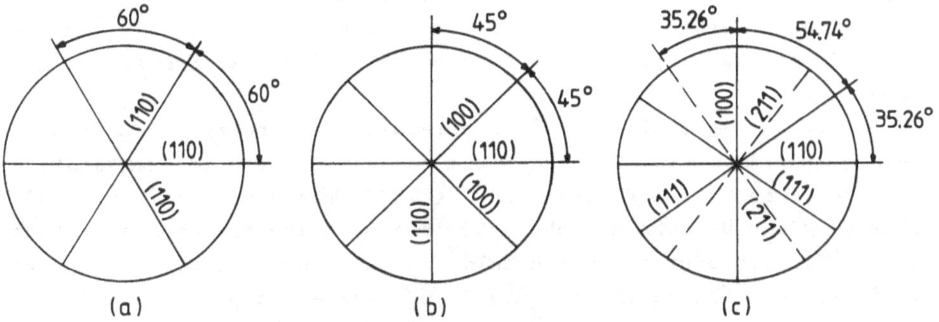

Fig.5.7. Orientation of the major planes around the three principal low-index axes <111> (a), <100> (b), and <110> (c) in cubic crystals (from Ref.[118])

yield. If a cubic crystal is rotated about the <110> axis, pronounced minima are detected for the (111) and (110) planes. In bcc crystals the (211) planar minima are pronounced (dashed lines in Fig.5.7c) but are not as great as the (111) minima in diamond or fcc structures.

From the preceding explanations it is obvious that the RBS channeling technique is a useful method for the alignment of crystalline samples with respect to the analyzing ion beam. But for most practical applications, additional "random" spectra are necessary, i.e. backscattering spectra that coincide with those from amorphous samples. Such spectra can be obtained if the crystal is oriented so that the direction of the incident ion beam does not coincide with major crystallographic axes or planes. However, the choice of such a direction is not a trivial task, because the probability for the beam to be incident on high-index planes or to a shoulder of an axial or planar dip is not negligible. Thus, commonly the crystal is tilted so that the ion beam is incident on the crystal well away from the crystallographic axis and continuously rotated while recording the spectrum.

5.2. IMPERFECT CRYSTALS

By means of backscattering spectrometry utilizing the effect of channeling, both qualitative and quantitative analysis of various kinds of lattice defects in monocrystalline solids can be performed. The specific information on the type and the density of defects under investigation results from the study of dechanneling due to these defects. It appears that it is possible to reduce the description of the different defect-induced dechanneling processes to only a few basic mechanisms (Sect.1.3.3.2) . Ion dechanneling can be caused by uncorrelated multiple scattering processes (in the case of point defects, defect clusters, amorphous regions, and grain boundaries), due to bent channels (within the strain fields around dislocations), or due to displaced strings or planes (in layers containing stacking faults and twin lamellae).

To identify the special kind(s) of defects, the energy dependence of the measured minimum yield represents a decisive tool, in combination with further methods of analysis (TEM, electron diffraction). Quantitatively the defects are characterized by means of specific defect densities. Two procedures for the calculation of depth distributions of defect densities (discontinuous model and two-beam approximation) are applied, considering typical examples of each kind of defects. The availability of both models is discussed.

In the case of twins, the investigation is restricted to two extreme cases, thin twin lamellae and thick twinned regions. Polycrystalline layers are a subject of analysis too, if both the size of the crystallites and the spread of their orientations still permit channeling. From the measurement of angular scans at different energies, this spread of the orientations of crystallites or twinned regions can be determined. Angular scans, temperature-dependent measurements, and measurements for different crystal axes are used for the lattice location of point defects (or impurity atoms) that occupy preferred positions in the lattice cell.

The examples discussed in this section usually refer to axial channeling. In special cases (dislocations, stacking faults, point defects with preferred positions) it is advantageous to involve also planar channeling measurements. To analyze the experimental results, the equations and procedures described in Sects.1.3.3.2, 1.3.4, and 1.3.5 are used. So, in this section equations to be applied to the analysis are represented only if further specifications are necessary (thick twinned regions and polycrystalline layers).

5.2.1. Point defects, clusters, and amorphized layers

Regarding their interaction with channeled ions (dechanneling and direct backscattering of the ions), point defects, defect clusters, and amorphous regions in monocrystals or completely amorphized layers of monocrystals can be treated as uncorrelated displaced lattice atoms. Particularly large cluster structures ("platelets") formed by a condensation of mobile point defects cannot be included. Those clusters which are rudiments of extended defects (dislocation loops) already show characteristics of correlated interaction with channeled ions.

Crystal defects consisting of uncorrelated displaced atoms with defect concentrations n_{pd} accessible by backscattering analysis ($0.01 \leq n_{pd} \leq 1$) can be produced by means of ion implantation in semiconductors (e.g. Si, GaAs; Refs.[68,119]) or monocrystalline insulators (e.g. quartz, $LiNbO_3$; Refs.[120,121,122]), provided that the implantation temperature does not generally exceed room temperature. Depending on the parameters of ion mass, implantation temperature T_i, and ion dose D, the whole spectrum of damage, ranging from single point defects to the total amorphization of the implanted layer, can be obtained.

5.2.1.1. Depth dependence of the minimum yield $\chi_{min}(z)$

Silicon and GaAs layers damaged by ion implantation are examples chosen for the analysis of such defects. In Fig.5.8, backscattering spectra $Y_{al}(E_2)$ for silicon implanted with ions of low mass (boron) at low temperatures ($T_i \approx 80$ K) and at room temperature are compared. The spectra are measured using He^+ with an energy E_o = 1.4 MeV (data from Ref.[123]).

These spectra show that only in the case of damage "frozen" by low temperature implantation a pronounced contribution of direct backscattering to the measured yield Y_{al} (peak in the energy region 550 keV $\leq E_2 \leq$ 750 keV) is achieved. On the other hand, defects formed

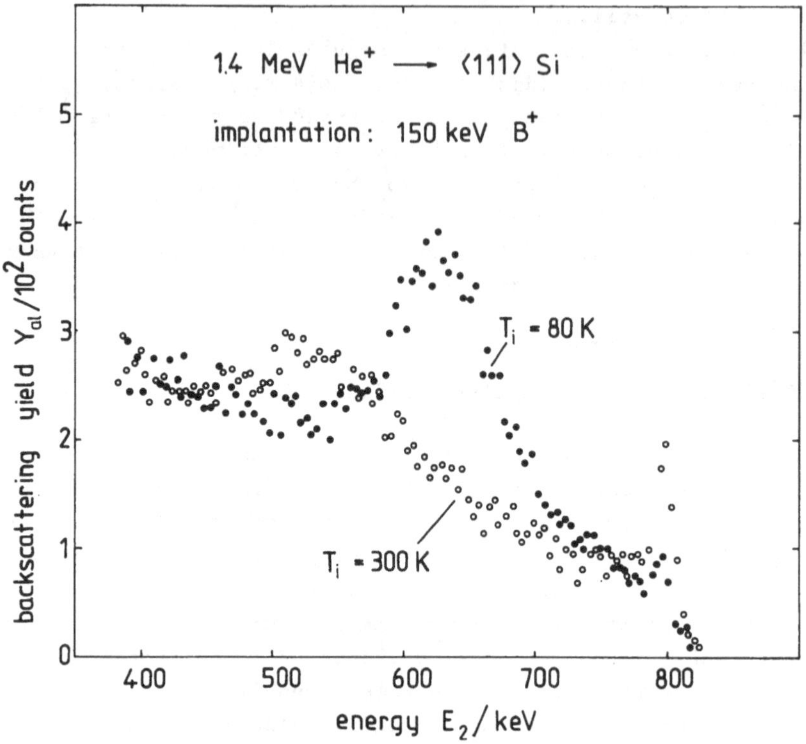

Fig.5.8. Backscattering spectra (aligned incidence) for silicon damaged by B^+ implantation at different temperatures: T_i = 80 K (D = 2.5*10^{14} cm^{-2}); T_i = 300 K (D = 1*10^{16} cm^{-2})

during boron implantation at room temperature preferentially cause an increase of Y_{al} with increasing depth z due to dechanneling. In Fig.5.9, for the case of implantation of different boron doses at $T_i \approx 80$ K, the measured backscattering spectra are represented (Ref.[124]). The fact that for doses D > 7*10^{14} cm^{-2} the random yield Y_r is obtained indicates the total amorphization of the depth region of maximum damage production. The existence of pronounced peaks for the implantation conditions $T_i \approx 80$ K and D ≥ 2*10^{14} cm^{-2} refers to displaced atoms which can be characterized by large distances r_a of their positions from the <111> strings. The concentration of defects giving rise to these peaks distinguish these layers as heavily damaged crystals (corresponding to the characterization in Sect.1.3.5.1) where preferred positions of the displaced atoms are improbable. Consequently, the displaced atoms can be assumed to be randomly distributed within the lattice cell.

The almost complete absence of a peak in the spectrum on the one hand and the predominant dechanneling due to the defects produced by boron implantation at $T_i \approx 300$ K (Fig.5.8) on the other hand indicate substantial deviations from the case of randomly distributed defects and the existence of preferentially small displacements r_a from the strings. The basis for the calculation of the depth distribution of defect

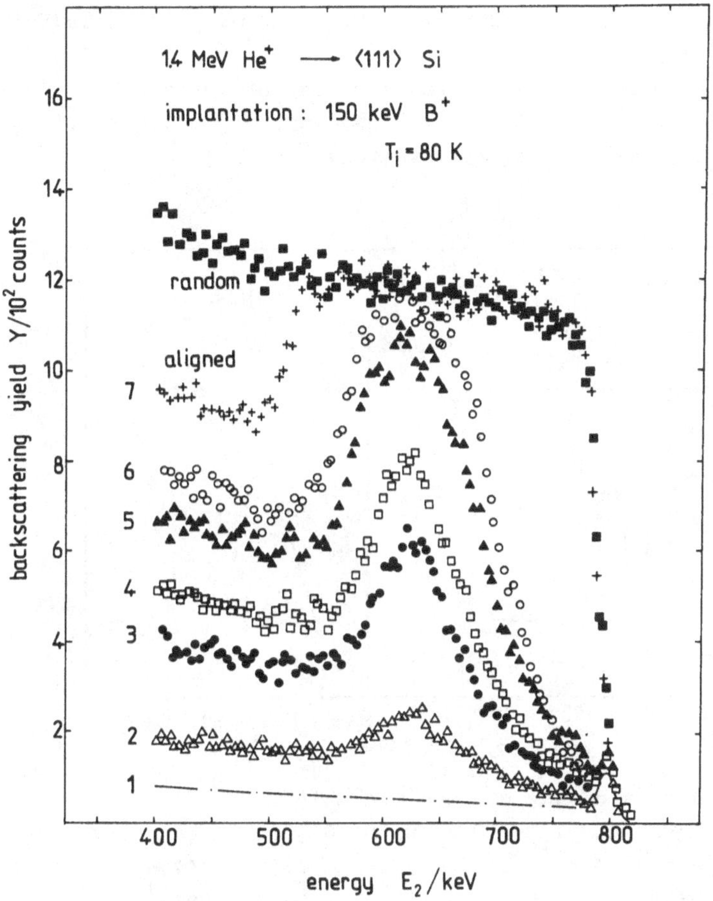

Fig.5.9. Backscattering spectra (aligned and random incidence) for silicon implanted with different doses D of 150 keV B^+ at a temperature T_i = 80 K: D = 2*10^{14} cm^{-2} (2), 3*10^{14} cm^{-2} (3), 3.5*10^{14} cm^{-2} (4), 5*10^{14} cm^{-2} (5), 7*10^{14} cm^{-2} (6), and 5*10^{15} cm^{-2} (7). The aligned spectrum (1) for virgin silicon is shown for comparison.

concentration $n_{pd}(z)$ from the measured backscattering spectra $Y_{al}(E_2)$ is the normalized (by the random yield Y_r) minimum yield of backscattered ions $\chi_{min}(z) = Y_{al}(z)/Y_r(z)$. This presumes the conversion of the energy scale of the spectrum into a depth scale $z = z(E_2)$ (see Sect.4.2.1). Fig.5.10a shows the dependence $\chi_{min}(z)$ obtained from the spectra in Fig.5.9 together with the dependence for the perfect crystal $\chi_{min,perf}(z) = Y_{al,perf}(z)/Y_r(z)$. In Fig.5.10b those results for $T_i \approx 300$ K are represented. Besides the case of light ion implantation (boron), an example of the implantation of heavier ions (silicon) is included here (Ref.[125]). It can be seen that with ions of higher masses, even during room temperature implantation heavy radiation damage is produced, characterized approximately by randomly distributed displaced atoms.

Fig.5.10c again comprises the two basic types of curves $\chi_{min}(z)$ described above for the case of implanted GaAs layers (no trace of a peak visible after room temperature implantation).

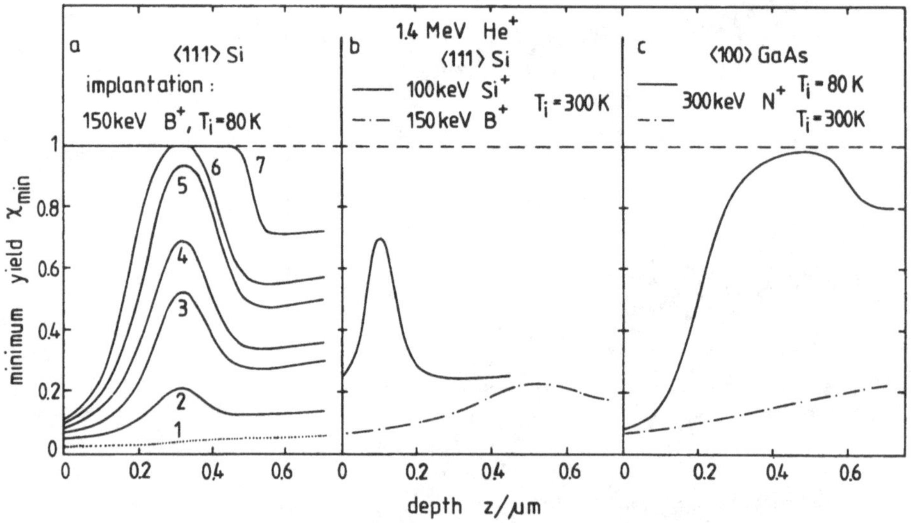

Fig.5.10. Depth dependences of the minimum yield for
(a) silicon implanted at T_i = 80 K with B^+ (derived from the backscattering spectra in Fig.5.9) ; implantation doses D = 2*10^{14} cm^{-2} (2), 3*10^{14} cm^{-2} (3), 3.5*10^{14} cm^{-2} (4), 5*10^{14} cm^{-2} (5), 7*10^{14} cm^{-2} (6), and 5*10^{15} cm^{-2} (7). The dependence for a perfect (virgin) silicon layer (1) is shown for comparison.
(b) silicon implanted at T_i = 300 K with Si^+ (D = 5*10^{14} cm^{-2}, solid line) and B^+ (D = 1*10^{16} cm^{-2}, dash-dotted line), respectively
(c) GaAs implanted with N^+ at T_i = 80 K (D = 3*10^{14} cm^{-2}, solid line) and T_i = 300 K (D = 1*10^{15} cm^{-2}, dash-dotted line), respectively

5.2.1.2. Energy dependence of the minimum yield

As shown in Sect.1.3.5.1 the difference of the minimum yields $\Delta\chi_{min} = \chi_{min} - \chi_{min,perf}$ for uncorrelated defects shows dependences on the incidence energy E_0 (in the MeV region) in the range of E_0^{κ} with $-0.65 \geq \kappa \geq -1$. The concrete E_0-dependence is determined both by the position r_a and by the concentration n_{pd} of the displaced atoms.

Therefore the energy dependence is not an unambiguous expedient for the identification of the defects. A better method for the evaluation of the position r_a of point defects is to measure the temperature dependence of $\Delta\chi_{min}$ (see Sects.1.3.5.1.2 and 5.3.3.2).

However, to test whether the assumptions about the defects derived from the shape of the backscattering spectra (Sect.5.2.1.1) are consistent with the measured energy dependences of the minimum yield, in Fig.5.11 the dependences $\Delta\chi_{min}(E_o)$ at $z = z_o$ (Ref.[125]) for two extremely different cases are compared (z_o is the depth corresponding to the thickness of the damaged layer). In both cases dependences $\Delta\chi_{min} \sim E_o^{\kappa}$ with $\kappa < 0$ are obtained, but with considerable differences of the value of κ. The result $\kappa \approx -0.8$ for the implantation of the heavier ions (Si^+) in silicon supports the assumption of randomly distributed displaced atoms derived from the typical shape of the corresponding curve $\chi_{min}(z)$ (pronounced peak; Fig.5.10b). Such values of κ are usually expected for dechanneling due to plural scattering by randomly distributed scattering centers of the concentrations given here.

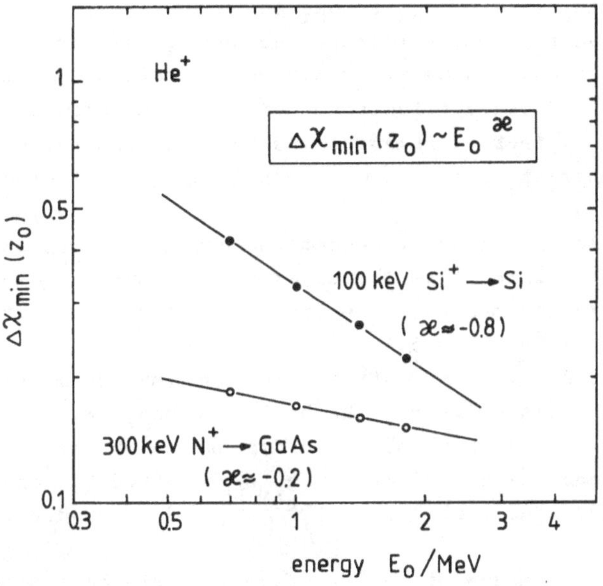

Fig.5.11. Difference of the minimum yields for implanted and unimplanted crystals as a function of the energy of the incident He ions; implantation (T_i = 300 K): Si^+ in silicon ($D = 5*10^{14}$ cm^{-2}) and N^+ in GaAs ($D = 1*10^{15}$ cm^{-2}), respectively

For implantation of light ions (N^+) in GaAs, the measured exponent $\kappa \approx -0.2$ indicates that the radiation damage produced deviates substantially from the case of randomly distributed displaced atoms. A predominance of small displacements r_a in this case has been concluded from the typical shape of the measured spectrum $\chi_{min}(z)$ (Fig.5.10c). But the value $|\kappa| \approx 0.2$ is even smaller than those calculated for uncorrelated displaced atoms with small displacements $r_a \geq a_{TF}$ and concentrations giving rise to a yield $\chi_{min}(z)$ similar to that in Fig.5.10c ($|\kappa| \approx 0.65$; Ref.[59]). This could be explained by the assumption that defect complexes have been formed here, causing lattice distortions (bent channels) which give rise to dechanneling contributions by <u>correlated</u> displaced atoms. For such contributions values $\kappa > 0$ have to be expected (see Sects.1.3.3.2 and 5.2.2).

5.2.1.3. Depth dependence of defect concentration $n_{pd}(z)$

The calculation of the defect concentration n_{pd} by means of the experimental data $\chi_{min}(z)$ in the case of point defects of low concentrations and with preferred positions r_a can be carried out using the procedure based on the "discontinuous model" which is described in Sect.1.3.5.1. The depth distribution $n_{pd}(z)$ has to be determined in <u>combination</u> with the position r_a, which presupposes the measurement of $\chi_{min}(z)$ at <u>different</u> temperatures. In Sect.5.3.3.2 which deals with the lattice location of point defects applications of this method are presented in detail. Here defect configurations and concentration regions are considered, characterized above as randomly distributed displaced atoms.

In Fig.5.12 the depth distributions of defect concentration $n_{pd}(z)$ are illustrated which were calculated from the experimental data $\chi_{min}(z)$ in Fig.5.10a by means of iteration procedures using two different models of defect analysis described in Sect.1.3.5.1.

(i) The curves calculated by means of the discontinuous model (Ref.[59]) are marked by dash-dotted lines. The basis for the iterative calculation of $n_{pd}(z)$ is Eq.(1.276) with the depth dependence of the ion distribution $g_i(z)$ influenced by $n_{pd}(z)$ as given by Eqs.(1.277) and (1.278). Following Eq.(1.250) the probability $\Pi_i^{(pd)}(r_a)$ for randomly distributed r_a is $\Pi_i^{(pd)} = 1$.

(ii) The curves calculated by means of the two-beam approximation are marked by the solid lines. In this case, the iteration procedure is based on Eq.(1.274). Instead of the change of $g_i(z)$ with depth, the change of the random component of the ion beam $\chi_r(z) = \chi_r(N_{pd}(z) = \int_0^z dz' n_{pd}(z'))$ due to defect-induced dechanneling has to be calculated using Eq.(1.275). The dependence $\chi_r(z)$ is described by defect-induced plural scattering processes represented by the dechanneling function $P(\tilde{\psi}_c, \tau)$ which has been calculated numerically (Ref.[38], tabulated in Appendix 9; see also Fig.1.54). The parameter

$\tau \sim N_{pd}(z)$ is given by Eq.(1.259) and the reduced critical angle $\tilde{\Psi}_c$ determined by the experimental conditions ($He^+ \rightarrow$ <111> Si, E_o = 1.4 MeV) is given by Eq.(1.258). As described in Sect.1.3.5.1, the critical angle ψ_c in Eq.(1.258) has to be adjusted to achieve $n_{pd}(z > z_o) = 0$. Contrary to the discontinuous model, the analysis based on the two-beam

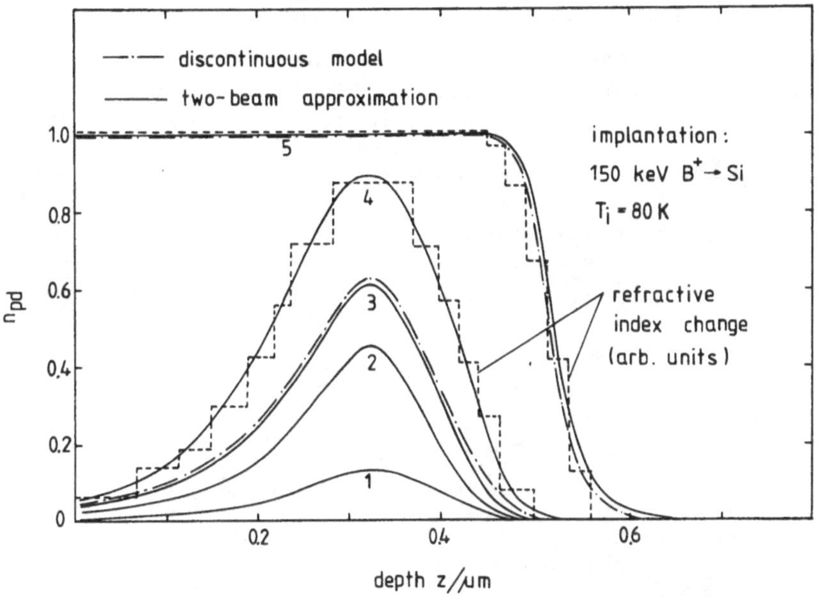

Fig.5.12. Depth distributions of defect concentration in B^+ implanted silicon: D = $2*10^{14}$ cm^{-2} (1), $3*10^{14}$ cm^{-2} (2), $3.5*10^{14}$ cm^{-2} (3), $5*10^{14}$ cm^{-2} (4), and $5*10^{15}$ cm^{-2} (5). The distributions were calculated from the data $\chi_{min}(z)$ in Fig.5.10a by means of the discontinuous model (dash-dotted lines) or the two-beam approximation (solid lines). Depth distributions of defect induced refractive index change (in arbitrary units) are shown for comparison.

approximation requires both the experimental data for the damaged <u>and</u> the perfect crystals, $\chi_{min}(z)$ and $\chi_{min,perf}(z)$, respectively. Comparing these curves, one can see that the application of the discontinuous model and of the two-beam approximation yields nearly identical results $n_{pd}(z)$. The formation of a completely amorphized layer is represented by n_{pd} = 1, i.e. the "displacement" of all the lattice atoms in this region.

To verify these results of RBS analysis, depth distributions of the damage-induced refractive index change $\Delta n(z)/n_{amorphous}$, measured by optical transmission and reflection spectroscopy for the same implanted layers (Ref.[68]) are included in Fig.5.12 (dashed lines). The depth distribution of randomly displaced atoms $n_{pd}(z)$ calculated from the backscattering spectra which characterizes the damage profile for high defect concentrations (defect clusters, amorphous zones, amorphized layers) proves to be in very good agreement with the corresponding distribution of refractive index change caused by these clusters or amorphous regions.

In Fig.5.13 depth distributions of defect concentration in a silicon implanted silicon layer ($D = 5*10^{14}\ cm^{-2}$) are compared, calculated by means of the two-beam approximation from He^+-backscattering spectra which were measured at different energies $1.0\ MeV \leq E_o \leq 1.8\ MeV$.

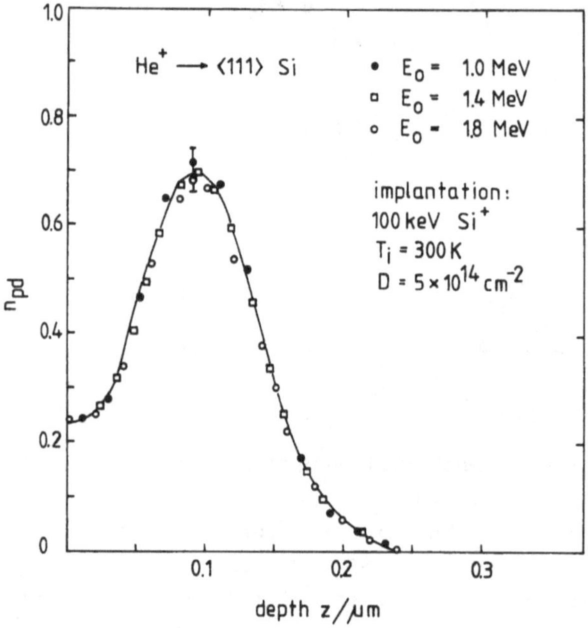

Fig.5.13. Depth distribution of defect concentration in Si^+ implanted silicon (comparison of the results obtained from RBS using He^+ of different energies E_o)

According to the shape of the spectra (Fig.5.10b) and the energy dependence $\Delta\chi_{min}(E_o)$ (Fig.5.11) the damage again corresponds to randomly distributed displaced atoms. The fact that (within the experimental error) for all energies identical profiles $n_{pd}(z)$ are obtained, is a good verification for the availability of the plural scattering model for the calculation of the dechanneling function $P(\tilde{\psi}_c, \tau)$ in the case of uncorrelated randomly distributed defects. By means of the E_o-dependence of $\tilde{\psi}_c$ in Eq.(1.258), the energy dependence of the measured spectra in question is described very well. Fig.5.14 demonstrates that the simple iteration procedure based on the two-beam approximation can be also used successfully to calculate $n_{pd}(z)$ in a compound material (GaAs), provided that the conditions for the application of the plural scattering model, as adduced above, are fulfilled.

In the same way damage profiles have been calculated for several compound monocrystals (e.g. SiO_2, Ref.[120]; $LiNbO_3$, Ref.[121,122]).

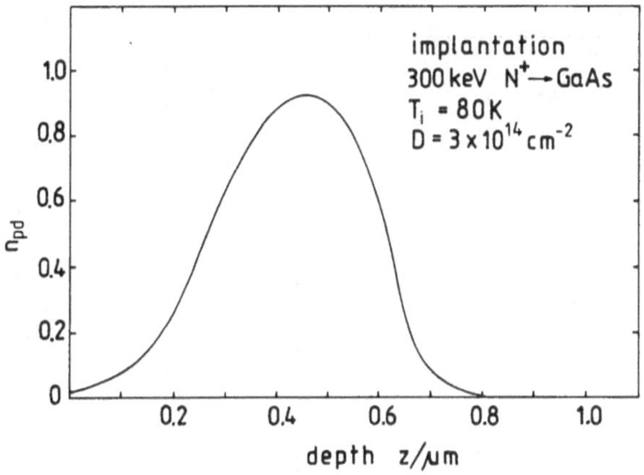

Fig.5.14. Depth distribution of defect concentration in N^+ implanted GaAs

5.2.2. Dislocations, stacking faults, and twin lamellae

5.2.2.1. Dislocations

Again ion implanted layers are chosen as examples of the analysis, because well-defined dislocation structures can be produced which are appropriate for demonstrating the accuracy of the quantitative analysis.

Dislocation structures in implanted monocrystals can be formed by the clustering of point defects or may result from lattice strain (e.g. due to high impurity concentrations). The formation of dislocation loops can be initiated by the bending of lines of point defects (e.g. vacancies) or by the nucleation of flat clusters (platelets) transformed into loops at sufficiently high annealing temperatures. In silicon, temperatures $T_a \geq 800...900$ K are necessary for loop formation, whereas in metals dislocation structures are already generated during implantation at room temperature (without post-annealing treatment). Due to interactions between the loops other types of dislocation structures (even irregular dislocation networks) can develop.

The dislocations analyzed in this section are produced by ion implantation into monocrystalline metal layers and by implantation and thermal annealing of monocrystalline silicon layers.

5.2.2.1.1. Depth dependence of the minimum yield

In Fig.5.15 the results of RBS measurements for two <110> Al single crystals implanted with 150 keV Zn^+ at room temperature ($D = 6*10^{15}$ cm^{-2} and $1.2*10^{16}$ cm^{-2}, respectively) are shown (Ref.[126]). The results are represented by the defect-induced increase of the minimum yield $\Delta\chi_{min}(z)$.

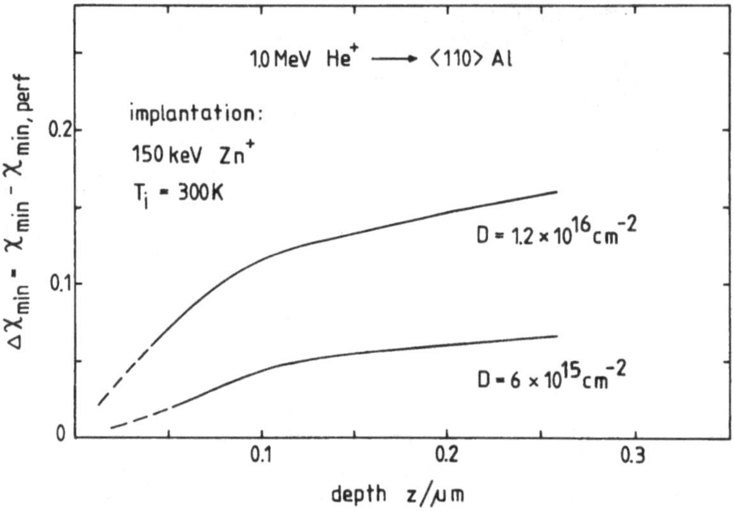

Fig.5.15. Depth dependence of $\Delta\chi_{min}$ for 1.0 MeV He ions incident on <110> Al implanted with different doses of Zn. The implantation of 150 keV Zn ions corresponds to a thickness of the implanted layers characterized by $R_p + \Delta R_p \approx 100$ nm.

The shape of the spectra is completely different from that for silicon implanted with heavy ions at room temperature (see for comparison Fig.5.10b in Sect.5.2.1). Within the surface layer damaged by the Zn^+ implantation (marked by $R_p + \Delta R_p \approx 100$ nm), $\Delta\chi_{min}$ increases with z, without any trace of a peak due to direct backscattering.

The results $\Delta\chi_{min}(z)$ for implanted and annealed silicon crystals are depicted in Fig.5.16. The curves correspond to the following implantation and annealing conditions:

(a,b) <100> Si (a) and <111> Si (b) implanted with 80 keV As^+ at $T_i \approx 300$ K, $D = 1*10^{16}$ cm^{-2}; annealed in a dry oxygen atmosphere at $T_a = 1220$ K for 960 min.

(c) <111> Si implanted with 200 keV P^+ at $T_i = 500$ K, $D = 1*10^{16}$ cm^{-2}; annealed in a nitrogen atmosphere at $T_a = 950$ K for 120 min.

The experimental data are taken from Ref.[67]. The solid, dashed, and dash-dotted lines are fitting curves for the calculation of the defect densities of these layers, N_{disl} and $n_{disl}(z)$, by means of the discontinuous model (demonstrated in this section).

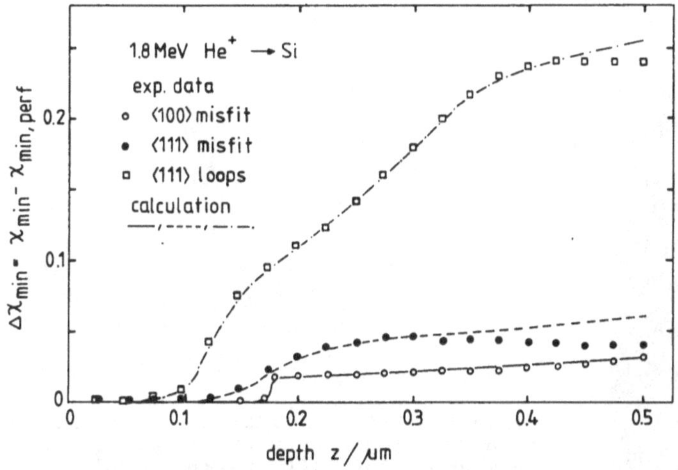

Fig.5.16. Depth dependences of $\Delta\chi_{min}$ for 1.8 MeV He^+ incident on <100> and <111> Si containing different dislocation structures (described in the text). The experimental data are fitted by curves calculated by means of the discontinuous model using the depth distribution $n_{disl}(z)$ in Fig.5.21.

A supposition for quantitative analysis is the identification of the type of defects and the evaluation of data characteristic of the special defect configuration (in the case of dislocation structures: Burgers vector $\vec{b}$ and the kind of the dislocation). This information can be derived from TEM analysis of the layers and from the measurement of the energy dependence of the minimum yield.

5.2.2.1.2. Results of TEM analysis

The TEM analysis of the Zn^+ implanted layers (Ref.[127]) showed a dense network of dislocations which appeared to be the predominant type of defects present after Zn^+ implantation in Al. For 200 keV Zn^+ and $D = 2.6*10^{16}$ cm^{-2}, an integral dislocation density (total projected length of the dislocation lines) $N_{disl} \approx 10^6$ cm^{-1} has been determined.

For the implanted Si layers (a), (b), and (c) TEM micrographs (Refs.[67,128]) exhibited the following defect configurations.

(a) The layer contains a regular two-dimensional misfit dislocation network of pure edge dislocations (i.e. all dislocations at the same depth) in the (100) plane parallel to the surface (<100> misfit for short; see Fig.5.17). The directions of the Burgers vectors are given by <011> and <01$\bar{1}$> representing dislocation lines in the <01$\bar{1}$> and <011> direction, respectively. A dislocation density $N_{disl} \approx (6.2 \pm 0.6)*10^4$ cm^{-1} has been measured.

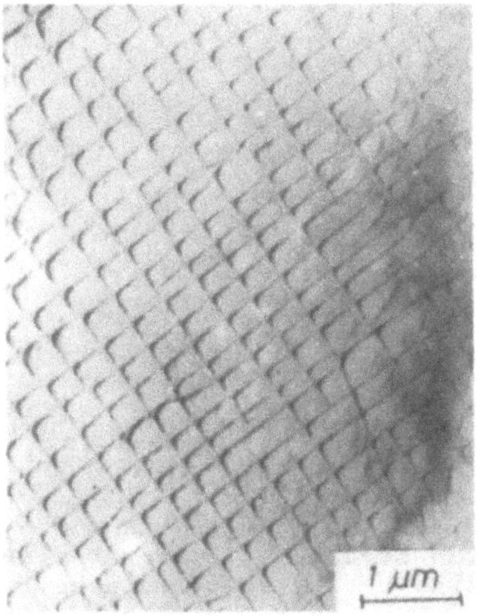

Fig.5.17. Bright-field TEM micrograph of a two-dimensional misfit dislocation network in the (100) plane parallel to the surface (dislocation structure produced by As^+ implantation and thermal annealing; parameters as described in the text).

(b) A misfit dislocation network parallel to the (111) plane (<111> misfit for short) was found. The directions of the Burgers vectors are given by <01$\bar{1}$>, <10$\bar{1}$>, and <1$\bar{1}$0> representing dislocation lines in the <$\bar{2}$11>, <1$\bar{2}$1>, and <11$\bar{2}$> directions; $N_{disl} = (2.0 \pm 0.4)*10^5$ cm^{-1}.

(c) These implantation and annealing conditions provided loops of perfect dislocations lying parallel to the (111) surface (<111> loops for short). The Burgers vectors are inclined with three possible directions: <110>, <101>, and <011>. The integral density amounts $N_{disl} = (2 \pm 1)*10^6$ cm^{-1}.

5.2.2.1.3. Energy dependence of the minimum yield

By measuring $\chi_{min}(E_o)$ at $z = z_o$ it is also possible to prove that the defects in the layers under investigation are dislocations. Particularly, it is possible to conclude from the energy dependence $\chi_{min}(E_o)$ to what extent further kinds of defects (which are not detected by TEM) yield contributions to the backscattering results used for quantitative defect analysis, and to what extent the measured $\Delta\chi_{min}(z)$ is caused only by dislocations.

Following the explanations in Sect.1.3.3.2.2 , all elements of the dechanneling matrix $Q_{ij}^{(disl)}$ have the same energy dependence $\sim E^{1/2}$, assuming that the distortion fields of the individual dislocations do not overlap. This is valid for dislocation densities $n_{disl} < 10^{11}$ cm^{-2} which is still fulfilled for the integrated densities $N_{disl} = \int_0^z dz' n_{disl}(z')$ determined by TEM in these cases.

If the simplified two-beam approximation is introduced (see Sect.1.3.4) corresponding to Eq.(1.255) for the total (integrated) dechanneling probability $\gamma(z)$, the same energy dependence as found for the dechanneling matrix is obtained,

$$\gamma_{disl} \sim E^{1/2} \quad . \tag{5.1}$$

Due to $\gamma_{disl}(z) = \lambda \int_0^z dz' n_{disl}(z')$, this dependence is also valid for the dechanneling diameter $\lambda \sim E^{1/2}$, in agreement with the results of Quéré (Sect.1.3.4, Eqs.(1.262, 1.263, 1.264), Ref.[60]). For considerably curved dislocation lines, the extension of the strain field of the dislocation is substantially reduced, giving rise to a weaker energy dependence $\sim E^{\kappa}$ with $\kappa \rightarrow 0$ for high energies E (if the oscillation length of the channeled ions is much larger than the width of the strain field). The results of Kudo [129] obtained from numerical calculations show that in the case of dislocation loops, such deviations from the $E^{1/2}$-dependence occur for small loop diameters. At energies in the range of ≈ 1 MeV for loop diameters $\geq$ some 10 nm, such deviations should be negligible. Consequently, for the loop structure considered here (<111> loops, sample (c)) the energy dependence typical for straight dislocation lines can be expected.

In Fig.5.18 for the case of the dislocation networks in the implanted Al layers and in Fig.5.19 for the three types of dislocation structures in silicon (<100> and <111> misfit dislocations and <111> loops), the energy dependences of the integrated dechanneling probability

$$\gamma(z) = \ln \frac{1 - \chi_{min,perf}(z)}{1 - \chi_{min}(z)} \tag{5.2}$$

are represented for $z = z_0$, calculated from the measured $\chi_{min}(z)$ at different incident energies E_0. In this parameter γ (within the two-beam approximation) the dechanneling due to the perfect crystal is eliminated. As can be seen, the dependence $\gamma(z_0) \sim E_0^{1/2}$ is verified for all defect structures investigated, showing that dislocations are the unique defects responsible for the measured defect-induced dechanneling ($\Delta\chi_{min}(z)$).

The experimental data in Fig.5.19 are compared with theoretical data

$\gamma(z_o)$ calculated for different energies E_o by means of the discontinuous model. The depth dependences of the dislocation density $n_{disl}(z)$ determined in Sect.5.2.2.1.4 for the three kinds of dislocation structures in silicon considered here are used for the calculation. The calculation of $\chi_{min}(z)$ by means of $n_{disl}(z)$ is performed according to Eqs.(1.246) and (1.247) with the geometrical parameters of the dislocation structures given in Sect.5.2.2.1.4.

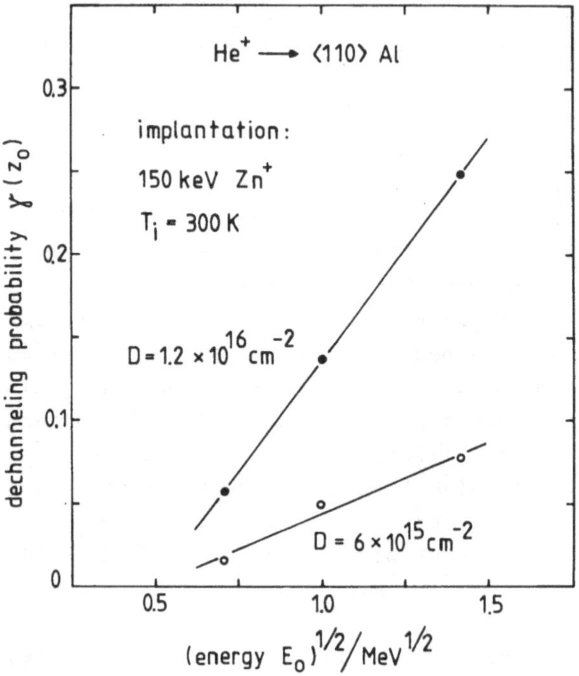

Fig.5.18. Energy dependence of the integrated dechanneling probability $\gamma(z_o)$ for He ions in <110> Al implanted with different doses of 150 keV Zn ions (experimental data taken from Ref.[126]). The depth of z_o corresponds to $R_p + \Delta R_p \approx 100$ nm.

Whereas the network of <100> misfit dislocations (identified by TEM to be only two-dimensional) provides the exact $E_o^{1/2}$-dependence, deviations are obtained in the case of the thick layer ($z_o = 0.45\,\mu m$) containing dislocation loops. For lower energies $E_o < 1$ MeV, a weaker dependence $\sim E_o^{\kappa}$ with $\kappa < 1/2$ was calculated in accordance with the computer simulation results of van Vliet [72]. This suggests the conclusion that for broad distributions of dislocations, the influence of dechanneling due to electronic and nuclear scattering ($\sim E_o^{-1}$) becomes important.

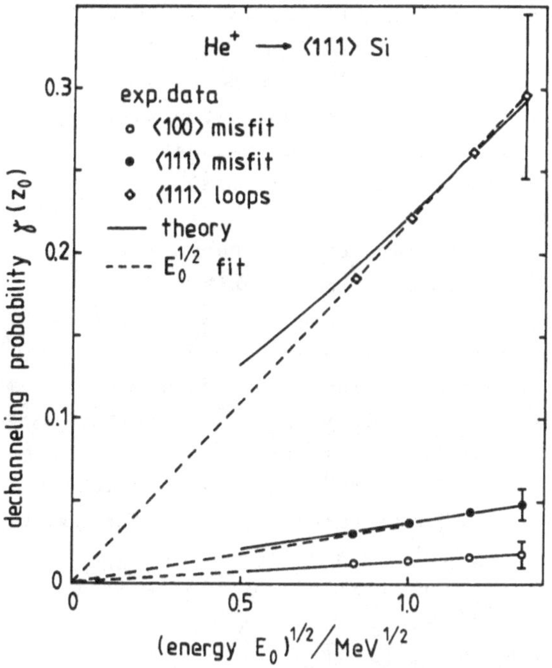

Fig.5.19. Energy dependence of the integrated dechanneling probability $\gamma(z_0)$ for He ions in Si; implantation and annealing parameters as described in the text (samples (a), (b), and (c)) with z_0 = 0.19 μm for <100> misfit dislocations, z_0 = 0.30 μm for <111> misfit dislocations, and z_0 = 0.45 μm for <111> dislocation loops

5.2.2.1.4. Integral density of dislocations N_{disl} and depth distribution of dislocation density $n_{disl}(z)$

The depth distribution of dislocations $n_{disl}(z)$ that can be calculated by means of the experimental data $\chi_{min}(z)$ is defined as the total projected length of the dislocations (perpendicular to the channel axis) per unit volume, and the integrated density $N_{disl} = \int_0^z n_{disl}(z')dz'$ consequently corresponds to a total projected length per unit area in agreement with the quantity derivable from TEM micrographs. Two models (the discontinuous model and the two-beam approximation) are again available. Analogous to the procedure for uncorrelated displaced atoms according to Sects.1.3.4 and 1.3.5.2 both models were employed here to determine dislocation densities by means of the experimental data $\Delta\chi_{min}(z)$ in Fig.5.16. The case of <u>axial dechanneling</u> is considered first.

(i) When the discontinuous model is applied, the calculation of $n_{disl}(z)$ has to be carried out by means of an iteration procedure based on Eqs.(1.282) and (1.283). The calculation starts from the first derivative of the measured minimum yield $d\chi_{min}(z)/dz$. As the type of the dislocations and the Burgers vector $\vec{b}$ are known from the TEM analysis (Sect.5.2.2.1.2), the dechanneling matrix $Q_{ij}^{(disl)}$ can be determined by Eqs.(1.236 - 1.241). The quantities x_d which are characteristic of the distortion field of a special type of dislocation are determined by Eqs.(1.232 - 1.234) for edge dislocations and by Eq.(1.235) for screw dislocations.

In Fig.5.20 the channel geometry together with the Burgers vector is mapped out for the dislocation structures in silicon. Both the <100> and the <111> misfit dislocations are pure edge dislocations with $\zeta = \pi/2$. An absolute value of the Burgers vector b = 0.384 nm and the Poisson ratio ν = 0.42 were used. The loops were treated in a simplified way by a stepwise fitting of the dislocation line, so that $Q_{ij}^{(loops)}$ resulted from both edge dislocations and screw dislocations (Ref.[63]). Such an approximation is justified only for loop diameters > 0.1 µm. The application of this approximation should be acceptable in the cases where an energy dependence $\gamma(E_o) \sim E_o^{1/2}$ can be measured (see Sect.5.2.2.1.3). N_{disl} then can be determined by integration of $n_{disl}(z)$.

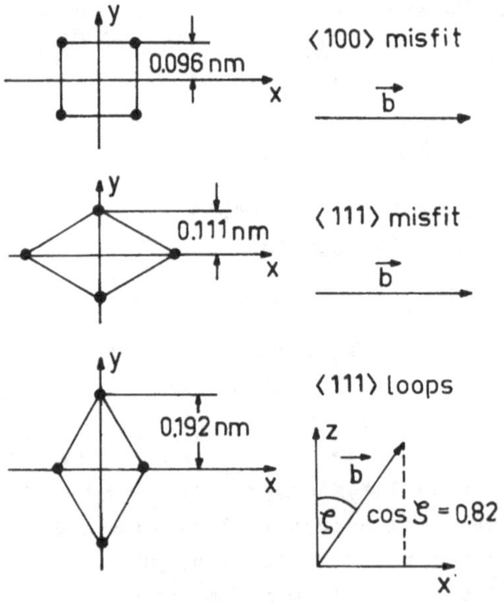

Fig.5.20. Channel geometry with respect to the Burgers vector $\vec{b}$ for different types of dislocation structures in silicon (described in the text). ζ is the angle between $\vec{b}$ and the channel axis.

(ii) The application of the two-beam approximation (i.e. the calculation of $n_{disl}(z)$ by means of the total dechanneling probability

$\gamma(z)$) presumes that the collective process of dechanneling in the distorted channels is restricted to a small depth interval. As for the description of these dechanneling events, a cross section (the dechanneling diameter λ) is introduced, this process within the framework of the two-beam approximation is characterized as a single scattering process (Sect.1.3.5.2). Single scattering can be assumed if the oscillation length of the ions is longer than the width of the deformation field of the dislocation, i.e. for $E_o > 100$ keV. A comparison of the defect densities with those determined by the discontinuous model and by means of TEM will provide a test (see also Sect.5.2.2.1.5), which can justify the description of dechanneling by means of a single parameter λ.

For the calculation of the dislocation densities the parameter $\gamma(z)$ is determined from the experimental data (Fig.5.16) by means of Eq.(1.279), and

$$\lambda_{ax} = \lambda_{o,ax}(E_o/\mathrm{MeV})^{1/2} \tag{5.3}$$

for axial channeling is given by Eq.(1.263) or by Eq.(1.265). The dechanneling width for the loops is approximately calculated using the mean value of λ_{ax} for pure edge and screw dislocations (Ref.[68]). The geometrical parameters β_{ax} necessary to calculate the $\lambda_{o,ax}$-values for the examples of analysis considered here are represented in Table 1.1. Values of $\lambda_{o,ax,Q}$ and $\lambda_{o,ax}$ are summarized in Table 5.3. The integral density of dislocations is obtained by the ratio $N_{disl} = \gamma(z_o)/\lambda_{ax}$. Similarly, the depth distribution $n_{disl}(z)$ is obtained by Eq.(1.280). As in the case of the discontinuous model, $n_{disl}(z)$ is determined by the derivative of the measured minimum yield $d\chi_{min}(z)/dz$ [1)].

The depth distributions of the dislocation density calculated by applying the discontinuous model (Ref.[63]; experimental results taken from Fig.5.16) are shown in Fig.5.21. In good agreement with the results of TEM (Sect.5.2.2.1.2) the distribution $n_{disl}(z)$ for the two-dimensional misfit dislocation network parallel to the (100) surface represents a sharp peak, whereas a broader distribution is obtained for the misfit dislocation network parallel to the (111) surface. A broad double-peaked distribution in the case of the loops was calculated with a surface layer $z \leq 50$ nm nearly free of dislocations which has been confirmed by TEM too. The differences in minimum yield $\Delta\chi_{min}(z)$ corresponding to the calculated depth profiles in Fig.5.21 are depicted in addition to the experimental data in Fig.5.16. While the

1) For sufficiently low values of $\chi_{min}(z)$ the dechanneling probability $\gamma(z)$ is well approximated by $\Delta\chi_{min}(z)$.

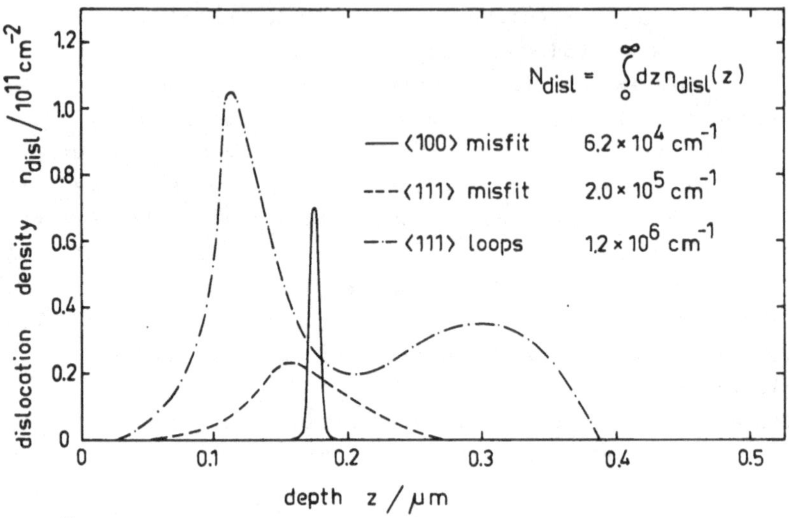

Fig.5.21. Depth dependence of dislocation density calculated from the curves $\Delta\chi_{min}(z)$ in Fig.5.16

calculated $\Delta\chi_{min}$ for larger depths $z > z_o$ (where $n_{disl} = 0$) is in good agreement with that measured for the <100> misfit, there is a deviation in the other cases. A constant or even decreasing $\Delta\chi_{min}(z)$ as it was measured (<111> misfit) is questionable and suggests small contributions of direct backscattering.

A comparison of the results obtained by the discontinuous model, the two-beam approximation, and by means of the TEM analysis is plotted in Table 5.2 for the <u>integral</u> densities N_{disl}.

TABLE 5.2
Integral density of dislocations N_{disl} obtained from the experimental data of Fig.5.16 by using the two-beam approximation (λ_{ax}-values due to Quéré and van Vliet, data from Ref.[68]), and the discontinuous model (data from Ref.[63]). The results of TEM (Sect.5.2.2.1.2) are shown for comparison.

dislocation structure	N_{disl}/cm^{-1}			
	two-beam approximation		dis-continuous model	TEM
	Quéré	v. Vliet		
<100> misfit	$1.6*10^4$	$7.0*10^4$	$6.2*10^4$	$(6.2 \pm 0.6)*10^4$
<111> misfit	$4.7*10^4$	$2.4*10^5$	$2.0*10^5$	$(2.0 \pm 0.4)*10^5$
<111> loops	$5.8*10^5$	$2.7*10^6$	$1.2*10^6$	$(2 \pm 1)*10^6$

While N_{disl} from Quéré's model (by means of $\lambda_{o,ax,Q}$) heavily underestimates the TEM results, the computer simulation data of van Vliet ($\lambda_{o,ax}$) as well as the discontinuous model provide results in agreement with TEM within the experimental error. This means that besides the discontinuous model, the two-beam approximation can also be used for quantitative analysis if an effective width $\lambda_{ax} = \lambda_{ax,eff} = \lambda_{o,ax}*(E_o/MeV)^{1/2}$ for dechanneling around a dislocation line (Eq.(1.265) is taken into account. The deviations of N_{disl} from the TEM results in the case of loops should not be surprising, because of the approximation used in both models and the relatively large error in the determination of N_{disl} from TEM measurements at these high densities.

From these results, conclusions can be derived concerning the detection limit of the RBS method for dislocations. In the case of narrow distributions $n_{disl}(z)$ comparable to the two-dimensional misfit dislocation network, the minimum density has been estimated to be $N_{disl}^{min} \approx 3...5*10^4\ cm^{-1}$ (which corresponds to $n_{disl}^{min} \approx 3...5*10^{10}\ cm^{-2}$). For broad distributions comparable to the <111> loops, the limit of detection is $N_{disl}^{min} \approx 10^5\ cm^{-1}$ ($n_{disl}^{min} \approx 2...5*10^9\ cm^{-2}$).

5.2.2.1.5. Availability of the dechanneling width λ as a parameter for quantitative analysis

The deviation of the energy dependence of the total dechanneling probability γ from $E_o^{1/2}$ for low energies (Sect.5.2.2.1.3) indicates that this parameter for dechanneling due to dislocations is also influenced by the dechanneling contribution from a perfect crystal. Therefore, in view of the application of the two-beam approximation to the quantitative analysis of dislocations, the question arises, if it is possible to describe dechanneling due to dislocations by only one parameter (γ and λ, respectively). This problem for axial channeling has been investigated in detail by Gärtner et al. [63] who calculated $\chi_{min}(z)$ in the case of the <111> misfit dislocations for many different distributions $n_{disl}(z)$ by means of the discontinuous model. With λ_{ax} determined from $\chi_{min}(z)$ by the Eqs.(1.279) and (1.261) the quantity of interest $\lambda_{o,ax} = \lambda_{ax}(E_o/MeV)^{-1/2}$ was derived. This reduced dechanneling diameter $\lambda_{o,ax}$ is expected to be constant if the two-beam approximation holds and the dechanneling probabilities P_{perf} and P_{def} are additive (Sect.1.3.4, Eq.(1.251)).

The results show that in the energy region 1 MeV < E_o < 2 MeV $\lambda_{o,ax}$ increases by about a factor 2, if (for $N_{disl} = 2*10^5\ cm^{-1}$) the width of $n_{disl}(z)$ is increased by an order of magnitude (≈30...300 nm FWHM). This effect is still more pronounced for lower energies E_o < 1 MeV. The reduced diameter $\lambda_{o,ax}$ also increases by about a factor 2 if (for fixed width: 80 nm FWHM) the integral density is increased by an order of magnitude ($10^5\ cm^{-1} \le N_{disl} \le 10^6\ cm^{-1}$). The fact that $\lambda_{o,ax}$ is not constant but depends on the shape and on the integral of the distribution $n_{disl}(z)$ indicates that there is no strict additivity of the dechanneling probabilities in the two-beam approximation for broader distributions, and that the dechanneling due to dislocations may not generally be described by only one parameter λ_{ax}. The efficiency of dechanneling by dislocations is also influenced by electronic and nuclear scattering. Therefore the enhancement of λ_o with increasing half width and integral of $n_{disl}(z)$ and with decreasing E_o is understood as the increasing influence of electronic and nuclear scattering. As the calculations showed, for large N_{disl} and small half widths $\lambda_{o,ax}$ becomes constant, because the dechanneling is then determined only by the dislocations. In these cases $n_{disl}(z)$ is easily and accurately provided by $\gamma(z)$ and λ_{ax}. Otherwise the dependence of λ_{ax} on N_{disl} and on the shape of $n_{disl}(z)$ has to be taken into account or the procedure of the discontinuous model may be used directly.

5.2.2.1.6. Analysis of dislocations by planar channeling

The typical dependence of dechanneling due to dislocations on the square root of the ion energy was also proved by the results of planar dechanneling measurements. For the calculation of N_{disl} or $n_{disl}(z)$ from planar dechanneling data the procedures are basically the same as described in Sect.5.2.2.1.4 for the axial case, but using the quantities for planar channeling in the equations considered. In the case of the two-beam approximation, this means that the axial dechanneling width λ_{ax} has to be replaced by a quantity $\lambda_{pl} = \lambda_{o,pl}*(E_o/MeV)^{1/2}$. The reduced dechanneling width $\lambda_{o,pl}$ for planar channeling can be calculated by means of Eq.(1.264) or Eq.(1.266).

In Table 5.3 for channeling in the (110) plane of silicon, values of $\lambda_{o,pl,Q}$ and $\lambda_{o,pl}$ calculated by means of Eq.(1.264) and Eq.(1.266), respectively, using the geometrical parameters β_{pl} given in Table 1.1, are compared to the corresponding values $\lambda_{o,ax,Q}$ and $\lambda_{o,ax}$ for the <111> axis.

The dechanneling widths $\lambda_{o,pl}$ prove to be more than one order of magnitude larger than the values $\lambda_{o,ax}$ for the axial case. Consequently, the sensitivity of the planar dechanneling measurements for the detection of dislocations is obviously higher than in the axial case.

The results of quantitative analysis by planar and axial dechanneling (based on the two-beam approximation) are compared for the three types of dislocation structures in silicon in terms of the integral densities N_{disl} in Table 5.4. The effective values λ_o (Table 5.3) are used in both cases. The angles ζ_{pl} for the special orientations of the dislocation

TABLE 5.3

Comparison of reduced dechanneling diameters for planar (110) and axial <111> dechanneling of He ions in silicon with different types of dislocations.

type of dislocation	$\lambda_{o,pl,Q}$/nm (110)	$\lambda_{o,ax,Q}$/nm <111>	$\lambda_{o,pl}$/nm (110)	$\lambda_{o,ax}$/nm <111>
edge dislocation $\vec{b} \perp$ (110), <111>	26	7.3	26	1.35
edge dislocation $\vec{b} \parallel$ (110), <111>	27	7.4	0	0
screw dislocation $\vec{b} \perp$ (110), <111>	16	4.45	16	0.8

lines (see Sect.5.2.2.1.2) had to be taken into account; e.g. the contributions to dechanneling in the (110) plane are completely different for the edge dislocations in the (100) plane (<100> misfit) lying in two different directions perpendicular to each other (see Eq.(1.266) and Table 5.3). Only those dislocations which are parallel to the channeling plane ($\vec{b} \perp$(110)) give rise to dechanneling. These amount to only one half of all the dislocations in this case.

TABLE 5.4

Comparison of the integral density of dislocations obtained from planar and axial dechanneling data and TEM, respectively. The angle between the incidence directions for (110) planar channeling and <100> or <111> axial channeling is $\geq 6^{o}$.

dislocation structure	N_{disl}/cm^{-1} planar (110)	axial	TEM
<100> misfit	1 *10^{5}	7 *10^{4}	(6.2 ± 0.6)*10^{4}
<111> misfit	1.7*10^{5}	2.4*10^{5}	(2.0 ± 0.4)*10^{5}
<111> loops	1.4*10^{5}	2.7*10^{6}	(2 ± 1)*10^{6}

The calculation of depth distributions of the dislocation density $n_{disl}(z)$ from measured $\chi_{min}(z)$ assumes that the effective strain field of the dislocation within a depth interval is <u>smaller</u> than the depth resolution of the RBS (20...30 nm for He^{+} in Si). Nevertheless, the effective range of the strain field parallel to the channeling direction causing dechanneling (approximately determined by λ) is much larger for the planar case (Table 5.3). With $\lambda_{pl} \approx 20...30$ nm (dependent on the incidence energy $E = E_0$) the effective strain field becomes larger than the depth resolution.

In Fig.5.22 for the two-dimensional <100> misfit dislocation network, the depth distributions obtained from planar and axial measurements (Ref.[128]) are compared considering the quantity $(d\gamma(z)/dz)E_0^{-1/2}$ for $E_0 = 1.0$ MeV and 1.8 MeV (the distribution obtained in the axial case is represented only by its width: hatched region). The results show that due to the large range of the strain field for dechanneling in the (110) plane, the extremely small depth distribution is substantially broadened, whereas a negligible broadening effect can be observed for <100> axial channeling (λ_{ax} in the order of 1 nm). Consequently, narrow distributions of dislocation concentration can be determined accurately only by axial channeling measurements.

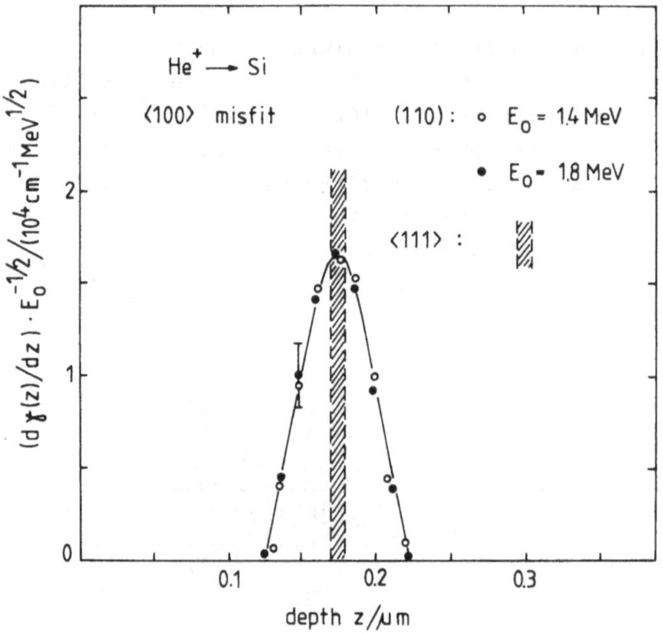

Fig.5.22. Depth distribution of dislocation density for the <100> misfit dislocation network characterized by the quantity $(d\gamma(z)/dz)E_0^{-1/2}$. The distribution is calculated from (110) planar channeling data; the distribution obtained from <111> axial channeling measurements is represented by its width (hatched region).

From the large values of λ_{pl} in the planar case, a second problem arises concerning the exact determination of depth distributions and integral densities. As has been proved for the high density of the <111> loops, the strain fields of the individual dislocations start to overlap already at $N_{disl} > 5*10^5...10^6$ cm^{-1} (Ref.[128]). This corresponds to $n_{disl} > 1...3*10^{10}$ cm^{-2} which is nearly one order of magnitude lower than in the axial case. The integral density $N_{disl} \approx 2*10^6$ cm^{-1} for the loops will consequently be underestimated in planar measurements (Table 5.4).

To avoid the effects of broadening of narrow distributions or of "degradation" and deformation of distributions for large concentrations, the results of planar channeling measurements should be analyzed together with results of axial channeling measurements.

5.2.2.2. Stacking faults

In silicon, stacking faults (deviations from the stacking sequence of (111) planes for the perfect diamond type lattice) can be formed due to the condensation of vacancies (intrinsic stacking faults) or interstitial atoms (extrinsic stacking faults). Such extended defects for example are observed after epitaxial growth of silicon layers. Stacking faults of limited extent are surrounded by dislocations. Often these defects are stable up to very high temperatures.

Two examples are chosen for qualitative and quantitative analysis of stacking defects:

(a) a thin surface layer of silicon monocrystal ($z_o \approx 0.3\,\mu m$) amorphized by ion implantation (phosphorus) and then epitaxially recrystallized by a thermal treatment at high temperatures (1300 K) (Ref.[128]) and

(b) a thick monocrystalline silicon layer ($z_o \approx 0.9\,\mu m$) grown on sapphire by heteroepitaxy (Ref.[66, 73]).

5.2.2.2.1. Depth dependence of the minimum yield

The dependence $\Delta\chi_{min}(z)$ measured for the two layers considered are represented in Fig.5.23 and Fig.5.24.

In the case of the recrystallized ion implanted layer (Fig.5.23) the experimental data for axial <111> and planar (110) channeling measurements are compared. In both cases a continual increase of $\Delta\chi_{min}$ can be observed without any trace of a peak due to direct backscattering. The thickness z_o of the layer containing defects is marked by a "knee" in the curves. The dechanneling for the planar case is larger than that for the axial one, indicating again a higher sensitivity of planar measurements to the detection of these extended defects.

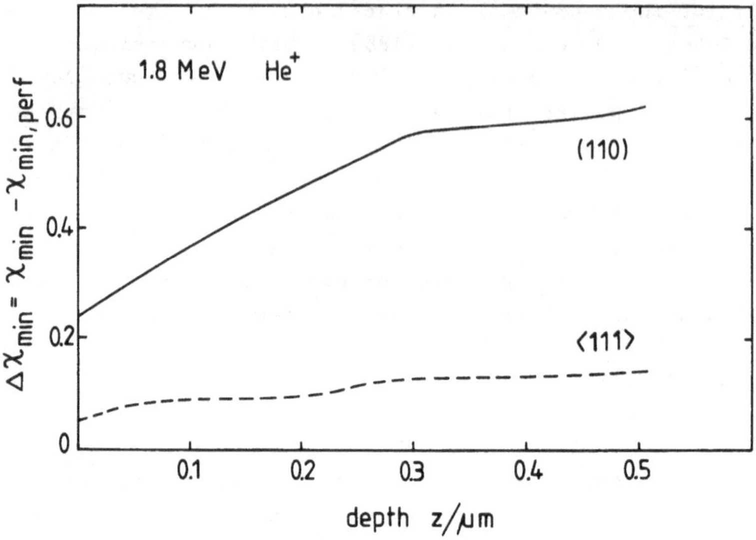

Fig.5.23. Depth dependence of $\Delta\chi_{min}$ for 1.8 MeV He^+ incident in the <111> and (110) direction of silicon containing stacking faults. The crystal was implanted with 200 keV P^+ ($D = 1*10^{16}$ cm^{-2}) at T_i = 80 K and annealed at T_a = 1300 K for 30 minutes in a nitrogen atmosphere.

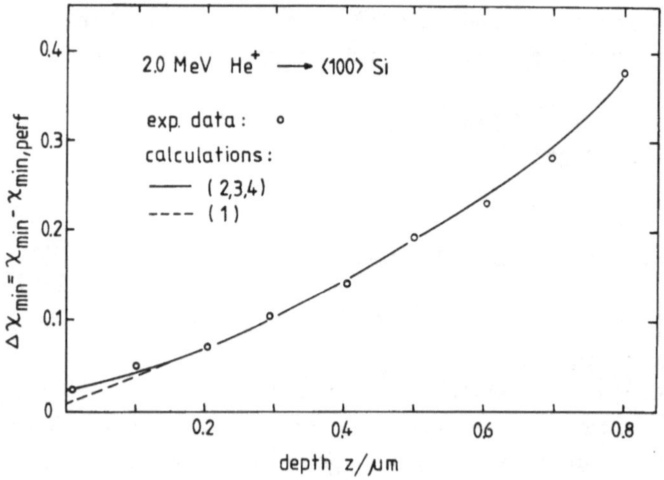

Fig.5.24. Depth dependence of $\Delta\chi_{min}$ for 2.0 MeV He^+ incident in the <100> direction of a silicon layer (thickness ≈ 0.9 µm) heteroepitaxially grown on sapphire. The experimental data (Ref.[73]) are fitted by a curve calculated on the assumption of different kinds of extended defects (Ref.[66]): stacking faults (1), twin lamellae (2), and a mixture of stacking faults and twin lamellae (3) which also can involve dislocations (4) (see Fig.5.29).

The curve $\Delta\chi_{min}(z)$ for axial (<100>) dechanneling of He^+ in the silicon layer epitaxially grown in the <100> direction on sapphire (Fig.5.24) shows a progressive increase of $\Delta\chi_{min}$ with z.

For both samples, but especially for the recrystallized implanted layer, already a clear defect-induced increase of the minimum yield $\Delta\chi_{min}(0)$ can immediately be measured at the surface.

As in the case of dislocations, information on the type and the structure of the defects is obtained by TEM analysis and by measuring the energy dependence of χ_{min}.

5.2.2.2.2. Results of TEM analysis

TEM micrographs provided the following information on the defects in the two layers.

(a) Recrystallized ion implanted silicon

A large number of stacking faults of different sizes is detected (Fig.5.25); other kinds of defects are not visible. The mean extension of the stacking faults amounts to ≈ 0.3...0.4 μm. The defects are equipartitioned across the four sets of (111) planes.

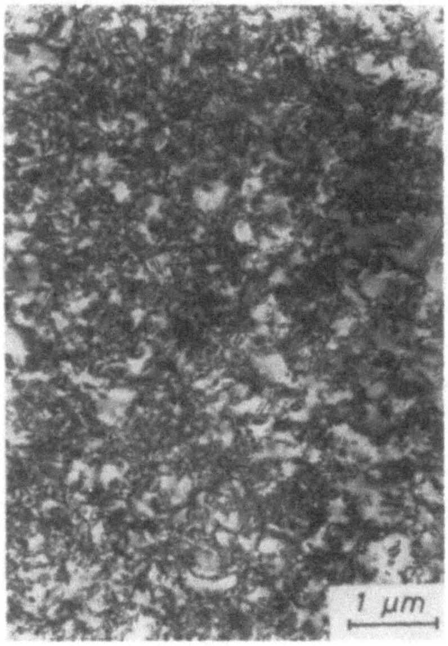

Fig.5.25. Bright-field TEM micrograph of implanted and annealed silicon containing stacking faults (data as given in Fig.5.23)

(b) Heteroepitaxial silicon layer on sapphire

The TEM micrographs (Ref.[73]) show a large number of stacking defects including both stacking faults and twin lamellae whose densities decrease from the sapphire/silicon interface to the layer surface. The mean extension of the stacking defects can be estimated to be in the order of $\approx 0.5\,\mu m$. The twin lamellae also occur on the different (111) planes (no information on the ratio of the stacking faults and the twin lamellae is given). The average thickness of the twin lamellae was determined to be ≈ 2 nm corresponding to about four atoms per <111> row.

In both cases the TEM analysis provided no quantitative results concerning the densities of the stacking defects.

5.2.2.2.3. Energy dependence of the minimum yield

As in the case of dislocations, a typical energy dependence of dechanneling is expected for stacking faults, again offering the possibility of identifying the defects by means of dechanneling measurements. As with dislocations, the dependence of the total dechanneling probability $\gamma(z = z_0)$ given by Eq.(1.279) on the incident energy E_0 is considered.

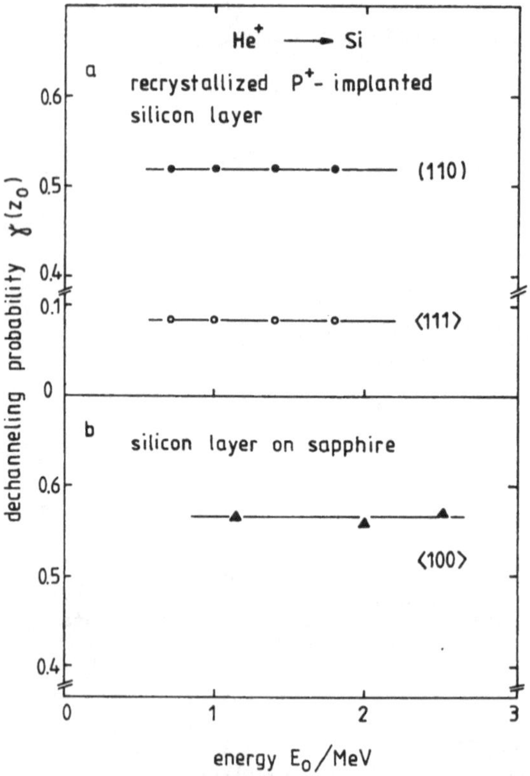

Fig.5.26. Energy dependence of the integrated dechanneling probability $\gamma(z_0)$ for He ions in silicon containing stacking defects; data for (a) as given in Fig.5.23 ($z_0 = 0.3\,\mu m$)

As dechanneling due to stacking faults is caused mainly by displaced strings (Sect.1.3.3.2) the dependence $\gamma(E_o)$ results from the energy dependence of the corresponding dechanneling matrix $Q_{ij}^{(dst)}$ (see for comparison Sect.5.2.2.1.3). The Eqs.(1.243) and (1.244) show that $Q_{ij}^{(dst)}$ is independent of the energy. Consequently, for stacking faults a dependence

$$\gamma(z_o) \sim E_o^o \tag{5.4}$$

can be assumed.

The experimental results $\gamma(E_o)$ for the recrystallized implanted silicon layer (a) and for the silicon layer on sapphire (b) are shown in Fig.5.26. In both cases an energy-independent defect-induced dechanneling is detected (for the recrystallized layer both by axial and planar measurements) which indicates that <u>displaced strings</u> (or displaced planes in the planar case) are the main defects responsible for the measured dechanneling ($\Delta\chi_{min}$). This information is based on the quantitative analysis described in the next section.

5.2.2.2.4. Density of displaced strings (n_{dst}, N_{dst})

The density of displaced strings $n_{dst}(z)$ represents an average number of strings which become displaced at the depth z within a unit volume. For short, $n_{dst}(z)$ is called the "depth distribution of displaced strings". The integral density $N_{dst} = \int_0^{z_o} dz' n_{dst}(z')$ then has to be understood as a total density of displaced strings per unit area.

The discontinuous model and the two-beam approximation were again used to calculate these densities by means of the experimental data of Figs.5.23 and 5.24.

(i) The procedure for calculation in the case of the discontinuous model is the same as for dislocations based on Eqs.(1.284) and (1.285). The calculation of the dechanneling matrix $Q_{ij}^{(dst)}$ is described in Sect.1.3.3.2.2. For strings displaced by a distance of δ_r, $Q_{ij}^{(dst)}$ is given by Eqs.(1.242 - 1.245). Values of δ_r for the <111> and the <100> strings in silicon are particularized in Sect.1.3.3.2.2. In the case of <100> strings, this displacement amounts to δ_r = 0.064 nm for stacking faults on all four sets of equivalent (111) planes; in the case of the <111> axis δ_r = 0.074 nm (see Eq.(1.223) and Fig.1.51a,b).

(ii) Within the framework of the two-beam approximation, the axial dechanneling due to stacking faults (because of the fact that the defect of the crystal symmetry is restricted to the stacking fault plane) is described by single scattering processes due to a new interior surface introducing an (energy-independent) dechanneling cross section σ_{dst} of one displaced string.

Calculating σ_{dst} by means of Eq.(1.270), n_{dst} and N_{dst} were provided by the total dechanneling probability γ using Eq.(1.281) in the same way as n_{disl} or N_{disl} were determined for dislocations. Values of γ again result from the measured minimum yields (Eq.(1.279)).

This procedure can be transcribed to the case of planar dechanneling too, introducing a temperature-dependent (and energy-independent) dechanneling cross section S_{dpl} per unit (projected) edge length of a displaced plane (because of the one-dimensional ion motion in the transverse plane, the planar dechanneling cross section being characterized by a length). The cross section S_{dpl} can be calculated using Eq.(1.272). As in the axial case, a density of displaced strings can be derived from the experimental data (i.e. $\gamma(z)$; Eq.(1.279)) by means of this cross section. The quantities characterizing the density of displaced strings in the planar case are the edge length of the displaced planes per unit volume, l_{dpl}, projected on the plane rectangular to the incident direction, and the total projected length per unit area, $L_{dpl} = \int_0^{z_o} dz' \, l_{dpl}(z')$. For a comparison of axial and planar channeling results, an approximate relation between n_{dst} and l_{dpl} can be used if the angle between the axial and planar incident directions is sufficiently small

$$n_{dst}(z) = l_{dpl}(z) \, N \, d \, d_p \quad . \tag{5.5}$$

If axial and planar results are compared by means of Eq.(5.5), it is necessary to consider the fact that for different channeling directions, the portion of stacking faults contributing to dechanneling is different. This is illustrated in Table 5.5.

TABLE 5.5

Portion of the four possible positions of stacking faults in silicon (corresponding to an equipartition across the four equivalent sets of (111) planes) giving rise to dechanneling for different incident directions

direction of incidence	portion of the stacking faults contributing to dechanneling
<100>	4/4
<111>	3/4
<110>	2/4
(110)	2/4

Consequently, the density n_{dst} obtained from measurements at (110) incidence by means of Eq.(5.5) has to be multiplied by a factor of 1.5 to compare it with n_{dst} for <111> incidence (angles $< 10^{o}$ between (110) and <111> incidence justify the use of Eq.(5.5)).

The densities of displaced strings $n_{dst}(z)$ calculated for the recrystallized ion implanted layer by means of the experimental results (for axial <111> and planar (110) incidence directions) in Fig.5.23 applying the two-beam approximation are shown in Fig.5.27a together with the depth distribution of the implanted phosphorus concentration, which is also measured by RBS.

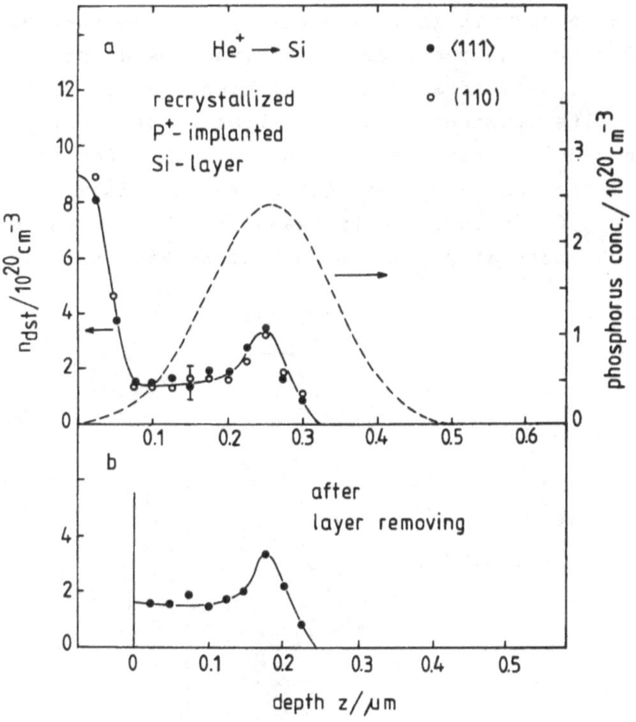

Fig.5.27. (a) Depth distribution of displaced atomic strings determined from the <111> and (110) channeling results shown in Fig.5.23 (full and open circles, respectively) and depth profile of the implanted P-atoms (dashed curve) (Ref.[128])
(b) Depth distribution of displaced strings after removing a surface layer of about 75 nm thickness

The results for axial and planar channeling agree well within the experimental error. Besides a small peak of $n_{dst}(z)$ near the maximum of the phosphorus profile, an extremely high density of displaced strings at the crystal surface ($n_{dst}/N > 0.01$ for $z < 0.1\,\mu m$) is obtained which corresponds to a somewhat stronger increase of $\Delta\chi_{min}(z)$ for small depths in Fig.5.23. To examine whether this high surface density is real, a layer of about 75 nm thickness was removed and $n_{dst}(z)$ was redetermined. The result shown in Fig.5.27b agrees with the depth distribution $n_{dst}(z)$ before the layer removal (Fig.5.27a). No peak is obtained at the surface now, indicating that surface effects are not responsible for this peak. Strong interactions of the stacking faults near the surface can be expected when stacking faults start to grow during epitaxial recrystallization of amorphous layers at the amorphous layer - crystal interface and the extension of the stacking faults increases with increasing distance from this original interface. It might be possible that due to this interaction a small portion of defects is formed, causing direct backscattering near the surface. Such defects - the density of which is too low to contribute to the energy dependence of dechanneling (Sect.5.2.2.2.3) by the procedure of analysis used here - were interpreted in terms of an additional density of displaced strings.

A comparison of the results obtained by the two-beam approximation and by the discontinuous model is represented in Fig.5.28 for the heteroepitaxial silicon layer on sapphire (experimental data in Fig.5.24). The fact that $n_{dst}(z)$ calculated by means of the two-beam approximation increases with increasing z is in agreement with the TEM

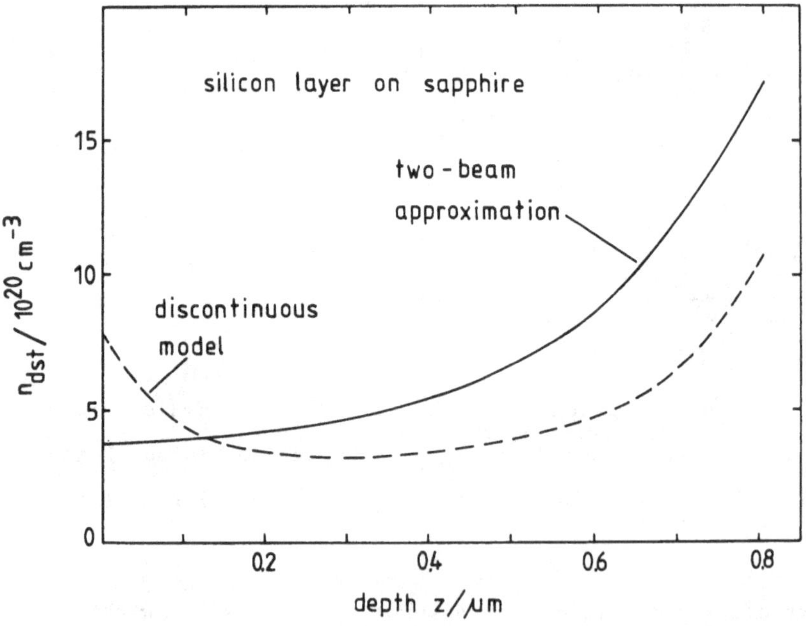

Fig.5.28. Depth distributions of displaced atomic strings determined from the experimental data in Fig.5.24 by means of the discontinuous model (Ref.[66]) and by means of the two-beam approximation (Ref.[73]), respectively

results (Sect.5.2.2.2.2), but a quantitative comparison with TEM results is not available. Between the two curves there are two main differences.

(1) A peak near the surface caused by the experimental data $\Delta\chi_{min}(0) > 0$, is obtained by using the discontinuous model. Assuming only dechanneling due to displaced strings, apart from the contribution from the head atoms of the displaced strings, there is no direct backscattering contribution to χ_{min}. Therefore, the measured $\Delta\chi_{min}(z)$ is reached only by enhanced dechanneling near the surface (stronger increase of the dashed curve in Fig.5.24 for $z < 0.1\,\mu m$).

No peak occurs when the two-beam approximation is used, because only the first derivative of the measured dependence $\Delta\chi_{min}(z)$ (or $d\gamma(z)/dz$) is of importance and not $\Delta\chi_{min}(0)$.

(2) With increasing depth, the densities of displaced strings $n_{dst}(z)$ provided by the two methods differ more and more up to a factor of 2. This indicates that dechanneling is underestimated by the two-beam approximation (model of new interior surfaces). The underestimation exists because the dechanneling probability of a new interior surface should be represented by $\chi_{min,perf}(0)$ (or the dechanneling due to a displaced string by a constant cross section σ_{dst} given by Eqs.(1.268 – 1.270)) only for small depths. Because the distribution of the transverse energy of the ions changes with increasing depth, the new interior surfaces become more effective for larger depths. The availability of the two-beam approximation using the dechanneling cross section σ_{dst} (or S_{dpl}) is considered in more detail in the next section.

No realistic results are attained (as indicated by the artifical peak of $n_{dst}(z)$ near the surface) when only displaced strings are considered in the description of dechanneling in the heteroepitaxial silicon layer. In Sect.5.2.2.3, the results $n_{dst}(z)$ are discussed for the same layer, if dechanneling due to twin lamellae is considered.

5.2.2.2.5. Availability of the dechanneling cross section as a parameter for quantitative analysis

The dechanneling cross sections σ_{dst} or S_{dpl}, used to calculate densities of displaced strings from experimental values for the total dechanneling probability γ based on the model of new interior surfaces (two-beam approximation), depend on the direction of incidence and on the temperature during the measurement (Eqs.(1.268, 1.269, 1.270, 1.272, 1.273)). The cross sections are independent of the displacement δ_r of the strings.

In order to examine the influence of δ_r on dechanneling $\Delta\chi_{min}(z)$ was calculated separately for the <100> axis of silicon as a function of δ_r (artifically varied) using the discontinuous model and data $n_{dst}(z)$ for the silicon layer on sapphire (Ref.[66]). The results showed that within a region $0.05\ nm \le \delta_r \le 0.1\ nm$, $\Delta\chi_{min}$ increases by about 50% with increasing δ_r. As this effect is neglected in the model of new interior surfaces, errors of the quantitative analysis and in some cases inconsistency of the $n_{dst}(z)$ attained for different crystal axes (Ref.[73]) arise.

5.2.2.3. Twin lamellae

The insufficient results obtained for $n_{dst}(z)$ of the silicon layer on sapphire (Fig.5.28), when taking only dechanneling due to displaced strings into account for the calculations based on the discontinuous model, suggest that dechanneling due to the twin lamellae (which were detected by TEM, Sect.5.2.2.2.2) yields a substantial contribution to the measured $\Delta\chi_{min}(z)$ (Fig.5.24).

As has been pointed out in Sect.1.3.3.2, the dechanneling due to twin lamellae (twinned regions of a very small average thickness d_{lam}) can be described by means of dechanneling due to a density $n_{dst}(z)$ of displaced strings and an additional dechanneling contribution due to randomly distributed point defects within the twin lamellae. The relative density n_{pd} of the point defects for a fixed value d_{lam} results from the density of the displaced strings and is given by Eq.(1.226). For the silicon layer on sapphire considered here by means of TEM analysis a value $d_{lam} \approx 2$ nm was estimated corresponding to about 4 atoms per string within the lamellae, which yields $n_{pd}(z) = 6\, n_{dst}(z)/N$. In this way, only one independent function $n_{dst}(z)$ is obtained that can be determined from the measured $\Delta\chi_{min}(z)$ in Fig.5.24 (experimental data fitted by the solid line) as in the case of the stacking faults, but using now Eqs.(1.288 - 1.290) in Sect.1.3.5.3 for $n_{disl} = 0$. The probability for hitting the randomly distributed point defects according to Eqs.(1.249) and (1.250) is $\Pi_i^{(pd)} = 1$.

Assuming that the silicon layer in question contains only pure twin lamellae with $d_{lam} = 2$ nm the dotted curve (2) in Fig.5.29 was calculated by this procedure for $n_{dst}(z)$. In comparison to the case of pure stacking faults (only displaced strings; curve (1)) on the average a lower density of displaced strings is obtained because of the additional point defects. In contrast to curve (1), $n_{dst}(z)$ now has a minimum at the surface. This results from a considerable direct backscattering contribution from the point defects. A more realistic description in agreement with the TEM results should therefore start with a mixture of stacking faults and twin lamellae (Sect.5.2.2.4).

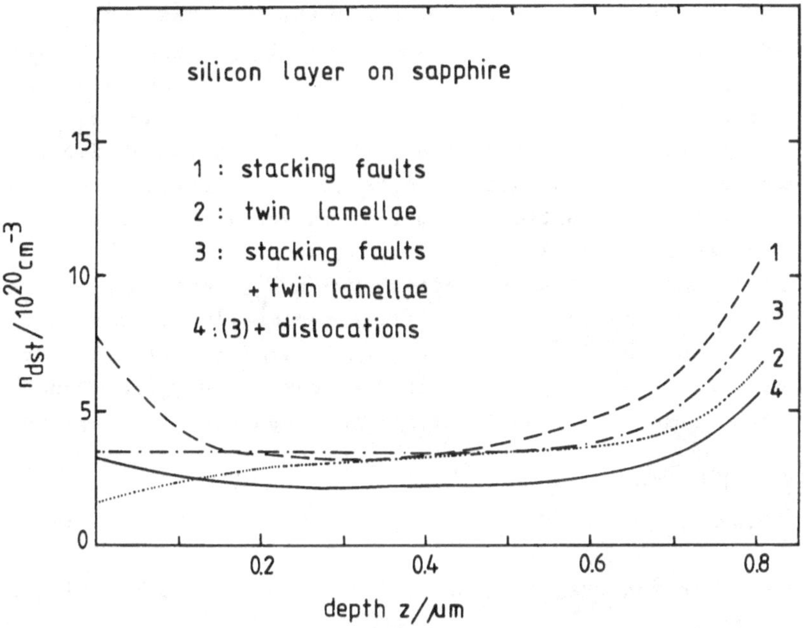

Fig.5.29. Depth distribution of displaced atomic strings determined from the experimental data in Fig.5.24 assuming different kinds of extended defects (Ref.[66])

5.2.2.4. Mixture of stacking faults, twin lamellae, and dislocations

The results of Sect.5.2.2.2.4 and Sect.5.2.2.3 represented by the curves (1) and (2) in Fig.5.29, indicate that the dechanneling in the silicon layer on sapphire cannot be satisfactorily described either by pure stacking faults or by pure twins. The TEM analysis (Sect.5.2.2.2.2) showed that the layer contains both kinds of defects, but the concentration ratio is unknown. On the other hand, it is known that stacking faults of finite extension are always surrounded by dislocations giving a specific contribution to dechanneling (Sect.5.2.2.1). Consequently, in this section mixtures of these defects are taken into account, which must involve a consideration of the energy dependence of dechanneling.

5.2.2.4.1. Determination of the defect densities

(i) First, we assume a mixture of stacking faults and twin lamellae with about half of the displaced strings caused by the twins. The relative concentration of point defects (within the twin lamellae) is now related to the density n_{dst} of displaced strings by $n_{pd}(z) = 3\, n_{dst}(z)/N$. The calculation of $n_{dst}(z)$ from $\Delta\chi_{min}(z)$ in

Fig.5.24 (fit of the data represented by the solid line according to Eqs.(1.288 – 1.290)) provides the dash-dotted curve (3) in Fig.5.29. As expected, the densities $n_{dst}(z)$ represented by curve(3) are between the two limits (1) and (2). The assumed ratio between $n_{pd}(z)$ and $n_{dst}(z)$ seems to be rather realistic, because the artifical peak (1) or dip (2) near the surface disappears (no indication is given of either by TEM analysis). Only near the silicon-sapphire interface $n_{dst}(z)$ does increases with increasing depth z.

It is possible to determine the ratio between n_{pd} and n_{dst} more accurately, if $\Delta\chi_{min}$ is measured more precisely especially for small z.

(ii) To make allowance for the fact, that stacking faults are surrounded by dislocations, the density of dislocations $n_{disl}(z)$ can be assumed to be approximately proportional to the density of displaced strings

$$n_{disl}(z) = \mu\, n_{dst}(z) \quad . \tag{5.6}$$

The ratio μ depends on the shape and the extension of the fault planes and can be expressed by

$$\mu = L_{stf} \frac{A_o}{A_{stf}} \tag{5.7}$$

where L_{stf} denotes the average circumference of a finite fault plane (length of the dislocation line) and A_{stf} its average area; A_o is the area of one string. In this way, μ can be estimated by TEM. Another possibility to estimate μ is by measuring the energy dependence of $\Delta\chi_{min}$. A value $\mu = 1.8{*}10^{-4}$ nm has been estimated from the independence of $\Delta\chi_{min}$ of E_o (Sect.5.2.2.4.2). This value of μ is not unrealistic, because the belonging extension of the fault planes ($\approx 0.5\,\mu$m corresponding to Eq.(5.7)) is compatible with the TEM results.

Using the values for n_{disl} resulting from μ, and considering dislocations of the Frank type approximated by pure edge dislocations with a Burgers vector $\vec{b}_F = (1,1,1)a/3$ ($b_F = 0.313$ nm, $\cos\zeta = 0.578$; see Sect.5.2.2.1.4) $n_{dst}(z)$ was again calculated as shown by curve (4) in Fig.5.29.

If this result, based on a realistic description of dechanneling (discontinuous model), and taking into account displaced strings as well as point defects and dislocations, is compared with the result of the simple description of dechanneling due only to displaced strings within the framework of the two-beam approximation (Fig.5.28), an essential underestimation of dechanneling can be concluded in the latter case, especially for larger depths.

5.2.2.4.2. Energy dependence of the minimum yield

Because the energy dependence of the specific contribution to the dechanneling matrix due to defects (and thus the energy dependence of $\Delta\chi_{min}$) is different for different kinds of defects (Table 5.6), some information on the ratio of these kinds of defects in the layer under investigation is expected.

A comparison of experimental and theoretical results for $\Delta\chi_{min}(z)$ at different energies (1.13 MeV $\leq E_0 \leq$ 2.5 MeV) is represented in Fig.5.30 for the example of the heteroepitaxial silicon layer on sapphire (see Sect.5.2.2.2).

TABLE 5.6

Energy dependence of the defect-induced contribution to dechanneling (i.e. $Q_{ij}^{(def)}$, γ, and $\Delta\chi_{min}$, respectively) for different kinds of defects.

kind of defects	energy dependence of dechanneling
uncorrelated displaced atoms (point defects, clusters, amorphous regions)	E^{-1} ... $E^{-1/2}$ (Sects..1.3.5.1 and 5.2.1.2)
displaced strings	E^{0} (Sects.1.3.3.2 and 5.2.2.2.3)
dislocations	$E^{+1/2}$ (Sects.1.3.3.2 and 5.2.2.1.3)

The calculations were carried out considering the mixture of stacking faults and twin lamellae (half of n_{dst} caused by stacking faults).

The experimental data in Fig.5.30 do not show a dependence on the energy E_0 within the experimental errors. The energy dependence of the calculated $\Delta\chi_{min}(z)$ for the case considered here (i.e. without dislocations: $/u = 0$; curve (3) in Fig.5.29) is obviously too strong. This indicates that the dislocations accompanying the stacking defects must have a measurable influence on dechanneling in the layer. The dislocations partially compensate for the energy dependence of $\Delta\chi_{min}$ introduced by the point defects.

Calculations of $\Delta\chi_{min}(z)$ were repeated, including dislocations and considering different values for the density ratio $/u$. They showed that the experimental error of the data given in Fig.5.30 in this way permits only a rough estimation of the value for $/u$. The assumption of an

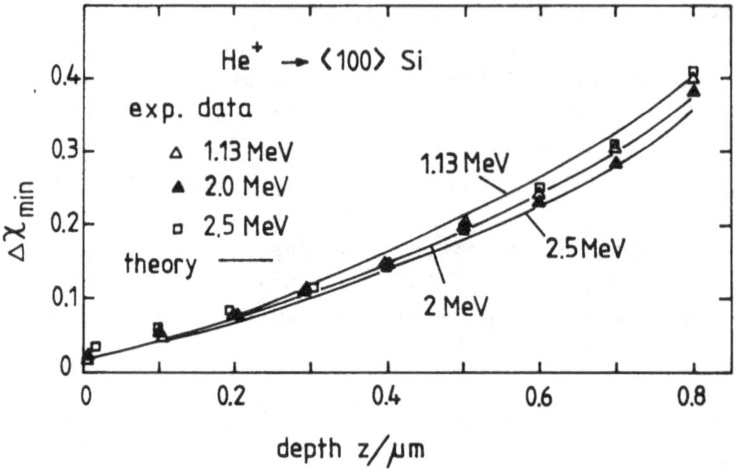

Fig.5.30. Depth dependence of $\Delta\chi_{min}$ for He^+ incident on <100> Si at different energies. Experimental data (Ref.[73]) and theoretical curves (Ref.[66]) assuming dechanneling both by stacking faults and by twin lamellae (and neglecting dislocations) are compared.

energy-independent $\Delta\chi_{min}(z)$ within 1.13 MeV $\leq E_0 \leq$ 2 MeV is fulfilled for a value $\mu = 1.8*10^4$ nm which is used for the quantitative analysis in Sect.5.2.2.4.1. In Fig.5.31 the calculated energy dependences of $\Delta\chi_{min}(z = 0.8\,\mu m)$ corresponding to the four assumptions for the defect structure in the silicon layer on sapphire (see Fig.5.29) are shown for comparison.

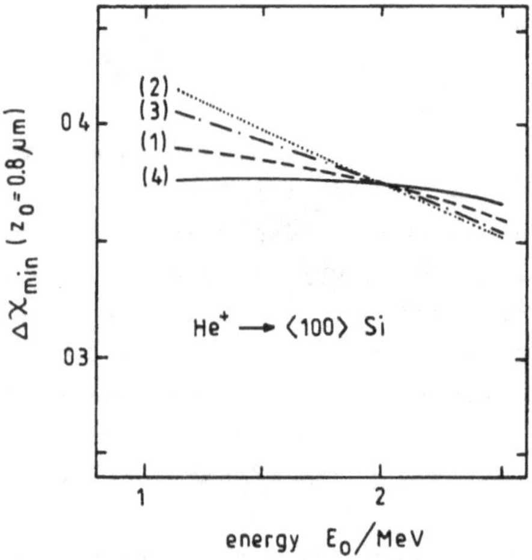

Fig.5.31. Energy dependence of $\Delta\chi_{min}$ at a fixed depth $z_0 = 0.8\,\mu m$ for He^+ incident on <100> Si. The curves (1-4) are calculated for different kinds of extended defects (described in the text) using the depth distributions represented in Fig.5.29.

Similar to the computer simulation results of Morgan and van Vliet [130] there is a weak decrease of $\Delta\chi_{min}$ with increasing energy E_o for pure intrinsic stacking faults, owing to the combined effect of the dechanneling from the perfect crystal ($\sim E_o^{-1}$) and intrinsic stacking faults (Table 5.6). Due to point defects, the decrease of $\Delta\chi_{min}$ becomes stronger for cases (2) and (3). Only the combined effect of displaced strings, point defects, and dislocations yields a nearly energy-independent $\Delta\chi_{min}$ in agreement with the experimental results.

5.2.3. Textured polycrystalline layers and layers consisting of thick twinned regions

The availability of channeling measurements in the case of layers containing crystallites or thick twinned regions of various orientations is considered in this section. The crystallites or the twinned regions considered should be sufficiently large, so that channeling within these regions (in contrast to thin twin lamellae) is possible. But within such a layer, a channeling effect will be not observable if the angular width of the orientation distribution of the single regions is larger than a critical value, where the measured backscattering yield approaches a value $\chi_{min} = 1$, characteristic of a random distribution of the crystalline regions or an amorphous substrate. So the polycrystalline or twinned layers suitable for analysis by RBS (determination of the angular spread of the orientations and proof of additional defects) should be characterized by small angular spreads and measured values of $\chi_{min} < 1$.

Both dechanneling and direct backscattering in polycrystalline layers or due to twinned regions result from crystalline regions (which additionally may contain displaced atoms) and from grain boundaries. With respect to their influence on channeled ions, the displaced atoms are assumed to be randomly distributed and the grain boundaries, because of the strong local lattice distortions, are considered to be amorphous regions. This means that the description of dechanneling and backscattering in such a layer can start from the simple equations provided by the two-beam approximation (Sect.1.3.4).

5.2.3.1. Calculation of the angular dependence of the backscattering yield $\chi^{(poly)}(\psi)$ and the minimum yield $\chi_{min}^{(poly)}$ (energy and Z_1-dependence)

5.2.3.1.1. Textured polycrystalline layers

Two assumptions have to be fulfilled to obtain a measurable diminution of the minimum yield $\chi_{min}^{(poly)}$ compared with that for amorphous material (Refs.[131, 132]):

- The size of the crystalline grains enables channeling within each crystallite (grain size > some 10 nm).
- The orientation distribution of the crystallites is characterized by a texture where the maximum angular spread with respect to a preferential orientation is in the order of the critical angle for channeling in monocrystals.

Two cases of polycrystalline layers on monocrystalline substrates will be treated here. In the first case (i) the orientation of the crystallites is well "aligned" with the substrate. In the second case (ii) the crystallite orientations are tilted to each other, described by an angular distribution $h(\varphi)$ with respect to the most preferential crystallite orientation.

(i) Assuming large grains (in the order of the layer thickness) with negligible defect densities compared to the amorphous-like grain boundaries (grain boundaries parallel to the surface normal), the minimum yield corresponding to the layer surface is given by

$$\chi_{min}^{(poly)}(0) = C_1\, \chi_{min,perf}(0) + C_2 \quad . \tag{5.8}$$

C_1 and $C_2 = 1 - C_1$ are the relative parts of the crystalline grains and the amorphous grain boundaries, respectively, and $\chi_{min,perf}$ is the minimum yield for the perfect crystal. Note that $\chi_{min,perf}(0)$ and therefore $\chi_{min}^{(poly)}(0)$ are independent of the ion energy E_o.

The minimum yield corresponding to the polycrystalline - monocrystalline interface $\chi_{min}^{(poly)}(z_o)$ consists of a contribution $\chi_{min}^{(gr)}$ due to dechanneling in the grains, a contribution $\chi_{min}^{(int)}$ due to dechanneling at the interface between the polycrystalline layer and the substrate (which is treated as a new interior surface similar to dechanneling by stacking faults; Sect.5.2.2.2) and a contribution $\chi_{min}^{(gr\ b)}$ due to the grain boundary regions.

$$\chi_{min}^{(poly)}(z_o) = \chi_{min}^{(gr)} + \chi_{min}^{(int)} + \chi_{min}^{(gr\ b)} \tag{5.9}$$

with

$$\chi_{min}^{(gr)} = C_1\, \chi_{min,perf}(z_o) \quad ,$$

$$\chi_{min}^{(int)} = C_1\, \chi_{min,perf}(0) \quad ,$$

and

$$\chi_{min}^{(gr\ b)} = C_2 \left[\chi_{min,perf}(z_o) + (1-\chi_{min,perf}(z_o))\right] \tilde{P}(\tilde{\psi}_c,\tau) \quad .$$

The contribution $\chi_{min}^{(gr\ b)}$ is described by plural scattering represented by the dechanneling function $\tilde{P}(\tilde{\psi}_c,\tau)$ (Sect.1.3.4) with $\tau = \pi a_{TF}^2 N z_o$.

(ii) In the case of tilted grains, the distribution of the crystallite orientations $h(\varphi)$ is assumed to be cylindrically symmetric and approximately Gaussian

$$h(\varphi) = (\pi\Delta^2)^{-1/2}\ e^{-\varphi^2/\Delta^2} \quad . \tag{5.10}$$

The quantity Δ is the standard deviation of this distribution. The angular dependence of the backscattering yield ("channeling angular dip") $\chi(\Phi)$ from a crystallite is also assumed to be a cylindrically symmetric Gaussian like distribution (Ref.[132])

$$\chi(\Phi) = 1 - (1 - \chi_{min} - s\ \frac{\Phi^2}{\psi_s^2})\ e^{-\Phi^2/\psi_s^2} \quad , \tag{5.11}$$

where Φ designates the angle between the ion beam and the axis of the crystallite, χ_{min} is the minimum yield $\chi(\Phi = 0)$ (for large crystallites > some 100 nm a contribution of the grain boundary and defects can be neglected and χ_{min} has to be subseded by $\chi_{min,perf}$), and ψ_s is the standard deviation of $\chi(\Phi)$ which is proportional to the critical angle for channeling (i.e. ~ ψ_1). The parameter s represents the magnitude of the "shoulder" of the dip.

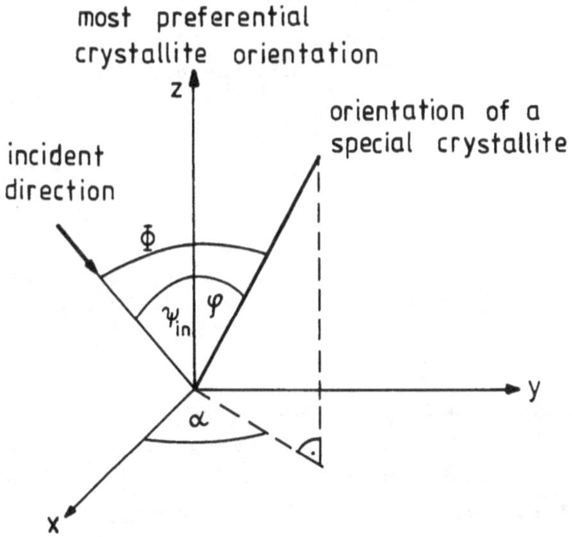

Fig.5.32. Geometric relation between the ion beam direction and the orientation of a crystallite (beam direction in the x-y plane)

The yield from a polycrystalline layer $\chi^{(poly)}(\psi)$ for an ion beam incident at an angle ψ against the most preferential crystallite orientation (z-axis, Fig.5.32) then is given by

$$\chi^{(poly)}(\psi) = \int_0^{2\pi} \int_0^{\pi} \chi(\Phi)\, h(\varphi) \sin\varphi \, d\varphi \, d\alpha \quad , \tag{5.12}$$

where α is the azimuthal angle, and the angle Φ between the incident direction and the orientation of a special crystallite can be expressed as

$$\cos\Phi = \cos\varphi \cos\psi + \sin\varphi \sin\psi \cos\alpha \quad . \tag{5.13}$$

Under the assumption Δ, $\psi_s << 1$ Eq.(5.12) can be analytically integrated. As with medium energies in an azimuthally averaged angular dip $\chi(\Phi)$, no remarkable shoulder will exist s = 0 can be assumed as a good approximation in Eq.(5.11). In this case the integration provides (Ref.[133])

$$\chi^{(poly)}(\psi) = 1 - \frac{1 - \chi_{min}}{1 + \Delta^2/\psi_s^2}\, e^{-(\psi^2/(\psi_s{}^2 + \Delta^2))} \tag{5.14}$$

By means of Eq.(5.14), the influence of the angular spread Δ of the crystallites on the minimum yield can be discussed. For $\Delta = 0$, the yield of monocrystalline material χ_{min} is obtained, and for $\Delta > \psi_s$ the yield $\chi_{min}^{(poly)}$ approaches the value 1. That means that a polycrystalline layer with nearly randomly distributed crystallites (Δ large) cannot be distinguished from an amorphous layer by means of RBS.

From Eq.(5.14), the following expressions for the dip depth $1 - \chi_{min}^{(poly)}$, the half width of the dip $\psi_{1/2}^{(poly)}$, and the dip volume V, are derived

$$1 - \chi_{min}^{(poly)} = \frac{1 - \chi_{min}}{1 + \Delta^2/\psi_s^2} \quad , \tag{5.15}$$

$$\left[\psi_{1/2}^{(poly)}\right]^2 = \ln 2 \; (\psi_s^2 + \Delta^2) \quad , \quad \text{and} \tag{5.16}$$

$$V = \pi \; (1 - \chi_{min}) \; \psi_s^2 \quad . \tag{5.17}$$

It can be concluded from Eqs.(5.15) and (5.16) that for polycrystals both the depth and the width of the dip vary with the spread Δ of the orientations. By means of Eq.(5.16) the value of Δ can be determined experimentally if the dependence of $\psi_{1/2}^{(poly)}$ on the energy E_o is measured. Because of $\psi_s^2 \simeq \psi_1^2$ and $\psi_1^2 \sim Z_1/E_o$ an extrapolation of the measured dependence $\left[\psi_{1/2}^{(poly)}\right]^2$ on Z_1/E_o for $Z_1/E_o \to 0$ yields $\left[\psi_{1/2}^{(poly)}\right]^2 = \ln 2\, \Delta^2$.

Eq.(5.17) shows that V is independent of Δ. This equation provides a method to estimate χ_{min} experimentally from the dependence $V(Z_1/E_o)$. As χ_{min} is determined only by the scattering centers and not by the orientation spread, it appears to be a measure of the defect density in the polycrystalline layer (mainly represented by the grain boundaries). A relationship between $\chi_{min}^{(poly)} = \chi_{min}$ (i.e. without contributions due to tilted crystallites) and the volume fraction C_2 of the grain boundaries is given by Eqs.(5.8) and (5.9).

Another simplified method to obtain a measure of Δ and χ_{min} can be derived from Eq.(5.14) for $\Delta < \psi_s$, considering the minimum yield from the layer surface (Ref.[125]). For $\Delta < \psi_s$, Eq.(5.14) can be approximated by

$$\chi_{min}^{(poly)}(0) = \chi_{min}(0) + [1 - \chi_{min}(0)] \frac{4\pi\varepsilon_o d\, \Delta^2}{2Z_1Z_2e^2} E_o \quad . \tag{5.18}$$

Eq.(5.18) shows a linear energy dependence of the minimum yield on the surface of a polycrystalline layer for <u>tilted</u> crystallites ($\Delta > 0$) in contrast to an energy-independent $\chi_{min}^{(poly)}(0)$ for crystallites of <u>uniform orientation</u> ($\Delta = 0$; see also Eq.(5.8)). Using Eq.(5.18), the measurement of $\chi_{min}^{(poly)}(z=0,E_o)$ provides an estimate of the angular spread of the orientations (up to the order of ψ_s) and of χ_{min} (which contains information about defects).

5.2.3.1.2. Thick twinned regions

In a twinned layer, two different axes are alternating (<111> and <115> in Si; Sect.1.3.3.2). Again we distinguish between the two cases of (i) untilted and (ii) tilted twinned regions.

(i) If the layer is perfectly twinned, all twinned regions have a uniform orientation with their <115> axes parallel to the <111> axes of the untwinned regions. The transition of channeled ions into a twinned region causes substantial dechanneling because of the strong change of the atomic spacing d .

If the twinned regions are very <u>thin</u> twin lamellae, dechanneling can be described in a way similar to stacking faults, considering the contribution of the atoms within the twins to be a contribution of uncorrelated randomly distributed displaced atoms (no channeling effect within the lamellae; Sect.1.3.3.2). A description of the dechanneling due to transitions from untwinned into <u>thick</u> twinned regions (permitting channeling) is more difficult, especially within the framework of the simple two-beam approximation. Thus, for simplification, equipartitioned thick twinned and untwinned regions are presumed to be arranged mainly "side by side". If the extension of these regions is small compared with the ion beam spot, the minimum yield for <111> incidence is given by

$$\chi_{min}^{(tw)}(z) = C_1' \chi_{min,perf}^{<111>}(z) + C_2' \chi_{min,perf}^{<115>}(z) \quad . \tag{5.19}$$

C_1' and $C_2' = 1 - C_1'$ are the relative volume fractions of <111> oriented and <115> oriented (twinned) regions. Additional contributions of displaced atoms are again taken into account in terms of a dechanneling function resulting from the sum of contributions from both regions (Ref.[134])

$$\tilde{P}^{(tw)}(\tilde{\psi}_c^{(tw)},\tau) = C_1' \tilde{P}(\tilde{\psi}_c^{<111>},\tau) + C_2' \tilde{P}(\tilde{\psi}_c^{<115>},\tau) \quad . \tag{5.20}$$

Because of $\tilde{\psi}_c^{<115>} < \tilde{\psi}_c^{<111>}$, the dechanneling probability in a twinned layer is higher than in untwinned material. The half width of the dip $\psi_{1/2}^{(tw)}$ measured for a twinned layer should be between those of the <111> axis and the <115> axis, $\psi_{1/2}^{<111>}$ and $\psi_{1/2}^{<115>}$, respectively. An enhanced dechanneling and a decreased critical angle in combination with TEM results can be used to characterize twinned regions.

(ii) For describing tilted twins with sufficiently small Δ, Eq.(5.14) can be extended (<111> incidence), introducing again the standard deviations $\psi_s^{<111>} = \psi_{1/2}^{<111>}/(\ln 2)^{1/2}$ and $\psi_s^{<115>} = \psi_{1/2}^{<115>}/(\ln 2)^{1/2}$

$$\begin{aligned} \chi^{(tw)}(\psi) = {} & C_1'\left[1-(1-\chi_{min,perf}^{<111>})\right] e^{-\psi^2/(\psi_s^{<111>})^2} \\ & + C_2'\left[1 - \frac{1-\chi_{min,perf}^{<115>}}{1+\Delta^2/(\psi_s^{<115>})^2}\right] e^{-\psi^2/[(\psi_s^{<115>})^2+\Delta^2]} \quad . \end{aligned} \tag{5.21}$$

For $\psi = 0$ and $z = 0$, Eq.(5.21) provides

$$\chi_{min}^{(tw)}(0) = \chi_{min,perf}^{<111>}(0) + C_2'\left[1-\chi_{min,perf}^{<111>}(0) - \frac{1-\chi_{min,perf}^{<115>}(0)}{1+\Delta^2/(\psi_s^{<115>})^2}\right] \quad . \tag{5.22}$$

The standard deviation Δ of the orientation spread of the twinned regions, can be determined as in the case of tilted crystallites (Sect.5.2.3.1.1) by the measurement of $\psi_{1/2}^{(tw)}$ as a function of Z_1/E_0 (Eq.(5.16)). Using this value of Δ, the volume fraction C_2' of twinned regions can be determined by means of Eq.(5.22).

5.2.3.2. Determination of the angular spread of the crystallite and twin orientations

The following examples are chosen for a dechanneling analysis.

(a) Textured polycrystalline layers were prepared by high-vacuum evaporation of Pd (50...100 nm) onto <111> Si wafers. The evaporated samples were heat treated for 1...2 h in an argon atmosphere at 670 K $\leq T_a \leq$ 920 K.

(b) A layer containing twinned regions was prepared by implantation of 150 keV Si^+ into <111> Si at room temperature (dose D = $1*10^{16}$ cm^{-2}). Then the sample was annealed at T_a = 920 K for 30 min in a nitrogen atmosphere.

In the first case, it has been proved by means of RBS that a uniform compound layer of Pd_2Si was formed (Ref.[132]). In the second case, a large fraction of twinned regions with an average lateral extension of ≈ 0.3 µm has been detected by means of TEM and electron diffraction. The twins are not twisted azimuthally and their average thickness is < 0.1 µm (Ref.[125]).

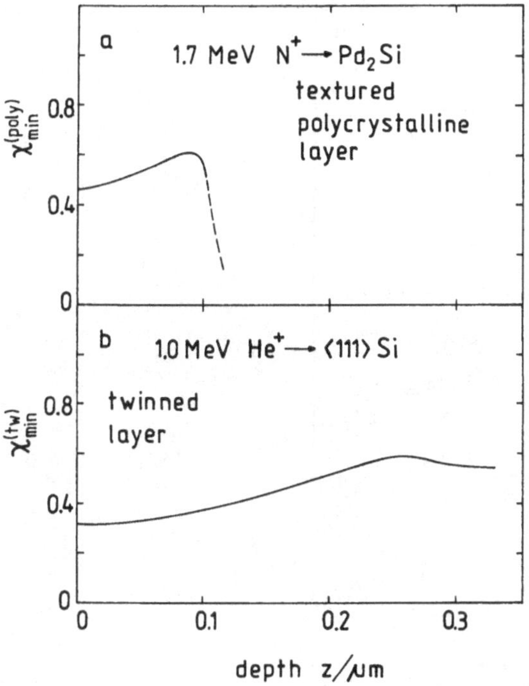

Fig.5.33. Depth dependences of the minimum yield (a) for 1.7 MeV N^+ incident on a textured polycrystalline Pd_2Si layer and (b) for 1.0 MeV He^+ incident on a twinned <111> silicon layer. The conditions of the preparation of the layers are described in the text (examples a,b).

In Fig.5.33, the depth dependences of the measured minimum yield for a polycrystalline Pd_2Si layer (a) and a twinned silicon layer (b) are shown. A minimum yield $\chi_{min}^{(poly)} < 1$ in Fig.5.33a indicates the presence of a textured polycrystal suitable for channeling analysis. In contrast to $\chi_{min}^{(poly)}$, which is largely independent of depth, $\chi_{min}^{(tw)}$ increases remarkably with depth (Fig.5.33b). This increase can be attributed to the fact that the thickness of the twinned regions (< 0.1 μm) is less than the thickness of the whole twinned layer (≈ 0.25 μm), so that dechanneling due to <111> - <115> transitions cannot be ignored. Consequently, for an analysis of the twinned layer without having to consider these transitions, only the backscattering yield from a near-surface layer is taken into account.

To characterize the spread of crystallite or twin orientations, the standard deviation Δ of the orientation distribution h(φ) (assumed to be

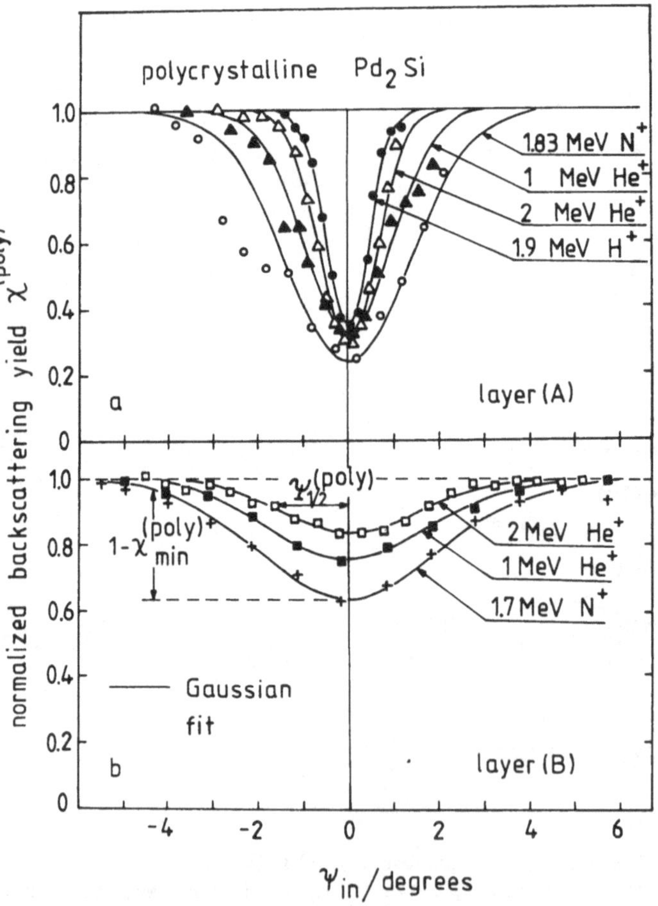

Fig.5.34. Angular dips for two textured Pd_2Si layers (A,B) corresponding to different ratios Z_1/E_o of the incident ions (Ref.[132]) fitted by Gaussian curves

Gaussian) has to be determined. As described in Sect.5.2.3.1, Δ can be derived from the measurement of the angular dip $\chi^{(poly)}(\psi)$ and $\chi^{(tw)}(\psi)$, respectively, for different energies E_o (or atomic numbers Z_1) of the incident ions.

In Fig.5.34, such dips are represented for two different textured polycrystalline Pd_2Si layers (A,B). Each dip corresponds to a special ratio of Z_1/E_o.
The solid lines show best-fit Gaussian curves. Considering a fixed value of Z_1 in Fig.5.34b, an increase of $\chi_{min}^{(poly)}$ as well as a decrease of the width of the dip with increasing E_o is observed, whereas the dip depth (i.e. $\chi_{min}^{(poly)}$) is almost constant in Fig.5.34a. If this behaviour is compared with the E_o-dependence of the minimum yield $\chi_{min}^{(poly)}(E_o)$ in Eq.(5.18), one can conclude that layer (B) contains tilted crystallites, and that the crystallites in layer (A) are almost untilted.

For the twinned layer, the E_o-dependence of the minimum yield from the surface $\chi_{min}^{(tw)}$ $(z = 0, E_o)$ is measured (Fig.5.35). The linear increase of $\chi_{min}^{(tw)}$ with increasing E_o can again be explained by slightly tilted twins.

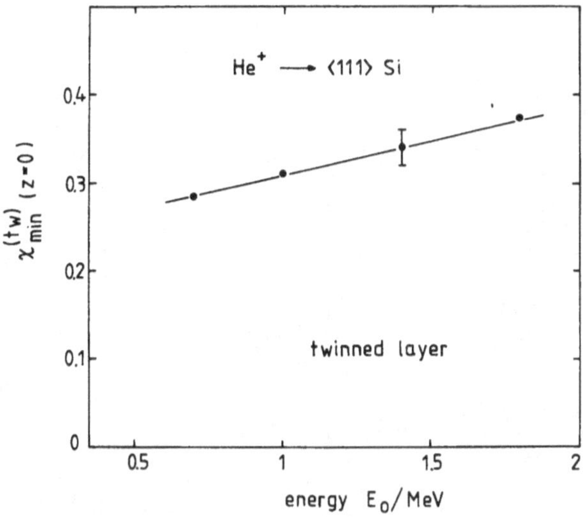

Fig.5.35. Dependence of the minimum yield of He^+ backscattered from the surface of a twinned silicon layer (example b) on the incident energy (Ref.[125])

The dependence of $\left[\psi_{1/2}^{(poly)}\right]^2$ and $\left[\psi_{1/2}^{(tw)}\right]^2$ on Z_1/E_o for the two Pd_2Si layers and the twinned silicon layer are shown in Fig.5.36a,b. The intercepts with the vertical axis yield the values $\ln 2\ \Delta^2$ and the gradient of each line represents the relation between $\psi_{1/2}^2 = \ln 2\ \psi_s^2$ and

Z_1/E_o. In Fig.5.36b the lines marked by <111> and <115>, respectively, represent the theoretical dependences of $\left[\psi_{1/2}^{<111>}\right]^2$ and $\left[\psi_{1/2}^{<115>}\right]^2$ on Z_1/E_o.

For the polycrystalline Pd_2Si layers (A) and (B), a value $\psi_s = 0.96\ (Z_1/E_o)^{1/2}$ is obtained (Ref.[132]) which is between the two extreme cases corresponding to Si and Pd rows (numerical factors 0.75 and 1.35, respectively). The standard deviations derived from Fig.5.36a are $\Delta = 0^o$ for layer (A) and $\Delta = 2^o$ for layer (B). The half width of the dip obtained for the twinned layer, due to a superposition of contributions from dechanneling in <115> (twinned regions) and in <111> (untwinned regions) is between $\psi_{1/2}^{<111>}$ and $\psi_{1/2}^{<115>}$. The standard

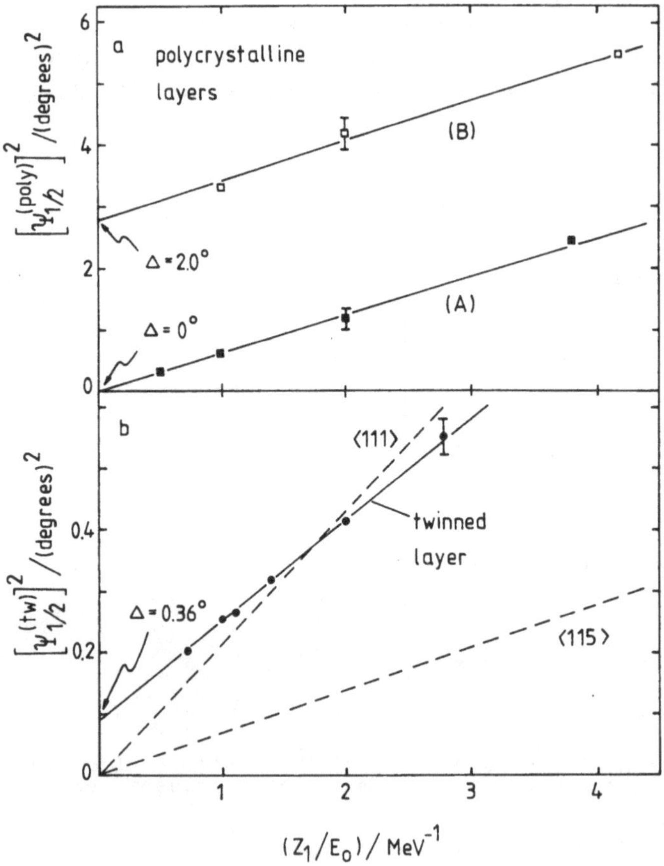

Fig.5.36. Dependences of the half widths of the measured angular dips on the ratio Z_1/E_o of the incident ions (Refs.[132,125]) (a) for the polycrystalline layers (A,B) and (b) for the twinned layer. The dashed lines in (b) represent the theoretical dependences for channeling in the <111> and in the <115> axis of silicon, respectively. The standard deviations Δ of the angular spread are derived from the intercepts of the measured linear dependences with the vertical axis.

deviation derived from Fig.5.36b is $\Delta = 0.36^{o}$. It is impossible to confirm these results for Δ by means of TEM, because a detection of orientation spreads by this method requires tilt angles > 5°.

5.2.3.3. Determination of the volume fractions of grain boundaries in polycrystals and of twins in twinned layers

The densities of scattering centers for the polycrystalline Pd_2Si layers have been determined (Ref.[132]) using the method based on Eq.(5.17) and described in Sect.5.2.3.1. For the different layers, a unique value $\chi_{min} \approx 0.2...0.3$ is obtained independent of Δ within the experimental error. Assuming χ_{min} to be caused mainly by amorphous grain boundaries, the volume fraction of the grain boundaries can be evaluated from χ_{min}. The results suggest that the average value of the grain size in these layers is independent of the orientation spread of the grains.

In the case of the twinned silicon layer, Eq.(5.22) has been applied to determine the volume fraction C_2' of twinned regions, using the value $\Delta = 0.36^{o}$ and the data $\chi_{min}^{(tw)}(0)$ from Fig.5.35. For all energies $0.7\ MeV \leq E_o \leq 1.8\ MeV$ within the experimental error, a unique value $C_2' \approx 0.6$ is determined. This is in agreement with the TEM results, which show that more than half of the layer is occupied by twinned regions. The agreement of the C_2'-values for different E_o also confirms that the dechanneling due to tilted crystallites at the layer surface is well characterized by the simple model used here.

5.2.3.4. Conclusions

(i) For point defects, clusters, amorphous regions or completely amorphized layers, which can be characterized as uncorrelated randomly distributed displaced lattice atoms, the application of both the discontinuous model and the two-beam approximation yields realistic depth distributions of defect concentration $n_{pd}(z)$, in monoelemental as well as in compound monocrystals. The results obtained using the much simpler procedure of two-beam approximation for these defect configurations do not differ from the results obtained using the more complicated (and generally more realistic) discontinuous model.

Information on the possibility of characterizing the displaced atoms as uncorrelated and "randomly distributed" can be derived from the shape of the measured backscattering spectra $\chi_{min}(z)$ (pronounced peak). For special cases (fixed defect concentrations), the assumptions about defects resulting from the shape of the spectrum can be confirmed by means of the dependence of the backscattering yield on the incidence energy $\chi_{min}(E_o, z_o)$. But it has to be taken into consideration that this energy dependence for uncorrelated displaced atoms proves to be <u>both</u> a function of their lattice position <u>and</u> of their concentration: $\Delta\chi_{min}(z_o) \sim E_o^{\kappa}$ with $-1/2 \geq \kappa(r_a, n_{pd}) \geq -1$. A better method to get information on r_a is to measure the temperature dependence of the minimum yield.

(ii) Dislocations and stacking faults in monocrystalline solids can be detected by RBS, based on an analysis of defect-induced dechanneling, assuming the defect densities to be larger than the minimum densities $N_{disl}^{(min)}$ and $N_{dst}^{(min)}$ for detection. Taking silicon layers of $\leq 0.5\,\mu m$ thickness as typical examples, these detection limits have been evaluated for axial dechanneling at <u>dislocations</u> to be $N_{disl}^{(min)} \approx$ some $10^4\ cm^{-1} \ldots 10^5\ cm^{-1}$ (or $n_{disl}^{(min)} \approx$ some $10^{10}\ cm^{-2} \ldots 10^9\ cm^{-2}$) depending on their depth distribution. The limit for detection of <u>stacking faults</u> is $N_{dst}^{(min)} \approx 5 * 10^{15}\ cm^{-2}$ (or $n_{dst}^{(min)} \approx 10^{20}\ cm^{-3}$).

These detection limits can still be reduced by a factor of $\approx 3 \ldots 5$ using planar measurements. In the case of stacking defects, it is preferable to measure in the <100> axis or the (100) plane, because stacking faults on all four sets of (111) planes then contribute to dechanneling.

To obtain clear evidence for stacking faults and dislocation loops, the extension (or diameter) of these defects should be larger than some 10 nm.

The backscattering spectra measured in the case of dislocations and stacking faults are characterized by the total absence of a peak due to direct backscattering.

The dependence of the measured defect-induced enhancement of the minimum yield on the incident energy $\Delta\chi_{min}(E_o)$ (or $\gamma(E_o)$) can be used in addition to TEM analysis to identify the kinds of defects in question. In the case of <u>dislocations</u>, an $E_o^{1/2}$-dependence is obtained for $E_o > 1$ MeV, assuming, the strain fields of individual dislocations do not overlap. Using axial channeling measurements, this is valid for $n_{disl} < 10^{11}\ cm^{-2}$.

Dechanneling by stacking defects is mainly caused by displaced strings. The dechanneling contribution due to displaced strings is independent of E_o. Dislocations also contribute to the measured energy dependence of <u>stacking faults</u>, in addition to the displaced strings. In the case of <u>twin lamellae</u>, randomly distributed point defects yield an additional contribution. In this way, information on the <u>ratio</u> of different kinds of defects can be obtained from $\Delta\chi_{min}(E_o, z)$, requiring high accuracy of the measured $\chi_{min}(z)$ (about 1%).

The calculation of the dislocation density or the density of displaced strings can be carried out by means of two different models, starting from the derivative of the measured yield, $d\chi_{min}(z)/dz$.

A very fast and simple procedure is provided by the <u>two-beam approximation</u>, assuming dechanneling by stacking faults or dislocations

to be due to single scattering processes (well fulfilled in the MeV-region) and using dechanneling cross sections.
Satisfactory results of quantitative analysis are obtained for dislocations by means of an effective dechanneling diameter λ when checking axial channeling. In some cases, a dependence of λ on the dislocation density and on the shape of the depth distribution of the dislocations has to be taken into account to achieve high accuracy.

For stacking faults corresponding to the model of new interior surfaces, a dechanneling cross section σ_{dst} is introduced which neglects both the influence of the change of the transverse energy distribution with depth and the influence of the absolute value for the displacement of the strings on dechanneling. So, especially for larger depths ($z \geq 0.4...0.5\,\mu m$) the calculated densities of displaced strings are too large, and inconsistent results are obtained for different incident directions.

By the more realistic (but also more complicated) description using the discontinuous model, quantitatively excellent results are obtained, both for the integral density and for the depth distribution. In contrast to the two-beam approximation, this model can be applied even if the layer under investigation contains a mixture of different defects.

For stacking faults, the results of planar dechanneling measurements agree with those of the axial case; also those of the two-beam approximation, if there is no substantial influence of different displacements of strings or planes in both cases.

Broadening of narrow distributions of dislocations (<< depth resolution of RBS) as well as "degradation" and deformation of distributions at larger densities (n_{disl} > some 10^{10} cm^{-2}) are measured for dislocations in the planar case because of the much larger width of the strain field.

(iii) Because of the sensitive dependence of the backscattering yield on the angle between an ion beam and a crystal axis, RBS proves to be a very fruitful method for the detection of small orientation spreads of crystallites in polycrystalline layers or twinned regions. Assuming that the average size of the crystallites or the thickness of the twins is greater than some 10 nm, tilts of such regions up to $\approx 3^{\circ}$ can be measured by this method, whereas tilt angles $> 5^{\circ}$ are necessary for detection by TEM.

The angular spread of the orientations of the grains or twins and the volume fractions of grain boundaries or twins can be estimated independently by means of energy-dependent (or additionally Z_1-dependent) measurements of angular channeling dips.

5.3. LATTICE LOCATION OF POINT DEFECTS

In the preceding sections, the possibilities for the identification of different defect types by means of the RBS analysis were discussed and procedures for the determination of the depth distributions of the defect densities for the different kinds of defects were described.

Now defects with preferred positions within the latice cell will be considered. For the determination of the depth distribution of such defects, additional experimental information is necessary which cannot be provided by the dependence of the difference of the minimum yield $\Delta\chi_{min}$ on the ion energy E_o.

Therefore, in contrast to the preceding considerations, the use of RBS channeling techniques for the location of point defects of small concentrations in crystals will be described. With the information obtained from such experiments, the depth distribution of the point defect concentration can be determined according to the procedure described in Sect.5.2.1.

Starting with some fundamental problems in the location of point defects, the procedures of locating impurity atoms by measuring the angular dependence of the backscattering yield for certain crystallographic axes and planes will be described. Furthermore, the application of this method for the location of self-interstitials is discussed and, finally, the use of temperature dependent dechannelling measurements for the determination of the distribution of slightly displaced lattice atoms is described.

However, the location of point defects, as well as the determination of their concentration, can be an experimentally difficult problem and also requires a careful theoretical analysis of the interaction processes of the ions with the material to be investigated (see Sect.1.3).

5.3.1. Basic considerations

Point defects in crystals may be impurity atoms on substitutional or interstitial sites and host lattice atoms on interstitial sites (self interstitials).

One method for determining the positions of such displaced atoms uses interaction processes with small impact parameters (see Sect.1.3). In such atom location experiments, the yield of collisions with small impact parameters is measured as a function of the angle of incidence of the analyzing ions with respect to a low index direction of the crystal.

The interaction yield may be

- backscattered ions,
- nuclear reaction products, or
- ion-induced X-rays.

For high-energy light ions incident on a crystal along a low index direction, displaced atoms act as scattering centers. The measured interaction yield depends on the position of the scattering center within the channel investigated. This is because the flux of ions is not uniformly distributed across the channels (flux peaking effect). In Sect.1.3.2.5 it was shown that where the incident ions interact with the atomic potential along a low index axial or planar direction, there is an enhanced flux density at the channel center and a reduced flux density near the atomic rows, if the ion beam is aligned with the corresponding crystal axis. Furthermore, the flux strongly depends on the angle ψ_{in} between the incident ion beam and the crystal axis to be considered (Fig.1.34), and on the position r_a of scattering centers within the channel (Fig.1.35). For axial channeling, the angular distribution of the flux density is most sensitive to changes of the position of scattering centers lying near the middle of the channel. On the other hand, in planar channels a variation of the position of the scattering centers within the channel is only connected to a shift of the pronounced peak in the angular distribution of the flux, corresponding to the variation of the position (Fig.1.38). Therefore, axial channeling experiments are more sensitive to locating scattering centers near the middle of the channel, whereas with planar channeling measurements, scattering centers lying between the middle of the channel and the atomic strings can be located. However, it should be noted again that the flux peaking effect is very sensitive to fluctuations of the angle of incidence (Figs.1.36 and 1.39) and is remarkably suppressed with increasing depth. This behaviour determines the experimental conditions for direct location analysis with respect to the beam collimation and recording depth of the interaction yield.

A further possibility for the location of point defects in weakly damaged crystals (no pronounced direct backscattering part of the minimum yield) is to analyse the dechanneling of energetic light ions. The basis for this method is the fact that the temperature dependence of the minimum yield for a crystal with point defects is characteristic of the distance r_a of these defects from the string (see Sect.1.3.4 and 1.3.4.2). Therefore it is possible to determine the defect concentration $n_{pd}(z)$ as well as the position r_a from the depth dependence of the minimum yield measured for different temperatures.

5.3.2. Impurity atoms

5.3.2.1. Detection of substitutional impurities

One important aspect of channeling experiments is the determination of the percentage of impurity atoms incorporated on substitutional sites. For this problem, consider the simple case of a crystal layer containing impurity atoms located partly on substitutional sites and partly on interstitial positions within the crystal lattice. In Fig.5.37a, this is illustrated schematically for a simplified two-dimensional lattice with substitutional impurities (solid circles) and interstitial impurities (open squares) at various positions. Information about the substitutionally incorporated part of impurities can be obtained if the

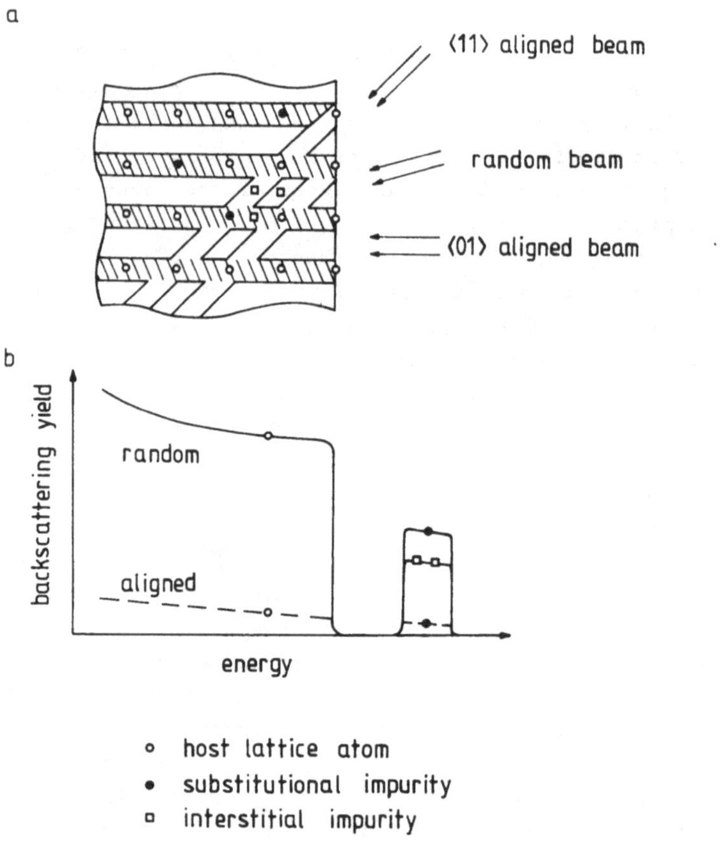

Fig.5.37. Schematic illustration of a simplified two-dimensional lattice with substitutional (•) and interstitial (▫) impurities (a) and backscattering spectra for random and aligned beams (b) of high energy light ions

yields of interactions with the impurity atoms for random and aligned incidence along a certain crystal direction are compared. In this simple procedure, the requirements with respect to beam collimation are small. Because the flux peaking effect in such experiments should be suppressed, the interaction yield must be recorded for a larger depth interval (commonly a depth region of ≈ 100 nm is sufficient to eliminate the influence of the flux peaking). Therefore, the procedure allows the determination of the depth distribution of the impurities displaced from the strings in a given crystal direction over the entire doped layer.

Generally, impurity location with this procedure can be done by measuring the interaction yields from nuclear reactions, PIXE, or directly backscattered light ions. However, the use of the RBS technique is only possible for impurities with masses sufficiently high compared to the mass of the host lattice atoms and provided that the thickness of the layer containing the impurities is sufficiently small. Under these conditions, an energy separation of the backscattering yields from lattice atoms and impurities is possible.

In the following, the application of the RBS technique for the location of impurities heavier than the lattice atoms incorporated in a thin surface layer of a crystalline target will be discussed in more detail. For random incidence of the ion beam, the crystalline structure of the material is not detected by the ions; the measured spectrum of backscattered ions corresponds to that of an amorphous target. The energy of the ions backscattered from the impurity atoms is higher than of those backscattered from the host lattice atoms. Therefore, a distinct peak above the high energy edge of the energy spectrum of the host lattice atoms is observed which is connected with the impurities (Fig.5.37b). When the ion beam enters the crystal along a low-index crystal axis or plane, between 95 and 98% of the incident ions are steered by the repulsive potentials of the atomic rows (axial channeling) or atomic planes (planar channeling). The channeled particles do not approach the axial rows more closely than about 0.01-0.02 nm, a distance three orders of magnitude greater than the impact parameter for wide-angle elastic scattering. As a consequence, the crystal atoms on lattice sites below the surface are "invisible" to the channeled particles. Thus, changing the beam-to-crystal orientation from random to aligned direction leads to a reduction of processes with small impact parameters by about two orders of magnitude (dashed line in Fig.5.37b).

When all the impurity atoms are located on substitutional sites (full circles in Fig.5.37) they are also shielded from direct interactions with the channeled part of the ion beam. Therefore, by tilting the crystal from random to aligned directions (see Fig.5.37b, right hand part), the backscattering yield from substitutional impurity atoms decreases in the same way as that from the host lattice atoms.

In practical cases, not all the impurity atoms will be located on well-defined substitutional lattice sites. For impurity atoms displaced from lattice sites, an interaction of the aligned ion beam with these atoms lying in the channels will take place. Thus, the decrease of the backscattering yield will be reduced as the orientation of the crystal with respect to the beam is changed. This behaviour is illustrated in Fig.5.37b (open squares) for the case of interstitial impurities on sites along the <01> channel and the analyzing ion beam incident along this direction. However, interstitial impurities positioned within the <01> atomic rows are shadowed by the host lattice atoms. They can be detected by ions incident parallel to the <11> direction.

From these considerations it is obvious that a determination of the percentage of substitutional impurities for a given crystallographic direction is possible by comparing the heights of the backscattering yield due to the impurities for random and aligned incidence of the ion beam. A quantitative analysis is possible in the framework of the two-beam approximation described in Sect.1.3.4. According to that the minimum yield connected with the impurities is given by

$$\chi_{min,imp}(z) = [1 - \chi_r(z)]\,(1 - f) + \chi_r(z) \qquad (5.23)$$

(see Eq.(1.256)). Here f is the percentage of impurities located within a distance of a_{TF} (Thomas-Fermi screening length) from the string, $\chi_r(z)$ given by Eq.(1.254) is the random beam interacting with all the impurities within the layer, and $[1 - \chi_r(z)]$ is the aligned beam which interacts with those impurities located at interstitial positions within the channels investigated. From that the percentage f of impurities at the depth z can be calculated by

$$f = \frac{1 - \chi_{min,imp}}{1 - \chi_r} \quad . \qquad (5.24)$$

The percentage of substitutional impurities f_s (located within a distance a_{TF} from a lattice site) can be estimated in the following way. Because of the lattice symmetry, the percentage $f-f_s$ of impurities positioned within a distance a_{TF} from the string but outside a distance a_{TF} from a lattice site must be equal for different equivalent crystal axes. For the number of different equivalent crystal axes n_{ea} the percentage $(n_{ea}-1)(f-f_s)$ of impurities contributes to the amount 1-f of impurities positioned beyond a distance a_{TF} from the string considered. Therefore $1-f > (n_{ea}-1)(f-f_s)$ must be true. On the other hand, $f_s < f$ is valid. Thus, we estimate the value of f_s by

$$(n_{ea}f-1)/(n_{ea}-1) < f_s < f \quad . \qquad (5.25)$$

The uncertainty of f_s decreases with increasing values of f and n_{ea}. For the <100> direction of a cubic crystal (n_{ea}= 3) we get $(3f-1)/2 < f_s < f$. However, the percentage of substitutional impurities f_s can be determined more precisely by measuring the value of f for different crystal axes.

This fast method is of particular interest for the investigation of substitutional incorporation of dopants in ion-implanted semiconductor layers caused by subsequent annealing. As an example, Fig.5.38 shows the backscattering yield of Te implanted into <100> GaAs (as implanted and annealed with different energy densities of a Q-switched Nd-glass laser, Ref.[135]). The measurements were carried out for the <100> direction. Using Eq.(5.24) and assuming $\chi_r \approx \chi_{min,perf}$ which is valid for near-surface regions, one finds 80% to 94% of the impurities located within a distance a_{TF} from the <100> string, depending on the laser energy density. Applying Eq.(5.25) to this case, it can be seen that for f=80%, between 70% and 80% of the impurities in most of the distribution are located on substitutional sites, for f=94% the concentration of substitutional impurities is between 91% and 94%. Furthermore, using the

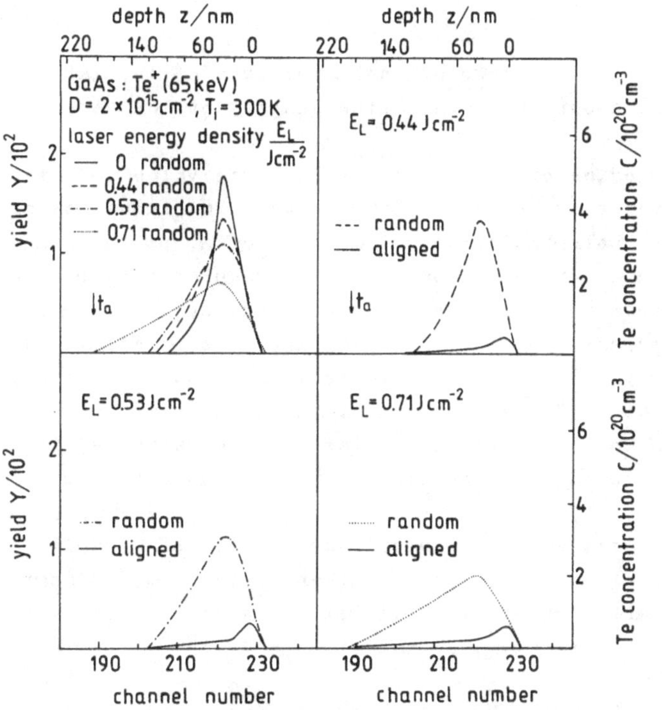

Fig.5.38. Depth distribution of Te along the <100> direction in GaAs annealed with laser pulses of different laser energy densities E_L from a Q-switched Nd-glass laser (wavelength λ = 1.06 μm, pulse duration $\tau \approx 40$ ns; from Ref.[135])

energy-to-depth conversion (described in Sect.4.2.1) and the measured $\chi_r(z)$, information about the depth distribution of these impurities can be achieved.

The preceding description starts with the assumption that all impurities are on substitutional sites or are located well away from substitutional sites at random positions. In some cases, especially for small point defect concentrations, the impurity atoms are only slightly displaced (in the order of the Thomas-Fermi screening radius a_{TF} = 0.01 to 0.02 nm) from substitutional lattice sites, or they are located on well-defined interstitial positions. In these cases, direct location by measuring the angular dependence of the interaction yield across different axial and/or planar channels is necessary.

5.3.2.2. Direct location by angular scan measurements

In direct location experiments, beyond the minimum yield χ_{min} the angular half-width $\psi_{1/2}$ of the measured angular scan, which is related to the characteristic angle ψ_1 by

$$\psi_{1/2} = c\psi_1 \qquad (5.26)$$

(c = 0.8...1.2, depending on the vibrational amplitude of the lattice atoms), gives information about the crystalline quality of the material investigated.

For the near-surface region of perfect crystals, the value of the minimum yield is in the range of 1 to $3*10^{-2}$ for low-index axial directions and the axial half-angle has values between about 0.4 and 1.2°. However, in Sect.1.3.3.1 it was shown that the angular dips become less pronounced with increasing depth due to dechanneling, i.e., for depths $z > 0$ χ_{min} increases and $\psi_{1/2}$ decreases (see Fig.1.43). Furthermore, the angular yield distributions are influenced by surface contaminations. From that and the fact that the flux distribution of ions in a channel depends on the depth and on the beam divergence, the following demands for the conditions in direct location experiments can be derived:

- The beam collimation angle should be about one order of magnitude lower than the critical angle ψ_1 for the given axial (or planar) direction of the crystal. Commonly, collimation angles $\leq 0.1°$ are required for axial channeling and angles $\leq 0.02°$ for planar channeling measurements.
- The interaction yield should be recorded in near surface regions. Otherwise, dechanneling and the depth dependence of the flux distribution must be taken into account in the analysis of the spectra.

- Surface contaminations should be prevented, to reduce scattering of the analyzing ions increasing the beam divergence (for example, a 2 nm thick oxide layer introduces an angular spread of 1 MeV He ions of 0.02° by scattering (Ref.[136])). Therefore, the experiments should be carried out in a vacuum chamber with a pressure lower than 10^{-5} Pa.

The practical procedure of location measurements consists of recording the yield of interactions with small impact parameters, corresponding to the host lattice and impurity atoms, respectively, versus the tilt angle for different crystallographic directions. The comparison of the angular scans for impurities and lattice atoms gives some first rough information about the incorporation of the impurities. This will be demonstrated for the case of two Si crystals doped with

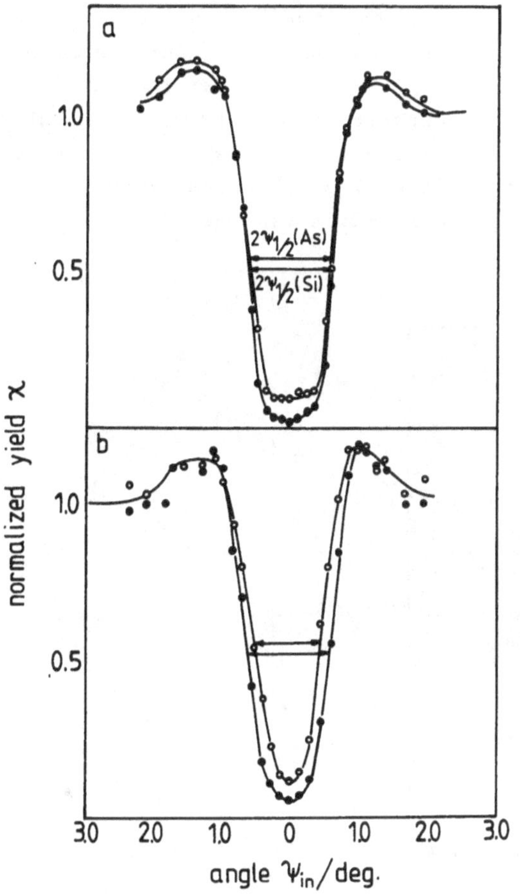

Fig.5.39. Angular yield profiles for tilts through the <110> axis for 1.8 MeV He ions incident on two crystals of Si (•) arsenic doped (o) with concentration of $6*10^{19}$ cm^{-3} (a) and (b) $1.5*10^{21}$ cm^{-3} (from Ref.[137])

different concentrations of arsenic (Ref.[137]). The analysis was carried out with 1.8 MeV He ions incident along the <110> axis of the Si crystals. Fig.5.39 shows the angular scans of backscattered ions. For the crystal with the lower arsenic concentration (Fig.5.39a) the measured values $\chi_{min} = 0.03$ and $\psi_{1/2} = 0.62^o$ of the host lattice are close to the experimental data for perfect Si, indicating the good quality of the silicon crystal. The only slightly smaller half-width of the angular profile for the As signal ($\psi_{1/2} = 0.6^o$) shows that the As atoms are predominantly on substitutional sites. From the measured minimum yields at $\psi_{in} = 0$ for backscattering at the Si and As atoms, the fraction of substitutional As can be determined using Eq.(5.24) in the preceding section. With $\chi_{min,imp} = 0.08$ and $\chi_{min} = 0.03$ for the Si scan, one finds that ≥ 95% of the As atoms are located on substitutional sites.

For the higher As concentration (Fig.5.39b) the half-width of the arsenic dip is remarkably narrower than that of the host lattice, $\psi_{1/2} = 0.5^o$. On the other hand, $\chi_{min,imp}$ is not greater than in the sample containing the lower impurity concentration (Fig.5.39a). This indicates that the As atoms in the latter case are slightly displaced from well-defined, crystallographic lattice sites. For such situations, accurate information about the lattice location can be obtained by

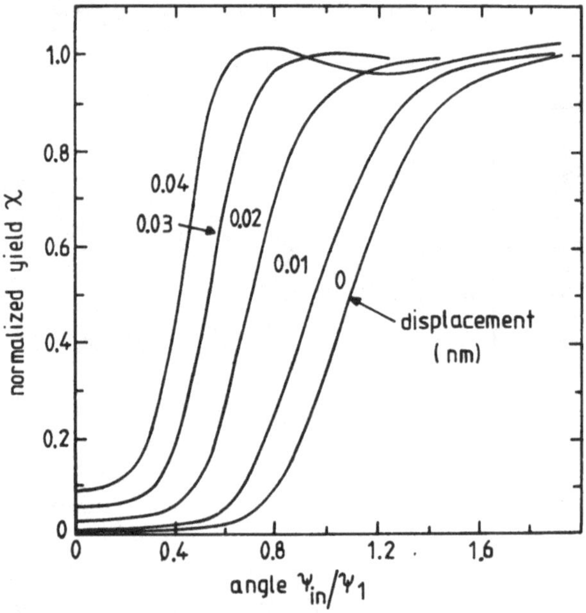

Fig.5.40. Calculated angular yield profiles of the signal from an impurity atom as a function of the displacement of the impurity from the <110> row for 1 MeV He ions incident on Si at 296 K. The angles are normalized to the characteristic angle ψ_1 (from Ref.[138]).

comparing the measured angular yield profiles with curves calculated for different possible positions of the impurities along atomic rows or planes. Calculated angular profiles for 1 MeV He ions as a function of the displacement distance of impurities from the <110> row of Si are shown in Fig.5.40 (Ref.[138]). As can be seen, a narrowing of the angular yield profile as well as a small increase of χ_{min} appears at displacements as small as 0.01 nm, and for greater displacement distances these effects become even more pronounced. The accuracy obtainable in location experiments, combined with computer simulations can also be seen with this simulation.

The identification of interstitial impurities located near the middle of crystallographic channels is somewhat easier than for slightly displaced impurity atoms. In this case, because the flux distribution depends on the position of the impurities in the channels, a significant modification of the angular yield profiles of the impurity atoms takes place, and in certain cases information about the lattice location can be obtained directly from the positions of the peaks of the measured angular scans (see e.g. Refs.[42, 136, 139-142]). In the following, as examples of the identification of interstitial impurity atoms, the location of B in W (Ref.[142]) and Br in Fe (Ref.[143]) will be discussed.

Boron as an atom with a low mass compared to tungsten can be located using the nuclear reaction $B^{11}(p,\alpha)\ Be^{8}$. In the experiment the angular profiles of the α-particles and of the backscattered protons were measured simultaneously, thus monitoring the interactions with the boron and the host lattice atoms. Expected interstitial sites in a bcc lattice as W are commonly the octahedral and the tetrahedral site, which are represented in Fig.5.41a. The corresponding projections of these sites along the three principal axes in a bcc lattice are illustrated in Fig.5.41b. The angular scans for these axes (Fig.5.41c) show pronounced peaks of the reaction yield from the B atoms in the middle of the axial channels. However, in the <100> and the <110> directions these peaks are superimposed on angular dips with a depth of about 1/3 of that of the dips of the host lattice atoms and angular widths comparable to those of the host lattice. This indicates that a remarkable number of impurities are located on interstitial sites within the <100> and <110> atomic rows. Although tetrahedral as well as octahedral sites would give peaks at $\psi_{in} \approx 0^o$ for the three axes, the experimental results are only consistent with the octahedral site. As can be seen in Fig.5.41b, in this case about 1/3 of the impurities lie within the <100> and <110> rows. Moreover, the high and narrow peak in the <100> axis agrees well with 2/3 of the boron being at the optimum peaking site in the center of the channel.

The location of impurities with sufficiently high masses compared to the mass of the host lattice atoms is possible by means of the RBS

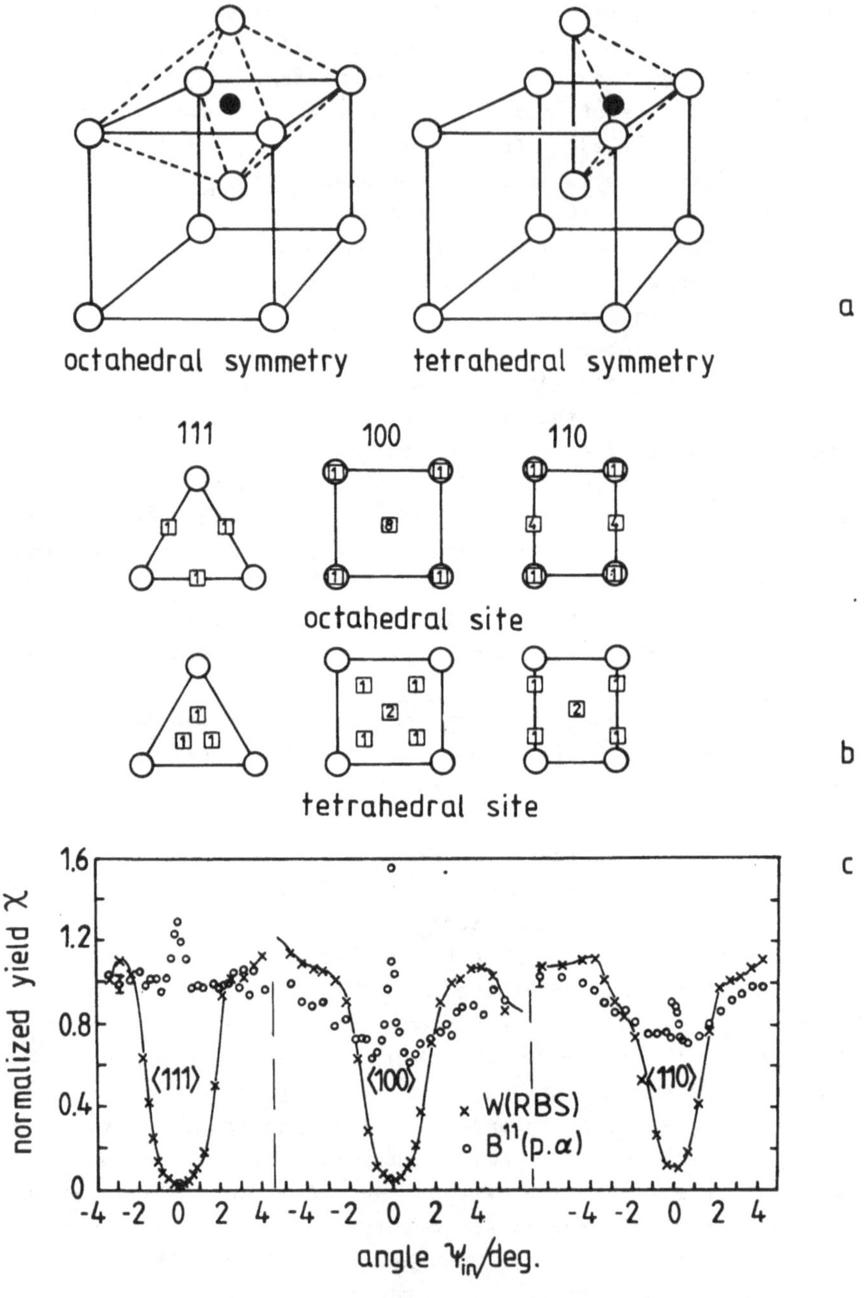

Fig.5.41. The two common interstitial sites in a bcc lattice (a), the position of octahedral and tetrahedral site with respect to the three principal axial channels in a bcc lattice (b), and backscattering yield as a function of crystal orientation for the W (x) and B (o) scanning through the <111>, <100>, and <110> axes (from Ref.[142])

technique. This will be demonstrated for the example of a Fe crystal implanted with $5*10^{14}$ Br ions cm^{-2} (Ref.[143]). The angular yield profiles of Br and Fe for different low index axial and planar directions measured with high energy He and N ions, respectively, are shown in Fig.5.42.

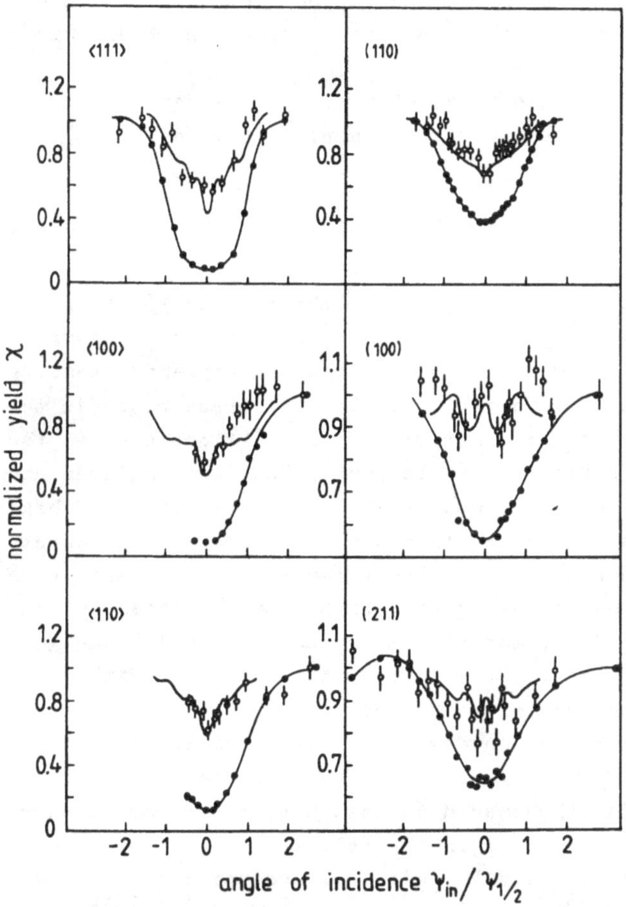

Fig.5.42. Angular scans across various low-index crystallographic axes and planes for Fe (•) and Br (o) in Br^+ implanted Fe. The smooth curves drawn through the Fe points are only to guide the eye. The experimental angular scans for the Br points are compared with the calculated yield curves (smooth curves) for a distribution of 60% Br in the X interstitial sites (see Fig.5.43) and 40% substitutional (Ref.[143]).

The following results can be derived from the experimental curves:

- For the three axial directions as well as for the (110) planar channels no peaks are observed in the angular profiles, indicating that there are no sites occupied at the center of <111>, <100>, or <110> axial channels or (110) planar channels.

- The peak in the (100) channel implies that a substantial fraction of the Br atoms are located at the center of this planar channel. The existence of a dip superimposed on the peak indicates that some Br atoms must occupy positions away from the channel center.
- The triple peak in the (211) scan indicates that a certain fraction of the Br atoms lies at the center of the channel, and that other atoms occupy sites between the middle of the channel and the planes. The dip underneath the triple peak points to a further fraction of Br atoms lying close to the channels walls.

The existence of two peaks for the (211) channel lets us determine the position of the interstitial sites in this channel. From Eq.(1.185) in Sect.1.3.2.5

$$E\psi_a^2 = Y(y_a) \tag{5.27}$$

using a Lindhard planar potential (Ref.[43])and the angular positions of the double peak $\psi_a = \pm(0.16^\circ \pm 0.03^\circ)$, one obtains for the distance from the plane $y_a = (0.42\pm0.04)\ d_p$, with the interplanar spacing $d_p = 0.0827$ nm for (211) in Fe. However, if all equivalent sites in the channel are taken into account, a distribution of positions across the channel could be responsible for the double peak. But the magnitude and the small angular width of the twin peaks indicate that a high concentration of atoms are located in a small number of discrete positions. Choosing sites of greatest planar symmetry consistent with $y_a = (0.42\pm0.04)\ d_p$, one obtains sites with high atomic concentrations at the center of the (211) plane, and on planes positioned at $\frac{3}{8}$ and $\frac{5}{8}$ of the distance across the channel (at $\frac{1}{8}\ d_p$ from the center). With this result and the other experimental facts for the different axes and planes, a systematic search for interstitial sites satisfying these constraints by using simple symmetry arguments as described in Ref.[143] can be made. As a result, for the case discussed here there remain three interstitial sites designated X, Y, Z in the following. The location of these interstitial sites in the Fe unit cell is illustrated in Fig.5.43; they occupy positions such as $\frac{3}{8}\frac{3}{8}\frac{1}{4}$ (X), $\frac{3}{4}\frac{3}{8}\frac{1}{2}$ (Y), and $\frac{1}{8}\frac{1}{4}\frac{1}{2}$ (Z). It should be noted that there exist 48 equivalent positions per unit cell for each site. Therefore, to obtain satisfactory quantitative information about the lattice location, calculations of the angular yield curves for atoms in these sites are required.

Theoretical angular-yield profiles for a particular site can be calculated from the flux distribution profiles (as a function of the angle of incidence) for each channel. For the following considerations we shall keep to the discussion of Monte Carlo computer simulations (Ref.[144]) based on a binary collision model. With this method, the angular scans measured for the Br atoms (open circles in Fig.5.42) were simulated for the three interstitial sites mentioned. The comparison of

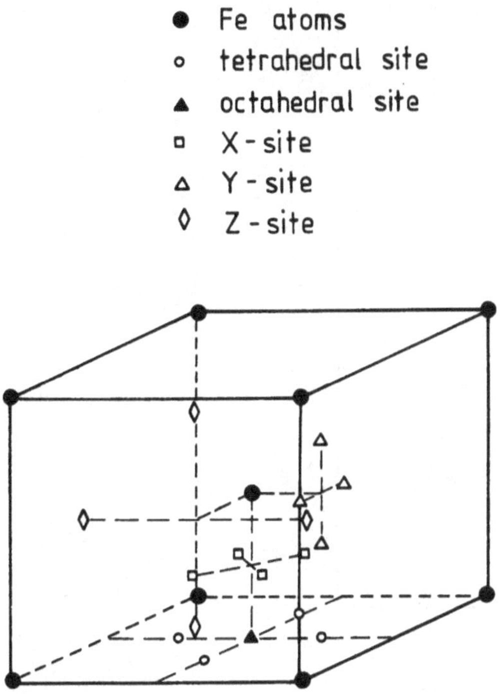

Fig.5.43. Location of the X, Y, and Z interstitial sites in the bcc Fe unit cell. The large filled circles represent Fe atoms. Only four of the 48 physically equivalent sites are shown in each case. The position of one of the three equivalent octahedral sites and four of the 12 equivalent tetrahedral sites are also shown (from Ref.[143]).

the simulated curves for different sites with the experimental results is illustrated in Fig.5.44 for the example of a scan across the (211) channel. For Br atoms in the X and Y interstitial site (dashed and dash-dotted curve in the figure, respectively) the triple peak observed in the experiment appears, but the calculated curves are too high compared with the experimental scan (this is also true for the other axial and planar directions). From this it was concluded that a substantial fraction of the Br atoms should occupy substitutional lattice sites. With this assumption, a remarkably better fit of the experimental data is obtained, as is illustrated in Fig.5.44 (dotted curve) for 60% Br in the X interstitial and 40% substitutional. A comparison of all simulations indicates that this case gives the best over-all agreement with the experimental data (Fig.5.42, solid lines). It was also shown that changes of only 5% in the percentage for the chosen sites agree even less, demonstrating the sensitivity of the calculation.

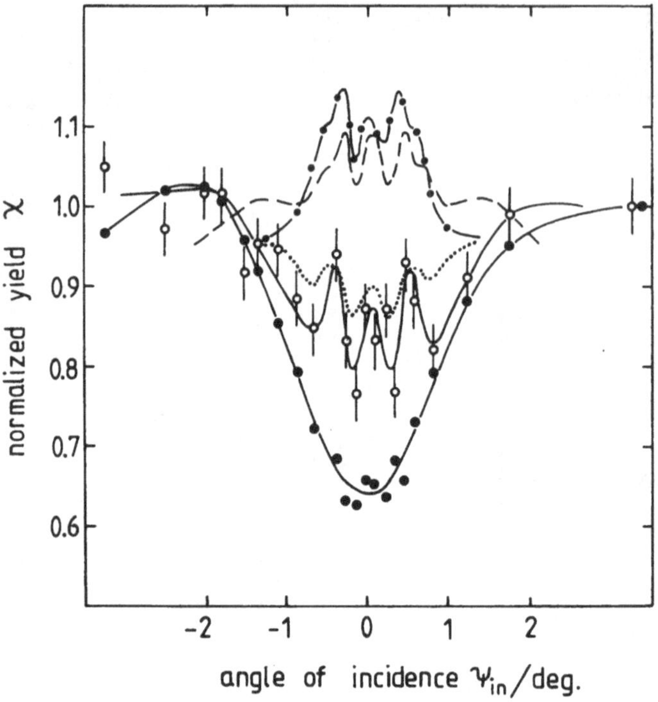

Fig.5.44. Comparison between the experimental angular scan for Fe (•) and Br (o) in Br implanted Fe across a (211) planar channel and the calculated yield curves for Br in the X interstitial sites (dashed curve), in the Y interstitial sites (dash-dotted curve), and for a distribution of 60% Br in the X interstitial sites and 40% substitutional (dotted curve; the lines through the experimental Fe points are only to guide the eye; from Ref.[143]).

5.3.3. Self interstitials

In damaged crystals, both point defects connected with the existence of impurities and displaced lattice atoms play an important role. In contrast to the location of impurity atoms, locating such self-interstitials is only possible by measuring the direct backscattering yield of energetic light ions. On principle, self-interstitials can be located, as can impurity atoms, by recording the angular yield profiles of backscattered ions for different low-index crystallographic axes and planes. Furthermore, especially in the case of self-interstitial concentrations down to $n_{pd} < 0.01$, information about lattice location and depth distribution of the defects may be obtained by investigating the dechanneling by defects as a function of the temperature.

5.3.3.1. Direct location by angular scan measurements

The direct locating of self-interstitials by recording angular yield profiles also makes use of the inhomogeneous distribution of the flux of ions at channeling conditions in low-index crystallographic axes and planes, respectively (compare Sects.1.3.2.5.1, 1.3.2.5.2). As in impurity location experiments (Sect.5.3.2.2), extremely good beam collimation and vacuum conditions are required. However, locating self-interstitials is significantly more difficult compared to the case of impurity atoms, because of the superposition of the signals from self-interstitials and regular lattice atoms. In order to resolve the additional structure of the angular dip resulting from the self-interstitials, high statistical accuracy as well as a very good beam collimation are necessary. This leads to very long measurement times (about 5 to 10 hours per crystallographic direction for a complete angular scan of the backscattering yield). Finally, self-interstitial location experiments require low target temperatures (near liquid nitrogen temperature) to minimize rearrangements of the defects.

Now, as an example, some problems of the location of self-interstitials in boron-implanted silicon are discussed (Refs.[49, 145]). The Si samples were implanted with $2*10^{14}$ B ions cm^{-2} at the temperature of 77 K. The RBS measurements were carried out with 1.4 MeV He ions with the samples cooled down to 120 K. Figs.5.45 and 5.46 show angular scans of the yield of backscattered He ions for different axial and planar directions recorded for the depth interval 0.225 μm to 0.485 μm (for comparison the angular dips for unimplanted Si measured at 300 K are shown). From the figures it is obvious that the implantation-induced defects cause an increase of the minimum yield as well as characteristic modifications in the angular distributions. Whereas the axial yield profiles do hardly exhibit a structure, in the planar scans, saddles and peaks are observed. In the case of the <111> direction, a significant narrowing of the dip is found for tilt angles $\leq \pm 0.5^{\circ}$ indicating backscattering of the He ions from several distinct positions outside the middle of the channel (near the atomic rows). The lack of a peak in the distribution indicates that interstitial sites near the channel center do not seem to be occupied by a great number of host lattice atoms; this is also the case for the <110> channel..The narrowing of the distribution for this direction is smaller than in the case of <111>; there is no hint at the existence of distinct positions outside the channel center. However, it was already mentioned above that the small change of the angular yield distributions due to the existence of self-interstitial atoms renders the location of small defect concentrations difficult. This is a problem especially for measurements in axial directions, because of the small sensitivity for interstitials lying outside the middle of the axial channels (see Sects.1.3.2.5, 5.3.1).

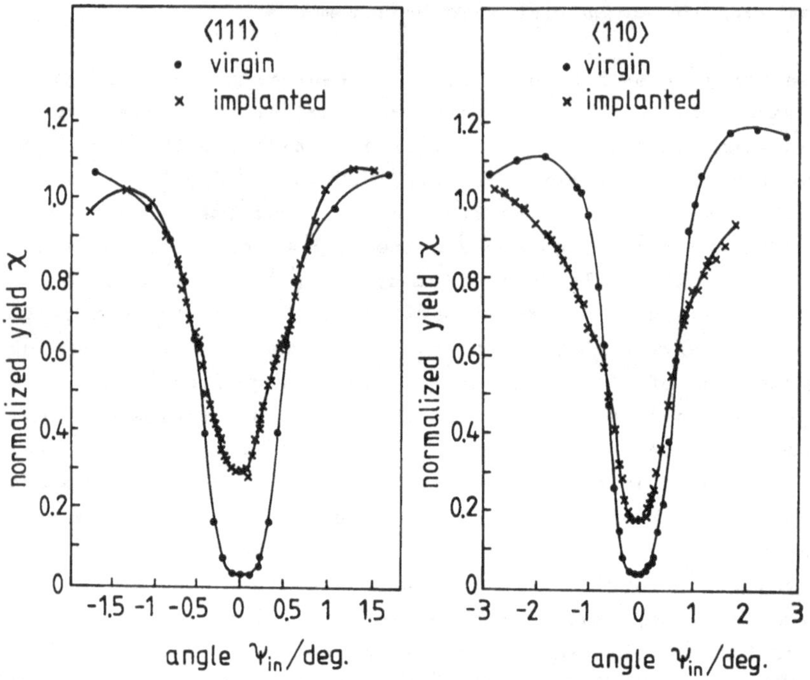

Fig.5.45. Angular scans across <111> and <110> axes (recorded in the depth interval 0.225 µm to 0.485 µm) for a Si sample implanted with $2*10^{14}$ B^+ cm^{-2} at 77 K. For comparison, the scans for virgin Si are shown (the solid lines through the measurement points are only to guide the eye; from Ref.[145]).

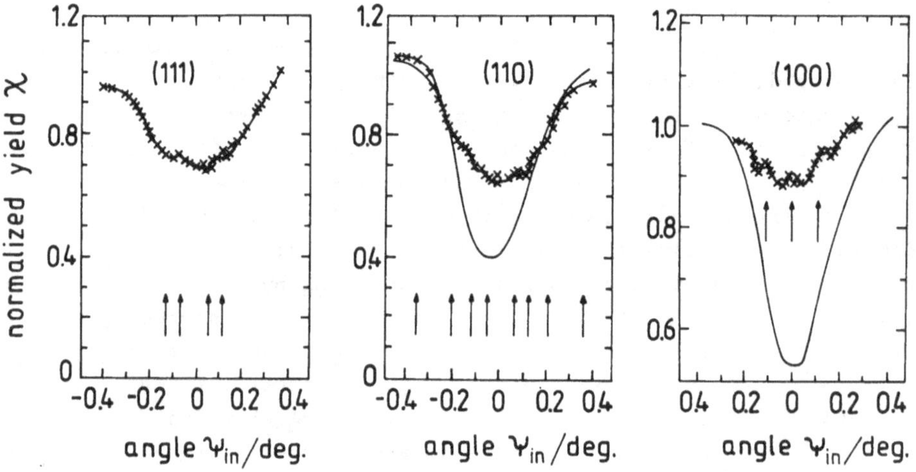

Fig.5.46. Angular scans across (111), (110), and (100) planes (recorded in the depth interval 0.225 µm to 0.485 µm) of the Si crystal (solid lines to guide the eye); the scans for virgin Si are indicated by solid lines (from Ref.[145]).

On the other hand, the flux distribution in planar channels favours the location of self-interstitial atoms by measuring the planar angular yield profiles. The results of such investigations are depicted in Fig.5.46. The most pronounced peaks can be seen in the yield profiles for the (100) plane at $\psi_a = \pm 0.12^\circ$ as well as in the center of the dip. For the (110) plane, peaks at $\psi_a = \pm 0.13^\circ$ are observed, near the channel center the yield is also increased. In the case of the (111) plane, broad peaks exceeding the normal yield by about 2 percent occur at $\psi_a \approx \pm 0.1^\circ$. This indicates the superposition of the scattering by two closely adjacent interstitial sites.

The occurrence of peaks and saddles in the planar angular scans lets us determine the corresponding intermediate planes according to the procedure described in Sect.5.3.2.2 for impurity atoms. Table 5.7 lists the angles ψ_a measured for the different peaks and the corresponding relative distances y_a/d_p from the atomic planes (Ref.[145]). The interstitial sites occupied by Si atoms, which are consistent with the measured angular scans, could be determined from the symmetry considerations described in Ref.[144] and using the results in Tab. 5.7 (for details see Ref.[145]).

The projections of the possible interstitial sites on the axial and planar channels considered are shown in Fig.5.47. The numbers in the circles correspond to the interstitial sites and give the weighting factors for the occupation of these sites (it should be noted that 12 interstitial sites per lattice atom, i.e. 96 interstitial sites for each axis and plane are possible). It can be seen that the distribution of interstitials in the <111> and <110> directions is consistent with the observed narrowing of the angular yield distributions (Fig.5.45).

TABLE 5.7

Relations between the measured characteristic angles ψ_a and the relative distance y_a/d_p from the plane.

	(111)	(110)			(100)	
d_p/nm	0.157 [1)]		0.096		0.1358	
ψ_a/deg.	0.100	0	0.070	0.130	0	0.120
y_a/d_p	0.220 [2)]	0.500	0.355	0.225	0.500	0.230

1) effective plane distance d_p = 0.157 nm
2) value related to the effective plane distance

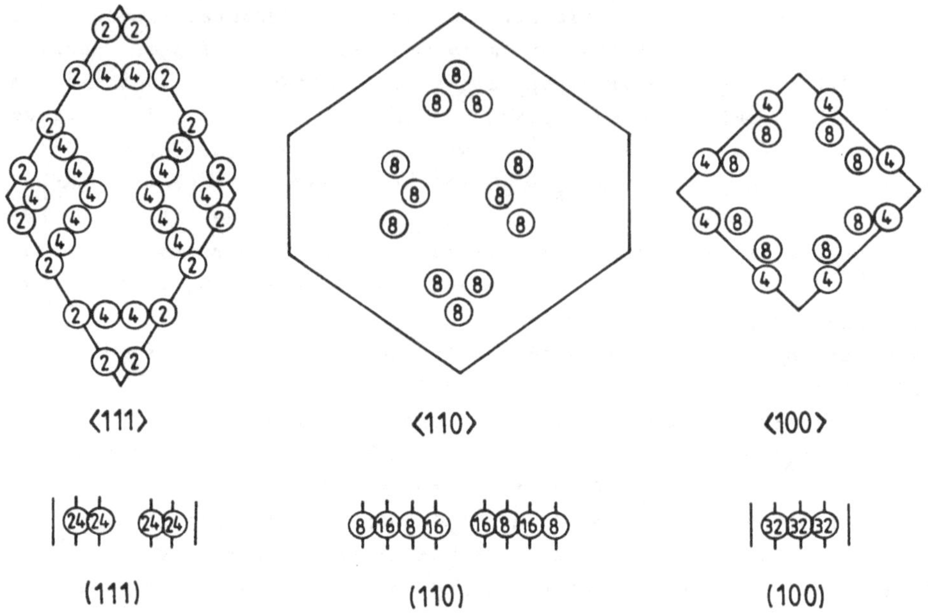

Fig.5.47. Projections of the possible interstitial sites of the <110> split configuration onto the major low-index crystallographic axes and planes (from Ref.[145])

Because of the higher resolution in planar channeling experiments as compared to the axial case, the (111) and (110) angular scans (Fig.5.46) exhibit a structure (saddles and peaks), which agrees with the projections of the interstitials in these channels. The three interstitial sites within the (100) channel lead to the three observed peaks in the corresponding angular yield profile.

From the given results, the interstitial atoms under the conditions of this experiment are assumed to form a <110> split configuration of orthorhombic symmetry. Such a split configuration is the symmetric arrangement of two interstitial atoms substituting a regular lattice atom, in this case parallel to the <110> direction. The distance between the two split atoms was determined to be 0.288 nm. Split configurations are characterized by a strong electronic binding of the displaced atoms with the neighbouring lattice atoms. Fig.5.48 represents a particular split configuration in the unit cell according to Ref.[146], which illustrates the represented results very well. The <111> projection of all equivalent split orientations shows a pronounced accumulation of displaced atoms in the vicinity of the string with an average distance $r_a = 0.061$ nm, whereas there is a more widespread distribution within the middle of the channel ($r_a = 0.118$ nm) in the <110> projection, which agrees with the distribution of interstitial positions across those channels shown in Fig.5.47.

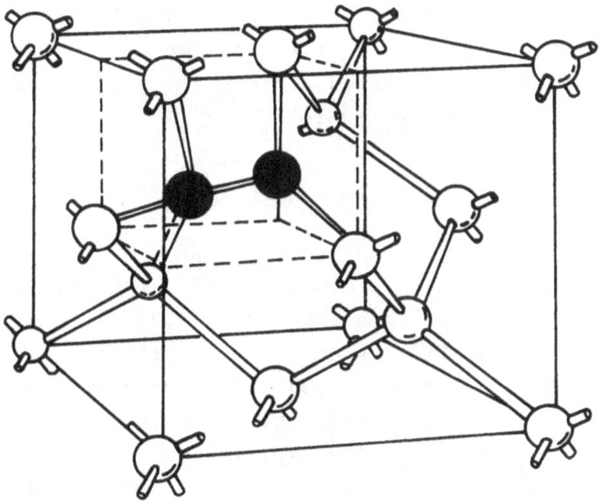

Fig.5.48. A special <110> split configuration of orthorhombic symmetry in the unit cell of silicon (o Si on regular lattice sites, • interstitial sites; from Ref.[146])

The method described above is especially successful for atoms positioned well away from the atomic strings and for defect concentrations of several percent. With this method, information is obtained about the lattice location in near-surface regions of the crystals investigated. If information about the depth distribution of the point defects is required, dechanneling must be taken into account or the measurements must be combined with successive etching techniques. Nevertheless, it can be said that the location of self-interstitials by angular scan measurements is an experimentally difficult and time-consuming procedure. Another method which overcomes these difficulties is described in the following section.

5.3.3.2. Dechanneling measurements

A further possibility for the lattice location of point defects in weakly damaged crystals (no pronounced direct backscattering part of the minimum yield) is to analyse the dechannelling of energetic light ions. The basis of this method is the fact that the temperature dependence of the minimum yield for a crystal with point defects is characteristic of the position r_a of these defects in the string. Therefore it is possible to determine the defect concentration $n_{pd}(z)$ as well as the position r_a from the depth dependence of the minimum yield, measured for different temperatures and crystallographic directions. The basic theoretical considerations and the procedure for determining the point defect concentration $n_{pd}(z)$ as well as the displacement r_a are described in Sects.1.3.4. The application of this method to the location of point defects in tantalum, weakly damaged ion implanted silicon (Ref.[59]) and gallium arsenide (Ref.[147]) is discussed in the following.

5.3.3.2.1. Tantalum

Fig.5.49 shows the depth dependence of the minimum yield of 2 MeV H ions incident on <111> tantalum (experimental data from Ref.[148]) for two different temperatures. A comparison with the calculations of $\chi_{min}(z)$ for a perfect crystal (dotted lines; the dash-dotted lines represent the theoretical curves from Ref.[53]) shows a significant difference, the theoretical curves are too low compared with the experimental results. This is a hint at the existence of defects responsible for this large difference.

In order to understand the experimental data, the existence of three different kinds of point defects characterized by different displacements r_a of lattice atoms from the atomic strings is assumed: r_a = 0.04 nm, r_a = 0.13 nm and randomly distributed atoms (the radius of the area belonging to one <111> string in tantalum is r_o = 0.142 nm).

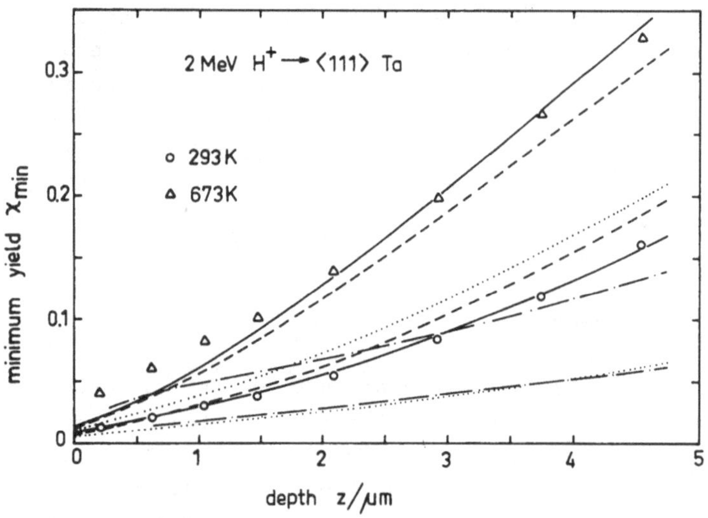

Fig.5.49. Depth dependence of the minimum yield of 2 MeV H ions incident on <111> tantalum with experimental data from Ref.[148]. The calculated minimum yields are given by dotted lines (perfect crystal), full lines (point defects with n_{pd} = 0.43% and r_a randomly distributed), and dashed lines (point defects with n_{pd} = 2.30% and r_a = 0.04 nm); the dash-dotted lines represent the diffusion model results according to Ref.[53] (from Ref.[59]).

From the measured minimum yields, the relative point defect concentrations $n_{pd}(z)$ for the two temperatures were calculated with these assumptions for r_a. The density $n_{pd}(z)$ was always found to be roughly constant with respect to the depth z. This is understandable, because the defects in question belong to the bulk material. In the case of randomly distributed r_a, the minimum yields at both temperatures are well described by the same point defect concentration $n_{pd} = 0.43\%$ (full lines in Fig.5.49). About the same result was obtained for 0.32% of point defects with fixed positions $r_a = 0.13$ nm (not shown in the figure). The situation changes for the third kind of defects ($r_a = 0.04$ nm, $n_{pd} = 2.30\%$, dashed lines in Fig.5.49). In this case, a significant difference is found in n_{pd} calculated at the two temperatures which is, of course, not possible. Therefore such point defects must be excluded. From the given results it is concluded that the minimum yield in tantalum for different temperatures can be understood by the existence of a constant concentration of point defects of 0.3-0.4% with randomly distributed r_a and/or preferred positions in the vicinity of the middle of the channel.

5.3.3.2.2. Silicon

In Sect.1.3.4.2.2 the procedure for determining the density $n_{pd}(z)$ and position r_a of point defects from the measured depth dependence of the minimum yield $\chi_{min}(z,T)$ for different temperatures is described, for the example of point defects in silicon generated by irradiation with 2 MeV He ions and subsequent thermal annealing. It was shown that the point defect density $n_{pd}(z)$ calculated from the measured $\chi_{min}(z,T_f)$ at a given temperature T_f, differs heavily for different positions r_a (see Fig.1.55). Therefore, for the determination of $n_{pd}(z)$ and r_a, the minimum yields for different temperatures must be considered. The theoretical calculations have shown that the temperature dependence of $\Delta\chi_{min}(z,T)$ as well as the second difference $\Delta^2\chi_{min}$ vary strongly with the position r_a (Fig.1.56). Thus, a comparison of the measured $\Delta\chi_{min}(z,T)$ or $\Delta^2\chi_{min}$ values with the calculated ones gives the position r_a. With the value r_a determined in this way, the defect density $n_{pd}(z)$ can be determined.

This method will be described in more detail for silicon crystals with point defects produced as mentioned above. The measurement of χ_{min} of the irradiated and perfect crystals was done with 1 MeV H ions incident along the <111> and <110> axes at sample temperatures of T = 280 K and 40 K. The measured differences in minimum yield are depicted as a function of the depth z in Figs.5.50 and 5.51 for the two directions (experimental data from Ref.[149]).

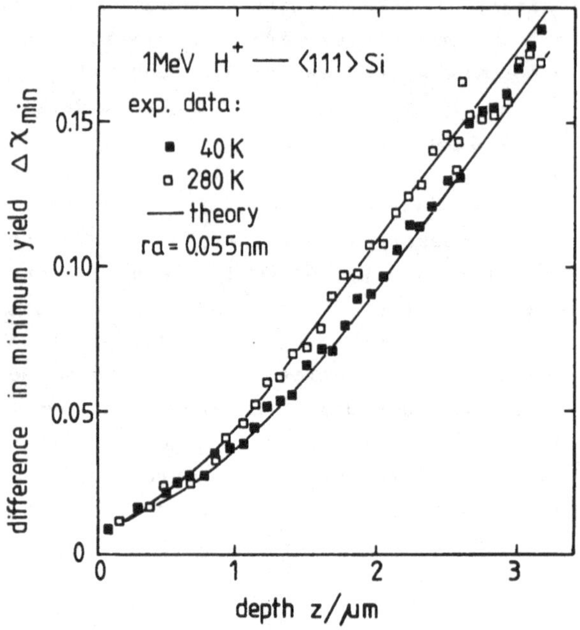

Fig.5.50. Depth dependence of $\Delta\chi_{min}$ for 1 MeV H ions incident on <111> silicon irradiated with $3.6*10^{16}$ He ions cm^{-2}. The irradiation with 2 MeV He ions at 40 K was followed by a 10 min anneal at 290 K (Ref.[149]). The full lines represent the theoretical results (from Ref.[59]).

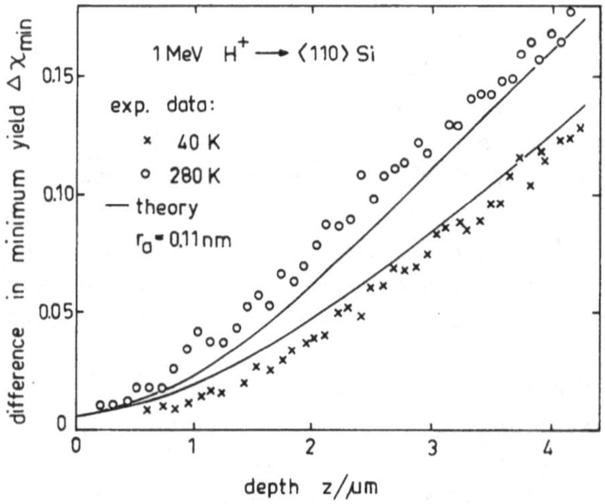

Fig.5.51. Depth dependence of $\Delta\chi_{min}$ for 1 MeV H ions incident on <110> silicon irradiated with $1.6*10^{16}$ He ions cm^{-2} at 40 K and annealed 15 min at 300 K (Ref.[149]) and results of theoretical calculation (from Ref.[59])

From the figures it can be seen that $\Delta\chi_{min}(z)$ increases more rapidly with the depth in the <111> direction than in the <110> direction. To prove that the assumption of preferred positions is responsible for the observed large difference of $\Delta\chi_{min}$ for the two axes, the second differences $\Delta^2\chi_{min} = \Delta\chi_{min}(280\ K) - \Delta\chi_{min}(40\ K)$ are calculated for different positions r_a. In this procedure, for each value of r_a the corresponding relative point defect concentrations $n_{pd}(z)$ are used, which were calculated from the experimental $\Delta\chi_{min}$ curves (for the <111> axis, $n_{pd}(z)$ is depicted in Fig.1.55). Fig.5.52 shows the calculated dependences $\Delta^2\chi_{min}(r_a)$ for fixed depths z. The strong dependence of $\Delta^2\chi_{min}$ on the position for small displacements up to $r_a \approx 0.07$ nm allows the determination of the displacement, particularly of atoms which occupy positions near the atomic rows. By comparing the experimental values of $\Delta^2\chi_{min}$ at the depth z = 2 µm (full dots in Fig.5.52) with the calculations, a fixed position with r_a = 0.045...0.073 nm in the <111> direction can be predicted (see Fig.5.52a). Point defects with large r_a and randomly distributed r_a can clearly be excluded. For the <110> direction the conclusion is not so clear. As is seen in Fig.5.52b, the maximum value of the calculated $\Delta^2\chi_{min}$ at depth z = 4 µm is smaller than the measured one, but within the experimental error. Thus, point defects with randomly distributed r_a are not strictly excluded by these results, but they are improbable because the <111> projection provides a fixed position. Therefore, for further consideration point defects with a

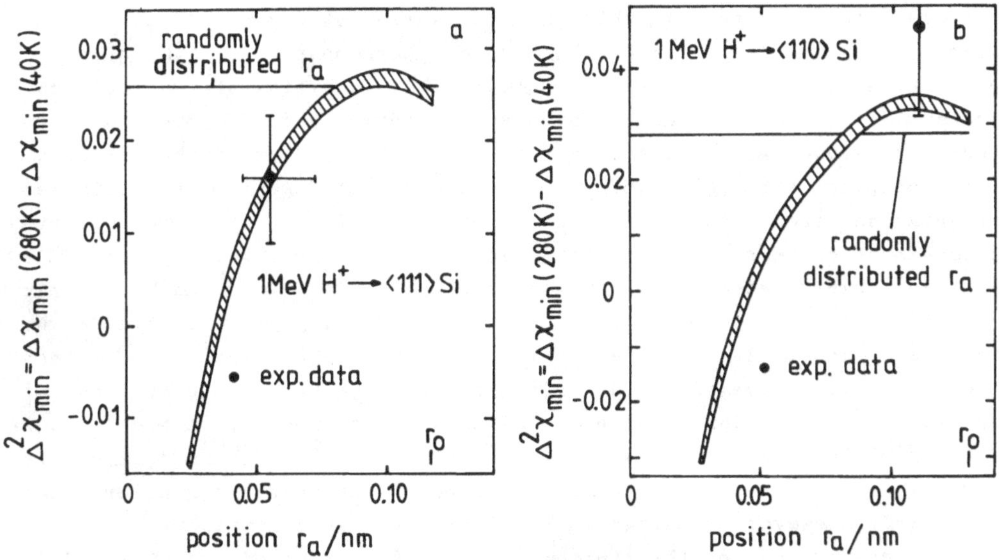

Fig.5.52. Calculated dependences of the second difference $\Delta^2\chi_{min}$ at depths z = 2 µm for <111> Si (a) and z = 4 µm for <110> Si (b) on the position of the point defects r_a. The experimental values for $\Delta^2\chi_{min}$(•) from Ref.[24] are included (from Ref.[59]).

position in the vicinity of r_a = 0.11 nm are assumed to be more realistic. For these positions, the calculated full depth dependences of $\Delta\chi_{min}$ for the two directions are added to the experimental data in Figs.5.51 and 5.52. There is good agreement for all temperatures and depths in the case of the <111> direction, but the temperature dependence of $\Delta\chi_{min}$ for <110> Si is somewhat underestimated for smaller depths. To summarize, the temperature dependence of $\Delta\chi_{min}$ in weakly damaged silicon may be understood by assuming point defects with preferred positions given by $r_a \approx 0.11$ nm in the <110> projection and by r_a = 0.045 ... 0.073 nm in the <111> projection. Such point defects were also found in silicon crystals irradiated with 150 keV B ions at 77 K ($n_{pd}^{max} \approx 20\%$) by measuring the angular dependence of the RBS yield as it has been discussed in Sect.5.3.3.1 (Refs.[145,150]). From these results, the existence of <110> split configurations of self-interstitials with a distance of 0.288 nm between the two split atoms was suggested. The mean displacement distances $\overline{r_a}$ determined for the projections of the equivalent split orientations to the <111> and <110> directions ($\overline{r_a}$ = 0.061 nm and 0.118 nm, respectively), agree with the results obtained from the dechanneling measurements.

5.3.3.2.3. Gallium arsenide

RBS investigations of GaAs have shown that under certain implantation conditions (light ions, sufficiently high implantation temperature), defects are produced in this material which strongly influence the dechanneling of light ions. These defects are connected with specific near edge optical properties. From the analysis of the optical properties it was concluded that high concentrations of point defects such as vacancies, anti-site defects and/or complexes of them exist in such samples (Refs.[151,152]). Therefore, to get more detailed information about the character of these defects, the temperature dependence of the minimum yield was measured (Ref.[147]). The GaAs samples investigated were implanted with $5*10^{15}$ N ions cm^{-2} (energy 280 keV) at room temperature. The backscattering yield of 1.4 MeV He ions was measured at the temperatures 293 K and 60 K for both perfect and implanted crystals. In order to avoid excessive generation of defects by the He ions in these preliminary experiments, a small total fluence of He ions equivalent to 1 μC (per 0.78 mm^2) was used. In this way, the increase of the minimum yield by the He ions was smaller than 0.01. This, however, resulted in a relatively high statistical error.

The difference of the minimum yields of implanted and virgin GaAs $\Delta\chi_{min}(z)$ is depicted in Fig.5.53 for the two temperatures (to show the statistical uncertainties, the experimental curves are not smoothed). In spite of the large statistical error, there is no doubt that the difference in minimum yield is larger for the lower temperature (Fig.5.53a). This is also true if we take into account the generation of defects during measurements, because the corresponding increase of $\Delta\chi_{min}$ is smaller than the difference of $\Delta\chi_{min}$ for 60 K and 293 K. The existence of such a negative temperature dependence for displaced atoms positioned near the strings was predicted from theoretical calculations (Ref.[59], see Fig.1.56). With the experimental data illustrated in Fig.5.53a the relative concentration of point defects $n_{pd}(z)$ was calculated from the smoothed experimental curve $\Delta\chi_{min}(z)$ at T_f = 293 K, assuming different fixed values of the distance r_a from the string and for randomly distributed r_a. The calculated profiles (Fig.5.53b) strongly depend on r_a, with the maximum defect concentration varying

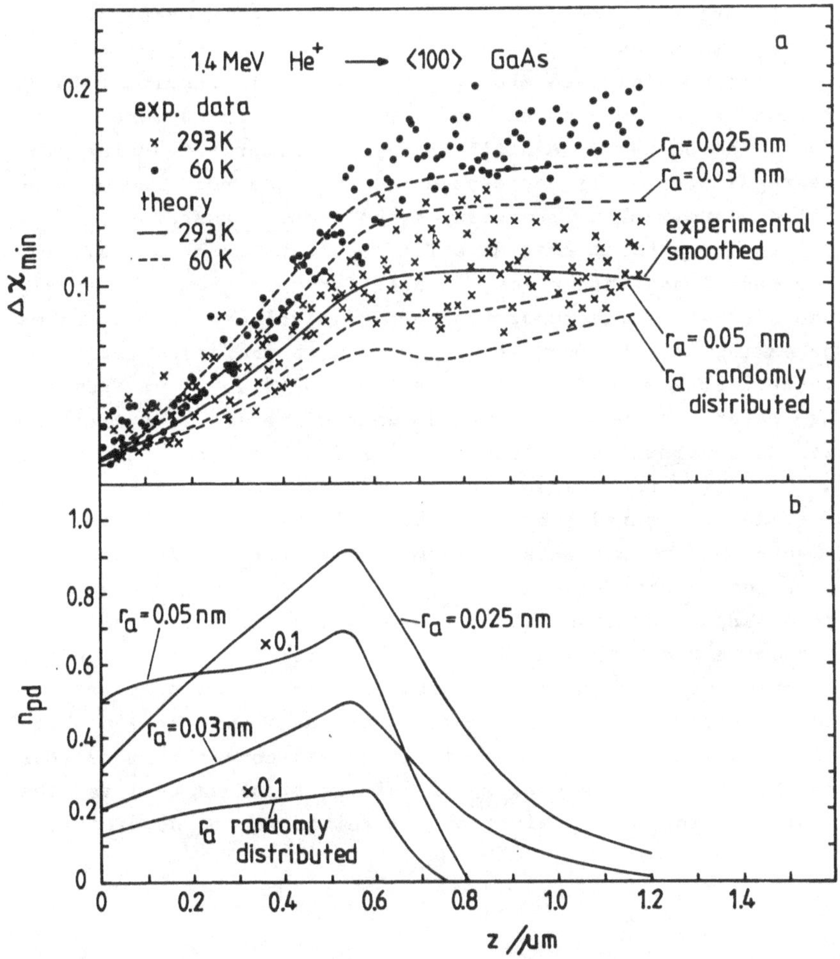

Fig.5.53. Depth dependence of $\Delta\chi_{min}$ (a) and of the relative concentrations of point defects $n_{pd}(z)$ (b) for 1.4 MeV He ions incident on <100> GaAs irradiated with $5*10^{15}$ N ions cm^{-2} at 293 K. Parameter is the distance from the <100> string (from Ref.[147]).

from about 2% for randomly distributed r_a to 90% for $r_a = 0.025$ nm. First of all one can state that the values $r_a = 0.025$ nm and $r_a = 0.03$ nm may be more probable because for $r_a = 0.05$ nm and randomly distributed r_a we get negative defect concentrations for depths $z \geq 0.8\,\mu m$ (omitted in the figure).

More accurate information about the most probable position r_a can be gained from the temperature dependence of $\Delta\chi_{min}$. For the defect distributiones in Fig.5.53b , the differences in the minimum yield $\Delta\chi_{min}(z)$ at 60 K were calculated with the corresponding r_a values. The results are shown in Fig.5.53a (dashed curves) as compared with the experimental data. The point defects with $r_a = 0.05$ nm and with randomly distributed r_a exhibit a positive temperature dependence of $\Delta\chi_{min}$, i.e. $\Delta\chi_{min}$ decreases with decreasing temperature (lower curves in Fig.5.53b). This case can be excluded because the temperature dependence of the measured $\Delta\chi_{min}$ is unambiguously a negative one. However, with respect to the experimental errors discussed above, one cannot distinguish between the two cases with $r_a = 0.025$ nm and $r_a = 0.03$ nm. Furthermore, with increasing depth the number of additional defects generated during the RBS measurements increases in correspondence with the profile of the energy deposited into nuclear processes by He ions. Therefore, more emphasis is put on the experimental data for depths up to about $0.5\,\mu m$. Comparing experimental and theoretical results for this depth interval, it can be concluded that high concentrations of displaced lattice atoms (values in the maximum of the profile 50...90%) with preferred positions in the lattice cell at a mean distance of about 0.025...0.03 nm from the atomic string exist in weakly damaged GaAs layers. The results distinctly exclude interstitials, because their distance from the string is larger than 0.03 nm, and suggest (in correspondence with the results of the optical measurements) that high concentrations of vacancies, anti-site defects and/or complexes of them are responsible for the high concentration of weakly displaced atoms.

From the preceding discussions it can be said that to locate slightly displaced lattice atoms, the application of temperature dependent dechanneling measurements seems to be favoured over the measurement of angular yield profiles. The first method in addition to its high sensitivity for the detection of small point defect concentrations has the advantage that the displacement r_a of lattice atoms as well as the depth distribution of the point defect concentration can be determined.

Appendices

APPENDIX 1 Atomic data (Ref.[2], mass density Ref.[153])

Z	element	isotopic mass [amu]	relative abundance	atomic weight [amu]	atomic density [cm^{-3}]	mass density [g/cm^3] [g/l]*
1	H	1.007825	0.9995	1.008	5.38 E 19	0.0899*
2	He	4.002603	1.0000	4.003	2.69 E 19	0.1787*
3	Li	6.015125	0.0756	6.940	4.60 E 22	0.53
		7.016004	0.9244			
4	Be	9.012186	1.0000	9.012	1.24 E 23	1.85
5	B	10.012939	0.1961	10.814	1.30 E 23	2.34
		11.009305	0.8039			
6	C	12.000000	0.9889	12.011	1.13 E 23	2.26
		13.003354	0.0111			
7	N	14.003074	0.9963	14.007	5.38 E 19	1.251*
		15.000108	0.0037			
8	O	15.994915	0.9976	15.999	5.38 E 19	1.429*
		16.999133	0.0004			
		17.999160	0.0020			
9	F	18.998405	1.0000	18.998	5.38 E 19	1.696*
10	Ne	19.992441	0.9092	20.171	2.69 E 19	0.901*
		20.993849	0.0026			
		21.991385	0.0882			
11	Na	22.989771	1.0000	22.990	2.54 E 22	0.97
12	Mg	23.985042	0.7870	24.310	4.31 E 22	1.74
		24.985839	0.1013			
		25.982593	0.1117			
13	Al	26.981539	1.0000	26.982	6.03 E 22	2.70
14	Si	27.976929	0.9221	28.086	5.00 E 22	2.33
		28.976496	0.0470			
		29.973763	0.0305			
15	P	30.973765	1.0000	30.974	3.54 E 22	1.82
16	S	31.972074	0.9500	32.061	3.89 E 22	2.07
		32.971462	0.0076			
		33.967865	0.0422			
		35.967090	0.0001			
17	Cl	34.968851	0.7577	35.453	5.38 E 19	3.17 *
		36.965899	0.2423			
18	Ar	35.967545	0.0034	39.948	2.69 E 19	1.784*
		37.962728	0.0006			
		39.962384	0.9960			
19	K	38.963710	0.9310	39.097	1.32 E 22	0.86
		39.964000	0.0001			
		40.961832	0.0688			

APPENDIX 1 continued

Z	element	isotopic mass [amu]	relative abundance	atomic weight [amu]	atomic density [cm^{-3}]	mass density [g/cm^3] [g/l]*
20	Ca	39.962589	0.9697	40.081	2.33 E 22	1.55
		41.958625	0.0064			
		42.958780	0.0015			
		43.955491	0.0206			
		47.952531	0.0019			
21	Sc	44.955919	1.0000	44.956	4.02 E 22	3.0
22	Ti	45.952632	0.0793	47.879	5.66 E 22	4.51
		46.951769	0.0728			
		47.947950	0.7394			
		48.947870	0.0551			
		49.944786	0.0534			
23	V	49.947164	0.0024	50.942	7.21 E 22	6.1
		50.943961	0.9976			
24	Cr	49.946055	0.0435	51.996	8.33 E 22	7.19
		51.940513	0.8376			
		52.940653	0.0951			
		53.938882	0.0238			
25	Mn	54.938050	1.0000	54.938	8.14 E 22	7.43
26	Fe	53.939617	0.0582	55.847	8.48 E 22	7.86
		55.934936	0.9166			
		56.935398	0.0219			
		57.933282	0.0033			
27	Co	58.933189	1.0000	58.933	9.09 E 22	8.90
28	Ni	57.935342	0.6788	58.728	9.13 E 22	8.90
		59.930787	0.2623			
		60.931056	0.0119			
		61.928342	0.0366			
		63.927958	0.0108			
29	Cu	62.929592	0.6917	63.546	8.49 E 22	8.96
		64.927786	0.3083			
30	Zn	63.929145	0.4889	65.387	6.58 E 22	7.14
		65.926052	0.2781			
		66.927145	0.0411			
		67.924857	0.1857			
		69.925334	0.0062			
31	Ga	68.925574	0.6040	69.717	5.10 E 22	5.91
		70.924706	0.3960			
32	Ge	69.924252	0.2052	72.638	4.41 E 22	5.32
		71.922082	0.2743			
		72.923463	0.0776			
		73.921181	0.3654			
		75.921405	0.0776			
33	As	74.921596	1.0000	74.922	4.45 E 22	5.72
34	Se	73.922476	0.0087	78.990	3.66 E 22	4.79
		75.019207	0.0902			
		76.919911	0.0758			
		77.917314	0.2352			
		79.916527	0.4982			
		81.916707	0.0919			
35	Br	78.918329	0.5069	79.904	2.35 E 22	3.12
		80.916292	0.4931			
36	Kr	77.920403	0.0035	83.801	2.69 E 19	3.74*
		79.916380	0.0227			
		81.913482	0.1156			
		82.914131	0.1155			
		83.911503	0.5690			
		85.910616	0.1737			

APPENDIX 1 continued

Z	element	isotopic mass [amu]	relative abundance	atomic weight [amu]	atomic density [cm^{-3}]	mass density [g/cm^3] [g/l]*
37	Rb	84.911800	0.7215	85.468	1.08 E 22	1.53
		86.909187	0.2785			
38	Sr	83.913430	0.0056	87.616	1.79 E 22	2.6
		85.909285	0.0986			
		86.908892	0.0702			
		87.905641	0.8256			
39	Y	88.905872	1.0000	88.906	3.03 E 22	4.47
40	Zr	89.904700	0.5146	91.224	4.28 E 22	6.49
		90.905642	0.1123			
		91.905031	0.1711			
		93.906313	0.1740			
		95.908286	0.0280			
41	Nb	92.906382	1.0000	92.906	5.44 E 22	8.4
42	Mo	91.906810	0.1584	95.890	6.41 E 22	10.2
		93.905090	0.0904			
		94.905839	0.1572			
		95.904674	0.1653			
		96.906022	0.0946			
		97.905409	0.2378			
		99.907475	0.0963			
43	Tc	98			7.04 E 22	11.5
44	Ru	95.907598	0.0551	101.046	7.27 E 22	12.2
		97.905289	0.0187			
		98.905936	0.1272			
		99.904218	0.1262			
		99.904218	0.1262			
		100.905577	0.1707			
		101.904348	0.3161			
		103.905430	0.1858			
45	Rh	102.905511	1.0000	102.906	7.26 E 22	12.4
46	Pd	101.905609	0.0096	106.441	6.80 E 22	12.0
		103.904011	0.1097			
		104.905064	0.2223			
		105.903479	0.2733			
		107.903891	0.2671			
		109.905164	0.1181			
47	Ag	106.905094	0.5183	107.868	5.86 E 22	10.5
		108.904756	0.4817			
48	Cd	105.906463	0.0122	112.434	4.63 E 22	8.65
		107.904187	0.0088			
		109.903012	0.1239			
		110.904188	0.1275			
		111.902763	0.2407			
		112.904409	0.1226			
		113.903360	0.2886			
		115.904762	0.0758			
49	In	112.904089	0.0428	114.818	3.83 E 22	7.31
		114.903871	0.9572			
50	Sn	111.904835	0.0096	118.734	3.62 E 22	7.30
		113.902773	0.0066			
		114.903346	0.0035			
		115.901745	0.1430			
		116.902958	0.0761			
		117.901606	0.2403			
		118.903313	0.0858			
		119.902198	0.3285			
		121.903441	0.0472			
		123.905272	0.0594			

APPENDIX 1 continued

Z	element	isotopic mass [amu]	relative abundance	atomic weight [amu]	atomic density [cm^{-3}]	mass density [g/cm^3] [g/l]*
51	Sb	120.903816	0.5725	121.755	3.27 E 22	6.62
		122.904213	0.4275			
52	Te	119.904023	0.0009	127.628	2.94 E 22	6.24
		121.903066	0.0246			
		122.904277	0.0087			
		123.902842	0.0461			
		124.904418	0,0695			
		125.903322	0.1871			
		127.904476	0.3179			
		129.906238	0.3448			
53	I	126.904470	1.0000	126.904	2.34 E 22	4.94
54	Xe	123.906120	0.0010	131.305	2.69 E 19	5.89*
		125.904288	0.0009			
		127.903540	0.0192			
		128.904784	0.2644			
		129.903509	0.0408			
		130.905085	0.2118			
		131.904161	0.2689			
		133.905397	0.1044			
		135.907221	0.0887			
55	Cs	132.905355	1.0000	132.905	8.61 E 21	1.90
56	Ba	129.906245	0.0010	137.327	1.53 E 22	3.5
		131.905120	0.0010			
		133.904612	0.0242			
		134.905550	0.0659			
		135.904300	0.0781			
		136.905500	0.1132			
		137.905000	0.7166			
57	La	137.906910	0.0009	138.905	2.67 E 22	6.17
		138.906140	0.9991			
58	Ce	135.907100	0.0019	140.101	2.87 E 22	6.67
		137.905830	0.0025			
		139.905392	0.8848			
		141.909140	0.1107			
59	Pr	140.907596	1.0000	140.908	2.89 E 22	6.77
60	Nd	141.907663	0.2711	144.241	2.92 E 22	7.0
		142.909779	0.1217			
		143.910039	0.2385			
		144.912538	0.0830			
		145.913086	0.1722			
		147.916869	0.0573			
		149.920915	0.0562			
61	Pm	147				
62	Sm	143.911989	0.0309	150.363	3.02 E 22	7.54
		146.914867	0.1497			
		147.914791	0.1124			
		148.917180	0.1383			
		149.917276	0.0744			
		151.919756	0.2672			
		153.922282	0.2271			
63	Eu	150.919838	0.4782	151.964	2.08 E 22	5.26
		152.921242	0.5218			

APPENDIX 1 continued

Z	element	isotopic mass [amu]	relative abundance	atomic weight [amu]	atomic density [cm^{-3}]	mass density [g/cm^3] [g/l]*
64	Gd	151.919794	0.0020	157.256	3.02 E 22	7.89
		153.920929	0.0215			
		154.922664	0.1473			
		155.922175	0.2047			
		156.924025	0.1568			
		157.924178	0.2487			
		159.927115	0.2190			
65	Tb	158.925351	1.0000	158.925	3.13 E 22	8.27
66	Dy	155.923930	0.0005	162.484	3.17 E 22	8.54
		157.924449	0.0009			
		159.925202	0.0229			
		160.926945	0.1888			
		161.926803	0.2553			
		162.928755	0.2497			
		163.929200	0.2818			
67	Ho	164.930421	1.0000	164.930	3.21 E 22	8.80
68	Er	161.928740	0.0014	167.261	3.26 E 22	9.05
		163.929287	0.0156			
		165.930307	0.3341			
		166.932060	0.2294			
		167.932383	0.2707			
		169.935560	0.1488			
69	Tm	168.934245	1.0000	168.934	3.33 E 22	9.33
70	Yb	167.934160	0.0014	173.036	2.43 E 22	6.98
		169.935020	0.0303			
		170.936430	0.1431			
		171.936360	0.2182			
		172.938060	0.1613			
		173.938740	0.3184			
		175.942680	0.1273			
71	Lu	174.940640	0.9741	174.967	3.39 E 22	9.84
		175.942660	0.0259			
72	Hf	173.940360	0.0018	178.509	4.42 E 22	13.1
		175.941570	0.0520			
		176.943400	0.1850			
		177.943880	0.2714			
		178.946030	0.1375			
		179.946820	0.3524			
73	Ta	179.947544	0.0001	180.948	5.53 E 22	16.6
		180.948007	0.9999			
74	W	179.947000	0.0014	183.842	6.32 E 22	19.3
		181.948301	0.2641			
		182.950324	0.1440			
		183.951025	0.3064			
		185.954440	0.2841			
75	Re	184.953059	0.3707	186.213	6.79 E 22	21.0
		186.955833	0.6293			
76	Os	183.952750	0.0002	190.333	7.15 E 22	22.6
		185.953870	0.0159			
		186.955832	0.0164			
		187.956081	0.1330			
		188.958300	0.1610			
		189.958630	0.2640			
		191.961450	0.4100			
77	Ir	190.960640	0.3730	192.216	7.05 E 22	22.5
		192.963012	0.6270			

APPENDIX 1 continued

Z	element	isotopic mass [amu]	relative abundance	atomic weight [amu]	atomic density [cm^{-3}]	mass density [g/cm^3] [g/l]*
78	Pt	189.959950	0.0001	195.081	6.61 E 22	21.4
		191.961150	0.0078			
		193.962725	0.3290			
		194.964813	0.3380			
		195.964967	0.2530			
		197.967895	0.0721			
79	Au	196.966541	1.0000	196.967	5.90 E 22	19.3
80	Hg	195.965820	0.0015	200.617	4.06 E 22	13.53
		197.966756	0.1002			
		198.968279	0.1684			
		199.968327	0.2313			
		200.970308	0.1322			
		201.970642	0.2980			
		203.973495	0.0685			
81	Tl	202.972353	0.2950	204.384	3.50 E 22	11.85
		204.974442	0.7050			
82	Pb	203.973044	0.0148	207.177	3.31 E 22	11.4
		205.974468	0.2360			
		206.975903	0.2260			
		207.976650	0.5230			
83	Bi	208.980394	1.0000	208.980	2.82 E 22	9.8

APPENDIX 2 Kinematic factor K for ^{1}H as projectile (Eq.(1.11))

Z_2	element	scattering angle θ									
		90°	100°	110°	120°	130°	140°	150°	160°	170°	180°
2	He	0.5977	0.5461	0.5004	0.4612	0.4285	0.4023	0.3823	0.3683	0.3600	0.3573
3	Li	0.7464	0.7093	0.6751	0.6446	0.6182	0.5964	0.5792	0.5669	0.5596	0.5571
4	Be	0.7988	0.7682	0.7397	0.7139	0.6913	0.6725	0.6576	0.6468	0.6403	0.6381
5	B	0.8295	0.8030	0.7781	0.7554	0.7355	0.7188	0.7055	0.6958	0.6900	0.6881
6	C	0.8452	0.8208	0.7979	0.7769	0.7585	0.7429	0.7306	0.7216	0.7161	0.7143
7	N	0.8658	0.8443	0.8241	0.8055	0.7891	0.7752	0.7641	0.7561	0.7512	0.7495
8	O	0.8815	0.8624	0.8442	0.8276	0.8128	0.8003	0.7902	0.7829	0.7785	0.7770
9	F	0.8992	0.8828	0.8672	0.8527	0.8399	0.8290	0.8202	0.8138	0.8099	0.8086
10	Ne	0.9048	0.8892	0.8744	0.8607	0.8485	0.8381	0.8297	0.8237	0.8200	0.8187
11	Na	0.9160	0.9022	0.8889	0.8767	0.8658	0.8565	0.8490	0.8435	0.8402	0.8391
12	Mg	0.9204	0.9072	0.8946	0.8830	0.8726	0.8637	0.8566	0.8514	0.8482	0.8471
13	Al	0.9280	0.9160	0.9046	0.8939	0.8845	0.8763	0.8698	0.8650	0.8621	0.8612
14	Si	0.9307	0.9192	0.9081	0.8979	0.8887	0.8809	0.8746	0.8700	0.8672	0.8662
15	P	0.9370	0.9264	0.9163	0.9070	0.8986	0.8914	0.8856	0.8814	0.8788	0.8779
16	S	0.9390	0.9288	0.9191	0.9100	0.9018	0.8949	0.8893	0.8852	0.8827	0.8818
17	Cl	0.9447	0.9354	0.9265	0.9182	0.9108	0.9044	0.8993	0.8956	0.8933	0.8925
18	Ar	0.9508	0.9425	0.9345	0.9271	0.9204	0.9147	0.9101	0.9067	0.9047	0.9040
19	K	0.9497	0.9413	0.9331	0.9256	0.9188	0.9130	0.9083	0.9048	0.9027	0.9020
20	Ca	0.9509	0.9427	0.9347	0.9273	0.9207	0.9150	0.9104	0.9070	0.9050	0.9043
21	Sc	0.9561	0.9487	0.9416	0.9349	0.9290	0.9239	0.9197	0.9167	0.9148	0.9142
22	Ti	0.9588	0.9518	0.9451	0.9388	0.9332	0.9283	0.9244	0.9216	0.9198	0.9192
23	V	0.9612	0.9546	0.9483	0.9424	0.9371	0.9325	0.9288	0.9261	0.9245	0.9239
24	Cr	0.9620	0.9555	0.9493	0.9435	0.9383	0.9338	0.9302	0.9276	0.9259	0.9254
25	Mn	0.9640	0.9578	0.9519	0.9464	0.9415	0.9373	0.9338	0.9313	0.9298	0.9292
26	Fe	0.9645	0.9585	0.9527	0.9473	0.9424	0.9382	0.9349	0.9324	0.9309	0.9304
27	Co	0.9664	0.9606	0.9551	0.9500	0.9454	0.9414	0.9382	0.9358	0.9344	0.9339
28	Ni	0.9663	0.9605	0.9550	0.9498	0.9452	0.9412	0.9380	0.9356	0.9341	0.9337
29	Cu	0.9688	0.9635	0.9583	0.9535	0.9492	0.9455	0.9425	0.9403	0.9390	0.9385
30	Zn	0.9696	0.9645	0.9595	0.9548	0.9506	0.9470	0.9441	0.9420	0.9406	0.9402
31	Ga	0.9715	0.9666	0.9619	0.9576	0.9536	0.9502	0.9475	0.9455	0.9442	0.9438
32	Ge	0.9726	0.9680	0.9634	0.9592	0.9554	0.9522	0.9495	0.9476	0.9464	0.9460
33	As	0.9735	0.9689	0.9645	0.9604	0.9568	0.9536	0.9510	0.9492	0.9480	0.9476
34	Se	0.9748	0.9705	0.9663	0.9624	0.9589	0.9559	0.9535	0.9517	0.9506	0.9502

APPENDIX 2 continued

Z_2	element	90°	100°	110°	120°	130°	140°	150°	160°	170°	180°
		scattering angle θ									
35	Br	0.9751	0.9709	0.9667	0.9629	0.9594	0.9565	0.9541	0.9523	0.9512	0.9509
36	Kr	0.9762	0.9722	0.9682	0.9646	0.9613	0.9584	0.9561	0.9544	0.9534	0.9530
37	Rb	0.9767	0.9727	0.9688	0.9652	0.9620	0.9592	0.9569	0.9553	0.9543	0.9539
38	Sr	0.9773	0.9734	0.9696	0.9661	0.9629	0.9602	0.9580	0.9564	0.9554	0.9550
39	Y	0.9776	0.9737	0.9700	0.9666	0.9634	0.9608	0.9586	0.9570	0.9560	0.9557
40	Zr	0.9782	0.9744	0.9708	0.9674	0.9644	0.9617	0.9596	0.9581	0.9571	0.9568
41	Nb	0.9785	0.9749	0.9713	0.9680	0.9650	0.9624	0.9603	0.9588	0.9579	0.9575
42	Mo	0.9792	0.9756	0.9722	0.9690	0.9661	0.9636	0.9615	0.9600	0.9591	0.9588
43	Tc	0.9796	0.9761	0.9728	0.9696	0.9668	0.9643	0.9623	0.9609	0.9600	0.9597
44	Ru	0.9802	0.9769	0.9736	0.9705	0.9678	0.9654	0.9635	0.9620	0.9612	0.9609
45	Rh	0.9806	0.9773	0.9741	0.9710	0.9683	0.9660	0.9641	0.9627	0.9619	0.9616
46	Pd	0.9812	0.9780	0.9749	0.9720	0.9694	0.9671	0.9653	0.9639	0.9631	0.9628
47	Ag	0.9815	0.9783	0.9752	0.9724	0.9698	0.9675	0.9657	0.9644	0.9636	0.9633
48	Cd	0.9822	0.9792	0.9762	0.9735	0.9710	0.9688	0.9671	0.9658	0.9650	0.9648
49	In	0.9826	0.9796	0.9767	0.9740	0.9716	0.9695	0.9678	0.9665	0.9658	0.9655
50	Sn	0.9832	0.9803	0.9775	0.9749	0.9725	0.9705	0.9688	0.9676	0.9669	0.9666
51	Sb	0.9836	0.9808	0.9780	0.9755	0.9732	0.9712	0.9696	0.9684	0.9677	0.9674
52	Te	0.9843	0.9816	0.9790	0.9766	0.9744	0.9725	0.9710	0.9698	0.9691	0.9689
53	I	0.9842	0.9815	0.9789	0.9765	0.9742	0.9723	0.9708	0.9697	0.9690	0.9687
54	Xe	0.9848	0.9821	0.9796	0.9772	0.9751	0.9733	0.9718	0.9707	0.9700	0.9698
55	Cs	0.9849	0.9824	0.9799	0.9775	0.9754	0.9736	0.9721	0.9710	0.9703	0.9701
56	Ba	0.9854	0.9829	0.9805	0.9782	0.9762	0.9744	0.9730	0.9719	0.9713	0.9711
57	La	0.9856	0.9831	0.9807	0.9785	0.9764	0.9747	0.9733	0.9722	0.9716	0.9714
58	Ce	0.9857	0.9833	0.9809	0.9786	0.9766	0.9749	0.9735	0.9725	0.9718	0.9716
59	Pr	0.9858	0.9834	0.9810	0.9788	0.9768	0.9751	0.9737	0.9726	0.9720	0.9718
60	Nd	0.9861	0.9837	0.9814	0.9793	0.9773	0.9756	0.9743	0.9733	0.9726	0.9724
61	Pm	0.9864	0.9840	0.9818	0.9796	0.9777	0.9761	0.9747	0.9738	0.9732	0.9729
62	Sm	0.9867	0.9844	0.9822	0.9801	0.9782	0.9766	0.9753	0.9743	0.9737	0.9735
63	Eu	0.9868	0.9846	0.9824	0.9803	0.9784	0.9768	0.9756	0.9746	0.9740	0.9738
64	Gd	0.9873	0.9851	0.9829	0.9810	0.9792	0.9776	0.9764	0.9754	0.9749	0.9747
65	Tb	0.9874	0.9852	0.9831	0.9812	0.9794	0.9778	0.9766	0.9757	0.9751	0.9750
66	Dy	0.9877	0.9855	0.9835	0.9816	0.9798	0.9783	0.9771	0.9762	0.9757	0.9755
67	Ho	0.9879	0.9858	0.9837	0.9818	0.9801	0.9786	0.9775	0.9766	0.9760	0.9759
68	Er	0.9880	0.9360	0.9840	0.9821	0.9804	0.9789	0.9778	0.9769	0.9764	0.9762
69	Tm	0.9881	0.9861	0.9841	0.9823	0.9806	0.9791	0.9780	0.9771	0.9766	0.9764

APPENDIX 2 continued

Z_2	element	scattering angle θ									
		90°	100°	110°	120°	130°	140°	150°	160°	170°	180°
70	Yb	0.9884	0.9864	0.9845	0.9827	0.9810	0.9796	0.9785	0.9777	0.9771	0.9770
71	Lu	0.9885	0.9866	0.9847	0.9829	0.9813	0.9799	0.9787	0.9779	0.9774	0.9772
72	Hf	0.9888	0.9868	0.9850	0.9832	0.9816	0.9803	0.9792	0.9783	0.9778	0.9777
73	Ta	0.9889	0.9870	0.9852	0.9834	0.9819	0.9805	0.9794	0.9786	0.9781	0.9780
74	W	0.9891	0.9872	0.9854	0.9837	0.9821	0.9808	0.9797	0.9790	0.9785	0.9783
75	Re	0.9892	0.9874	0.9856	0.9839	0.9824	0.9811	0.9800	0.9792	0.9787	0.9786
76	Os	0.9895	0.9876	0.9859	0.9842	0.9828	0.9815	0.9804	0.9797	0.9792	0.9790
77	Ir	0.9896	0.9878	0.9860	0.9844	0.9829	0.9817	0.9806	0.9799	0.9794	0.9792
78	Pt	0.9897	0.9879	0.9862	0.9846	0.9832	0.9819	0.9809	0.9802	0.9797	0.9795
79	Au	0.9898	0.9881	0.9864	0.9848	0.9833	0.9821	0.9811	0.9803	0.9799	0.9797
80	Hg	0.9900	0.9883	0.9866	0.9850	0.9836	0.9824	0.9814	0.9807	0.9803	0.9801
81	Tl	0.9902	0.9885	0.9869	0.9853	0.9839	0.9827	0.9818	0.9811	0.9806	0.9805
82	Pb	0.9903	0.9886	0.9870	0.9855	0.9841	0.9830	0.9820	0.9813	0.9809	0.9807
83	Bi	0.9904	0.9887	0.9871	0.9856	0.9843	0.9831	0.9822	0.9815	0.9810	0.9809
84	Po	0.9904	0.9888	0.9872	0.9857	0.9844	0.9832	0.9822	0.9816	0.9811	0.9810
85	At	0.9904	0.9888	0.9872	0.9857	0.9844	0.9832	0.9822	0.9816	0.9811	0.9810
86	Rn	0.9910	0.9894	0.9879	0.9865	0.9852	0.9841	0.9832	0.9825	0.9821	0.9820
87	Fr	0.9910	0.9894	0.9879	0.9865	0.9853	0.9842	0.9833	0.9826	0.9822	0.9821
88	Ra	0.9911	0.9896	0.9881	0.9867	0.9855	0.9844	0.9835	0.9828	0.9825	0.9823
89	Ac	0.9912	0.9896	0.9882	0.9868	0.9855	0.9844	0.9836	0.9829	0.9825	0.9824
90	Th	0.9914	0.9899	0.9884	0.9871	0.9858	0.9848	0.9839	0.9833	0.9829	0.9828
91	Pa	0.9913	0.9898	0.9884	0.9870	0.9858	0.9847	0.9838	0.9832	0.9828	0.9827
92	U	0.9916	0.9901	0.9887	0.9874	0.9862	0.9852	0.9843	0.9837	0.9833	0.9832

APPENDIX 3 Kinematic factor K for ^{4}He as projectile (Eq.(1.11))

Z_2	element	scattering angle θ									
		90°	100°	110°	120°	130°	140°	150°	160°	170°	180°
3	Li	0.2684	0.2102	0.1664	0.1344	0.1115	0.0954	0.0844	0.0773	0.0733	0.0721
4	Be	0.3849	0.3241	0.2747	0.2356	0.2056	0.1832	0.1671	0.1564	0.1502	0.1482
5	B	0.4597	0.4004	0.3503	0.3094	0.2770	0.2520	0.2336	0.2210	0.2137	0.2113
6	C	0.5001	0.4424	0.3929	0.3518	0.3187	0.2929	0.2736	0.2604	0.2526	0.2501
7	N	0.5555	0.5009	0.4532	0.4127	0.3795	0.3532	0.3333	0.3194	0.3113	0.3086
8	O	0.5998	0.5483	0.5027	0.4635	0.4309	0.4047	0.3848	0.3708	0.3625	0.3597
9	F	0.6520	0.6050	0.5627	0.5258	0.4946	0.4693	0.4498	0.4360	0.4278	0.4250
10	Ne	0.6688	0.6234	0.5824	0.5464	0.5159	0.4911	0.4718	0.4582	0.4501	0.4474
11	Na	0.7034	0.6615	0.6233	0.5896	0.5607	0.5370	0.5185	0.5053	0.4974	0.4948
12	Mg	0.7173	0.6769	0.6399	0.6071	0.5790	0.5558	0.5377	0.5248	0.5170	0.5145
13	Al	0.7416	0.7040	0.6693	0.6384	0.6117	0.5897	0.5724	0.5600	0.5525	0.5500
14	Si	0.7505	0.7139	0.6802	0.6500	0.6239	0.6022	0.5852	0.5731	0.5657	0.5633
15	P	0.7711	0.7370	0.7054	0.6770	0.6523	0.6318	0.6156	0.6040	0.5970	0.5946
16	S	0.7780	0.7448	0.7139	0.6861	0.6619	0.6418	0.6259	0.6145	0.6076	0.6053
17	Cl	0.7971	0.7663	0.7375	0.7115	0.6889	0.6699	0.6549	0.6441	0.6376	0.6354
18	Ar	0.8179	0.7898	0.7634	0.7395	0.7186	0.7010	0.6871	0.6770	0.6709	0.6689
19	K	0.8143	0.7857	0.7589	0.7347	0.7134	0.6956	0.6815	0.6713	0.6651	0.6630
20	Ca	0.8184	0.7904	0.7641	0.7403	0.7194	0.7019	0.6880	0.6779	0.6718	0.6698
21	Sc	0.8365	0.8109	0.7869	0.7650	0.7457	0.7295	0.7166	0.7073	0.7016	0.6997
22	Ti	0.8457	0.8214	0.7985	0.7777	0.7593	0.7438	0.7314	0.7225	0.7170	0.7152
23	V	0.8543	0.8312	0.8095	0.7896	0.7720	0.7572	0.7454	0.7368	0.7316	0.7298
24	Cr	0.8570	0.8344	0.8130	0.7934	0.7761	0.7615	0.7498	0.7414	0.7362	0.7345
25	Mn	0.8642	0.8425	0.8221	0.8033	0.7867	0.7727	0.7615	0.7534	0.7485	0.7468
26	Fe	0.8662	0.8449	0.8247	0.8062	0.7898	0.7760	0.7649	0.7569	0.7520	0.7504
27	Co	0.8728	0.8524	0.8331	0.8154	0.7997	0.7864	0.7758	0.7681	0.7634	0.7618
28	Ni	0.8724	0.8519	0.8326	0.8148	0.7991	0.7857	0.7751	0.7673	0.7626	0.7611
29	Cu	0.8815	0.8624	0.8442	0.8276	0.8128	0.8003	0.7903	0.7830	0.7785	0.7770
30	Zn	0.8846	0.8660	0.8483	0.8320	0.8176	0.8053	0.7955	0.7884	0.7840	0.7826
31	Ga	0.8914	0.8738	0.8570	0.8416	0.8279	0.8163	0.8069	0.8001	0.7960	0.7946
32	Ge	0.8955	0.8785	0.8624	0.8475	0.8342	0.8230	0.8139	0.8074	0.8034	0.8020
33	As	0.8986	0.8820	0.8663	0.8518	0.8389	0.8279	0.8191	0.8127	0.8087	0.8074
34	Se	0.9035	0.8878	0.8727	0.8588	0.8465	0.8360	0.8276	0.8214	0.8176	0.8164

APPENDIX 3 continued

Z_2	element	scattering angle θ									
		90°	100°	110°	120°	130°	140°	150°	160°	170°	180°
35	Br	0.9047	0.8891	0.8742	0.8605	0.8483	0.8379	0.8295	0.8234	0.8197	0.8185
36	Kr	0.9088	0.8939	0.8796	0.8664	0.8547	0.8446	0.8366	0.8307	0.8272	0.8260
37	Rb	0.9105	0.8958	0.8818	0.8688	0.8573	0.8474	0.8395	0.8338	0.8302	0.8291
38	Sr	0.9126	0.8982	0.8845	0.8718	0.8605	0.8509	0.8431	0.8375	0.8340	0.8329
39	Y	0.9138	0.8996	0.8861	0.8736	0.8624	0.8529	0.8452	0.8396	0.8362	0.8351
40	Zr	0.9160	0.9021	0.8889	0.8766	0.8657	0.3564	0.8489	0.8434	0.8401	0.8390
41	Nb	0.9174	0.9038	0.8907	0.8787	0.8679	0.8588	0.8514	0.8460	0.8427	0.8416
42	Mo	0.9199	0.9066	0.8939	0.8822	0.8718	0.8628	0.8557	0.8504	0.8472	0.8461
43	Tc	0.9215	0.9085	0.8961	0.8846	0.8743	0.8656	0.8585	0.8534	0.8503	0.8492
44	Ru	0.9238	0.9112	0.8991	0.8879	0.8779	0.8694	0.8625	0.8575	0.8544	0.8534
45	Rh	0.9251	0.9127	0.9008	0.8898	0.8800	0.8716	0.8648	0.8599	0.8569	0.8558
46	Pd	0.9275	0.9155	0.9039	0.8933	0.8837	0.8756	0.8690	0.8642	0.8613	0.8603
47	Ag	0.9284	0.9165	0.9052	0.8946	0.8852	0.8771	0.8706	0.8659	0.8630	0.8620
48	Cd	0.9312	0.9198	0.9088	0.8987	0.8896	0.8818	0.8755	0.8710	0.8682	0.8672
49	In	0.9326	0.9214	0.9106	0.9007	0.8917	0.8841	0.8780	0.8735	0.8707	0.8698
50	Sn	0.9348	0.9239	0.9135	0.9038	0.8951	0.8877	0.8817	0.8774	0.8747	0.8738
51	Sb	0.9363	0.9257	0.9155	0.9060	0.8976	0.8903	0.8845	0.8802	0.8776	0.8767
52	Te	0.9392	0.9290	0.9192	0.9102	0.9021	0.8951	0.8895	0.8854	0.8829	0.8821
53	I	0.9388	0.9286	0.9188	0.9097	0.9015	0.8945	0.8889	0.8848	0.8823	0.8814
54	Xe	0.9408	0.9309	0.9214	0.9126	0.9047	0.8979	0.8924	0.8884	0.8860	0.8852
55	Cs	0.9415	0.9317	0.9223	0.9136	0.9058	0.8991	0.8937	0.8897	0.8873	0.8865
56	Ba	0.9434	0.9339	0.9247	0.9162	0.9087	0.9021	0.8969	0.8931	0.8907	0.8899
57	La	0.9440	0.9346	0.9256	0.9172	0.9096	0.9032	0.8980	0.8942	0.8919	0.8911
58	Ce	0.9444	0.9351	0.9262	0.9178	0.9104	0.9040	0.8988	0.8951	0.8928	0.8920
59	Pr	0.9448	0.9355	0.9266	0.9183	0.9109	0.9045	0.8994	0.8956	0.8933	0.8926
60	Nd	0.9460	0.9369	0.9282	0.9201	0.9128	0.9066	0.9016	0.8979	0.8957	0.8949
61	Pm	0.9470	0.9381	0.9295	0.9215	0.9144	0.9083	0.9033	0.8997	0.8975	0.8968
62	Sm	0.9481	0.9394	0.9310	0.9232	0.9162	0.9102	0.9054	0.9019	0.8997	0.8990
63	Eu	0.9487	0.9400	0.9317	0.9240	0.9171	0.9111	0.9064	0.9028	0.9007	0.9000
64	Gd	0.9504	0.9420	0.9339	0.9265	0.9198	0.9140	0.9094	0.9060	0.9039	0.9032
65	Tb	0.9509	0.9426	0.9346	0.9272	0.9206	0.9149	0.9103	0.9069	0.9048	0.9041
66	Dy	0.9519	0.9438	0.9360	0.9287	0.9222	0.9167	0.9121	0.9088	0.9068	0.9061
67	Ho	0.9526	0.9446	0.9369	0.9298	0.9233	0.9178	0.9134	0.9101	0.9081	0.9075
68	Er	0.9533	0.9454	0.9378	0.9307	0.9244	0.9189	0.9145	0.9113	0.9094	0.9087
69	Tm	0.9537	0.9459	0.9384	0.9314	0.9251	0.9197	0.9154	0.9122	0.9102	0.9096

APPENDIX 3 continued

Z_2	element	scattering angle θ									
		90°	100°	110°	120°	130°	140°	150°	160°	170°	180°
70	Yb	0.9548	0.9471	0.9398	0.9329	0.9268	0.9215	0.9173	0.9142	0.9123	0.9116
71	Lu	0.9553	0.9477	0.9404	0.9337	0.9276	0.9224	0.9182	0.9151	0.9132	0.9125
72	Hf	0.9561	0.9487	0.9416	0.9349	0.9290	0.9238	0.9197	0.9167	0.9148	0.9142
73	Ta	0.9567	0.9494	0.9423	0.9358	0.9299	0.9248	0.9207	0.9178	0.9159	0.9153
74	W	0.9574	0.9502	0.9432	0.9368	0.9310	0.9260	0.9219	0.9190	0.9172	0.9166
75	Re	0.9579	0.9503	0.9439	0.9375	0.9318	0.9269	0.9229	0.9200	0.9182	0.9176
76	Os	0.9588	0.9518	0.9451	0.9389	0.9332	0.9284	0.9245	0.9216	0.9199	0.9193
77	Ir	0.9592	0.9523	0.9456	0.9394	0.9339	0.9291	0.9252	0.9224	0.9207	0.9201
78	Pt	0.9598	0.9530	0.9464	0.9403	0.9348	0.9301	0.9263	0.9235	0.9218	0.9212
79	Au	0.9602	0.9534	0.9469	0.9408	0.9354	0.9307	0.9270	0.9242	0.9225	0.9219
80	Hg	0.9609	0.9542	0.9479	0.9419	0.9365	0.9319	0.9282	0.9255	0.9238	0.9233
81	Tl	0.9616	0.9551	0.9488	0.9429	0.9377	0.9332	0.9295	0.9268	0.9252	0.9246
82	Pb	0.9621	0.9557	0.9495	0.9437	0.9385	0.9340	0.9304	0.9278	0.9262	0.9256
83	Bi	0.9624	0.9560	0.9499	0.9442	0.9390	0.9346	0.9310	0.9284	0.9268	0.9262
84	Po	0.9626	0.9562	0.9501	0.9444	0.9393	0.9349	0.9313	0.9287	0.9271	0.9266
85	At	0.9626	0.9562	0.9501	0.9444	0.9393	0.9349	0.9313	0.9287	0.9271	0.9266
86	Rn	0.9646	0.9586	0.9528	0.9473	0.9425	0.9383	0.9349	0.9324	0.9309	0.9304
87	Fr	0.9647	0.9587	0.9530	0.9476	0.9427	0.9386	0.9352	0.9327	0.9312	0.9307
88	Ra	0.9652	0.9593	0.9536	0.9482	0.9435	0.9394	0.9360	0.9336	0.9321	0.9316
89	Ac	0.9653	0.9595	0.9538	0.9485	0.9437	0.9396	0.9363	0.9339	0.9324	0.9319
90	Th	0.9661	0.9603	0.9548	0.9496	0.9449	0.9409	0.9376	0.9353	0.9338	0.9333
91	Pa	0.9659	0.9601	0.9546	0.9493	0.9447	0.9406	0.9374	0.9350	0.9335	0.9330
92	U	0.9669	0.9613	0.9559	0.9508	0.9462	0.9423	0.9392	0.9368	0.9354	0.9349

APPENDIX 4 Nuclear stopping cross section factor $\varepsilon_{n,o}$ and characteristic energies E_ε and E_η for H and He ions (Eqs.(1.72, 1.73, 1.85))

Z_2	element	H ions			He ions		
		$\varepsilon_{n,o}$	E_ε	E_η	$\varepsilon_{n,o}$	E_ε	E_η
1	H	1.498	0.087	25	4.210	0.492	100
2	He	1.060	0.124	50	4.758	0.438	200
3	Li	0.919	0.185	76	4.858	0.557	300
4	Be	0.909	0.256	101	5.146	0.720	400
5	B	0.912	0.333	126	5.391	0.894	500
6	C	0.949	0.415	151	5.750	1.087	600
7	N	0.923	0.498	176	5.758	1.267	700
8	O	0.899	0.584	201	5.742	1.453	800
9	F	0.833	0.672	227	5.460	1.629	900
10	Ne	0.849	0.767	252	5.624	1.839	1000
11	Na	0.803	0.861	277	5.410	2.029	1100
12	Mg	0.810	0.960	302	5.503	2.245	1200
13	Al	0.777	1.059	327	5.337	2.448	1300
14	Si	0.788	1.163	352	5.444	2.674	1400
15	P	0.753	1.267	378	5.255	2.884	1500
16	S	0.762	1.375	403	5.344	3.117	1600
17	Cl	0.722	1.483	428	5.106	3.330	1700
18	Ar	0.669	1.591	453	4.779	3.540	1800
19	K	0.710	1.707	478	5.069	3.799	1900
20	Ca	0.719	1.823	504	5.144	4.047	2000
21	Sc	0.665	1.937	529	4.799	4.264	2100
22	Ti	0.646	2.054	554	4.683	4.502	2200
23	V	0.627	2.173	579	4.567	4.743	2300
24	Cr	0.634	2.296	604	4.620	5.001	2400
25	Mn	0.618	2.418	629	4.520	5.249	2500
26	Fe	0.625	2.544	655	4.579	5.514	2600
27	Co	0.609	2.669	680	4.474	5.767	2700
28	Ni	0.626	2.799	705	4.607	6.044	2800
29	Cu	0.594	2.926	730	4.387	6.292	2900
30	Zn	0.591	3.056	755	4.375	6.562	3000
31	Ga	0.568	3.186	780	4.216	6.820	3100
32	Ge	0.558	3.319	806	4.148	7.089	3200
33	As	0.553	3.454	831	4.117	7.364	3300
34	Se	0.536	3.588	856	3.999	7.633	3399
35	Br	0.540	3.726	881	4.038	7.919	3499
36	Kr	0.526	3.863	906	3.937	8.194	3599
37	Rb	0.525	4.003	931	3.939	8.482	3699
38	Sr	0.522	4.144	957	3.920	8.770	3799
39	Y	0.524	4.286	982	3.937	9.063	3899
40	Zr	0.520	4.429	1007	3.911	9.355	3999
41	Nb	0.519	4.573	1032	3.910	9.651	4099
42	Mo	0.512	4.718	1057	3.859	9.944	4199
43	Tc	0.509	4.865	1083	3.842	10.243	4299
44	Ru	0.502	5.012	1108	3.792	10.540	4399
45	Rh	0.501	5.160	1133	3.785	10.844	4499
46	Pd	0.491	5.309	1157	3.722	11.144	4599
47	Ag	0.492	5.460	1183	3.730	11.453	4699
48	Cd	0.480	5.611	1208	3.637	11.753	4799
49	In	0.476	5.763	1234	3.616	12.063	4899
50	Sn	0.467	5.916	1259	3.550	12.370	4999
51	Sb	0.462	6.070	1284	3.513	12.682	5099
52	Te	0.447	6.225	1309	3.403	12.987	5199
53	I	0.455	6.382	1334	3.467	13.315	5299
54	Xe	0.446	6.539	1359	3.399	13.628	5399
55	Cs	0.446	6.698	1385	3.403	13.953	5499
56	Ba	0.437	6.856	1410	3.338	14.270	5599
57	La	0.438	7.017	1435	3.342	14.598	5699

APPENDIX 4 continued

Z_2	element	H ions $\varepsilon_{n,o}$	H ions E_ε	H ions E_η	He ions $\varepsilon_{n,o}$	He ions E_ε	He ions E_η
58	Ce	0.439	7.178	1460	3.355	14.928	5799
59	Pr	0.442	7.341	1485	3.376	15.262	5899
60	Nd	0.437	7.503	1511	3.339	15.589	5999
61	Pm	0.433	7.667	1536	3.316	15.920	6099
62	Sm	0.429	7.831	1561	3.281	16.251	6199
63	Eu	0.429	7.997	1586	3.284	16.589	6299
64	Gd	0.419	8.162	1611	3.211	16.919	6399
65	Tb	0.419	8.330	1636	3.213	17.259	6499
66	Dy	0.414	8.497	1662	3.178	17.597	6599
67	Ho	0.412	8.666	1687	3.165	17.939	6699
68	Er	0.411	8.836	1712	3.154	18.283	6799
69	Tm	0.411	9.007	1737	3.155	18.630	6899
70	Yb	0.405	9.177	1762	3.113	18.973	6999
71	Lu	0.405	9.349	1787	3.110	19.322	7099
72	Hf	0.400	9.522	1813	3.080	19.670	7199
73	Ta	0.399	9.695	1838	3.068	20.021	7299
74	W	0.396	9.870	1863	3.050	20.373	7399
75	Re	0.395	10.045	1888	3.040	20.728	7499
76	Os	0.390	10.220	1913	3.003	21.080	7599
77	Ir	0.390	10.397	1939	3.001	21.439	7699
78	Pt	0.387	10.574	1964	2.984	21.797	7799
79	Au	0.387	10.752	1989	2.983	22.158	7899
80	Hg	0.383	10.931	2014	2.955	22.517	7999
81	Tl	0.379	11.110	2039	2.927	22.878	8099
82	Pb	0.377	11.290	2064	2.913	23.242	8199
83	Bi	0.377	11.472	2090	2.912	23.609	8299
84	Po	0.379	11.654	2115	2.923	23.980	8399
85	At	0.382	11.837	2140	2.947	24.354	8499
86	Rn	0.364	12.017	2165	2.813	24.704	8599
87	Fr	0.365	12.201	2190	2.823	25.079	8699
88	Ra	0.363	12.386	2215	2.808	25.450	8799
89	Ac	0.364	12.571	2241	2.818	25.828	8899
90	Th	0.359	12.756	2266	2.779	26.198	8999
91	Pa	0.364	12.944	2291	2.813	26.582	9099
92	U	0.356	13.130	2316	2.752	26.951	9199

APPENDIX 5 ^{1}H stopping cross section ε [10^{-15}eVcm^{2}] (Ref.[30])

Z_2	element	energy of ^{1}H [keV] 200	400	600	800	1000	1200	1400	1600	1800	2000	2200	2400	2600
1	H	4.00	2.35	1.70	1.36	1.14	0.98	0.87	0.78	0.71	0.65	0.60	0.56	0.52
2	He	5.64	3.60	2.69	2.19	1.90	1.66	1.47	1.33	1.21	1.11	1.03	0.96	0.90
3	Li	6.99	4.77	3.70	3.06	2.64	2.31	2.07	1.87	1.72	1.58	1.47	1.38	1.30
4	Be	7.62	5.62	4.50	3.77	3.27	2.88	2.58	2.34	2.15	1.99	1.85	1.73	1.63
5	B	10.05	6.84	5.37	4.43	3.84	3.39	3.05	2.77	2.55	2.36	2.20	2.06	1.94
6	C	11.41	7.96	6.31	5.29	4.56	4.02	3.61	3.29	3.02	2.80	2.61	2.45	2.30
7	N	14.12	9.41	7.28	5.05	5.13	4.54	4.08	3.72	3.42	3.17	2.96	2.77	2.62
8	O	14.54	10.04	7.85	6.55	5.61	4.97	4.48	4.09	3.76	3.49	3.26	3.06	2.89
9	F	14.01	10.26	8.09	6.79	5.94	5.28	4.76	4.35	4.01	3.73	3.49	3.28	3.09
10	Ne	14.08	10.64	8.45	7.11	6.31	5.62	5.08	4.64	4.29	3.99	3.73	3.51	3.31
11	Na	16.23	11.66	9.19	7.73	6.77	6.03	5.46	5.00	4.61	4.29	4.02	3.78	3.57
12	Mg	18.68	12.66	9.97	8.40	7.25	6.46	5.85	5.36	4.95	4.61	4.32	4.06	3.84
13	Al	16.58	12.75	10.43	8.95	7.72	6.88	6.23	5.71	5.28	4.91	4.60	4.33	4.10
14	Si	19.23	13.63	11.02	9.40	8.26	7.37	6.67	6.11	5.65	5.26	4.93	4.64	4.39
15	P	20.66	14.85	11.74	9.91	8.70	7.77	7.04	6.46	5.97	5.56	5.21	4.91	4.64
16	S	21.23	15.24	12.15	10.31	9.12	8.15	7.39	6.78	6.27	5.85	5.48	5.16	4.88
17	Cl	24.75	16.70	13.16	11.11	9.56	8.55	7.76	7.12	6.59	6.14	5.75	5.42	5.13
18	Ar	24.58	16.84	13.49	11.43	9.95	8.90	8.08	7.42	6.87	6.40	6.01	5.66	5.36
19	K	24.55	17.29	13.97	11.94	10.56	9.45	8.57	7.86	7.28	6.79	6.36	6.00	5.68
20	Ca	26.00	18.19	14.66	12.52	11.09	9.91	9.00	8.25	7.64	7.12	6.68	6.29	5.96
21	Sc	25.84	18.15	14.66	12.55	11.11	9.96	9.06	8.32	7.72	7.20	6.76	6.38	6.04
22	Ti	25.68	18.29	14.84	12.74	11.35	10.19	9.28	8.53	7.92	7.39	6.94	6.55	6.21
23	V	27.37	19.46	15.51	13.20	11.66	10.48	9.54	8.78	8.15	7.62	7.16	6.76	6.40
24	Cr	27.43	20.04	15.85	13.41	11.68	10.52	9.60	8.85	8.22	7.69	7.23	6.84	6.48
25	Mn	25.83	19.80	15.94	13.58	11.92	10.74	9.81	9.05	8.42	7.88	7.41	7.01	6.65
26	Fe	26.01	19.85	15.99	13.66	12.18	10.99	10.04	9.27	8.62	8.07	7.60	7.18	6.82
27	Co	24.81	19.64	15.96	13.68	12.30	11.11	10.17	9.40	8.75	8.20	7.72	7.31	6.94
28	Ni	23.88	18.80	15.85	13.90	12.43	11.25	10.30	9.53	8.88	8.32	7.85	7.43	7.06
29	Cu	22.95	19.14	16.35	14.29	12.59	11.40	10.45	9.67	9.02	8.46	7.97	7.55	7.18
30	Zn	24.52	18.99	16.12	14.22	13.14	11.88	10.88	10.07	9.38	8.80	8.29	7.85	7.46
31	Ga	25.51	19.72	16.77	14.79	13.42	12.14	11.13	10.30	9.60	9.00	8.49	8.04	7.64
32	Ge	25.46	20.71	17.73	15.57	13.70	12.40	11.37	10.52	9.81	9.20	8.68	8.22	7.82
33	As	28.41	21.19	17.75	15.56	13.88	12.57	11.53	10.68	9.97	9.36	8.83	8.36	7.96
34	Se	26.61	21.35	18.32	16.16	14.28	12.93	11.86	10.98	10.25	9.62	9.08	8.60	8.18

APPENDIX 5 continued

Z_2	element	energy of ^1H [keV] 200	400	600	800	1000	1200	1400	1600	1800	2000	2200	2400	2600
35	Br	31.02	22.65	18.76	16.36	14.44	13.10	12.02	11.14	10.40	9.77	9.22	8.74	8.32
36	Kr	30.40	22.56	18.95	16.63	14.58	13.24	12.17	11.28	10.54	9.91	9.36	8.87	8.44
37	Rb	30.83	22.96	19.24	16.88	15.05	13.65	12.53	11.61	10.85	10.19	9.62	9.12	8.68
38	Sr	33.41	24.19	20.02	17.46	15.42	13.98	12.83	11.89	11.10	10.43	9.85	9.34	8.89
39	Y	34.62	24.85	20.45	17.80	15.62	14.17	13.01	12.06	11.26	10.59	10.00	9.48	9.02
40	Zr	36.42	25.74	21.00	18.21	15.90	14.41	13.23	12.27	11.46	10.77	10.17	9.65	9.18
41	Nb	37.81	26.42	21.40	18.49	16.00	14.50	13.31	12.34	11.53	10,84	10.24	9.72	9.26
42	Mo	37.45	26.44	21.41	18.50	15.97	14.47	13.29	12.34	11.53	10.85	10.26	9.74	9.28
43	Tc	39.45	27.49	22.08	19.00	16.30	14.77	13.57	12.59	11.77	11.07	10.46	9.93	9.47
44	Ru	38.22	27.26	21.98	18.95	16.42	14.89	13.68	12.70	11.88	11.18	10.57	10.04	9.57
45	Rh	38.33	27.45	22.06	18,99	16.36	14.86	13.68	12.71	11.90	11.20	10.60	10.08	9.61
46	Pd	36.44	26.93	21.80	18.82	16.31	14.83	13.67	12.72	11.91	11.23	10.63	10.11	9.64
47	Ag	32.33'	25.98	22.05	19.27	16.54	15.05	13.87	12.90	12.09	11.40	10.79	10.26	9.79
48	Cd	35.45	26.06	21.62	18.96	16.87	15.34	14.12	13.13	12.30	11.60	10.98	10.44	9.96
49	In	37.27	26.93	22.17	19.36	17.12	15.55	14.31	13.31	12.46	11.74	11.12	10.57	10.08
50	Sn	38.39	27.55	22.57	19.67	17.34	15.75	14.49	13.47	12.62	11.89	11.26	10.71	10.21
51	Sb	38.18	28.40	23.59	20.49	17.47	15.87	14.61	13.59	12.73	12.00	11.37	10.81	10.32
52	Te	40.82	28.85	23.49	20.41	17.80	16.16	14.88	13.84	12.96	12.22	11.58	11.01	10.51
53	I	43.26	29.95	24.21	20.96	17.68	16.07	14.81	13.78	12.92	12.18	11.55	10.99	10.49
54	Xe	45.00	30.83	24.82	21.45	18.30	16.61	15.30	14.23	13.33	12.57	11.91	11.33	10.82
55	Cs	43.00	30.47	24.82	21.57	19.00	17.22	15.84	14.72	13.78	12.98	12.29	11.69	11.15
56	Ba	45.17	31.62	25.64	22.23	19.60	17.75	16.31	15.14	14.17	13.35	12.63	12.01	11.45
57	La	46.02	32.22	26.13	22.65	20.07	18.17	16.69	15.48	14.49	13.64	12.90	12.26	11.69
58	Ce	44.79	31.84	25.93	22.53	20.18	18.28	16.80	15.60	14.60	13.75	13.01	12.36	11.79
59	Pr	44.49	31.78	25.90	22.52	20.25	18.37	16.89	15.70	14.69	13.84	13.10	12.45	11.88
60	Nd	44.18	31.71	25.87	22.51	20.25	18.39	16.93	15.74	14.74	13.89	13.15	12.51	11.93
61	Pm	43.83	31.62	25.81	22.47	20.26	18.43	16.98	15.80	14.80	13.96	13.22	12.57	12.00
62	Sm	43.49	31.57	25.81	22.49	20.24	18.43	17.00	15.82	14.84	13.99	13.26	12.62	12.04
63	Eu	43.14	31.50	25.78	22.48	20.24	18.46	17.04	15.87	14.89	14.05	13.31	12.67	12.10
64	Gd	44.13	32.09	26.20	22.83	20.68	18.85	17.40	16.20	15.19	14.33	13.58	12.92	12.34
65	Tb	42.40	31.38	25.76	22.50	20.42	18.66	17.25	16.08	15.10	14.26	13.52	12.88	12.30
66	Dy	39.80	30.35	25.17	22.11	20.31	18.59	17.20	16.06	15.09	14.26	13.53	12.89	12.32
67	Ho	38.66	29.85	24.83	21.85	20.14	18.46	17.12	16.00	15.05	14.23	13.51	12.88	12.32
68	Er	38.18	29.68	24.72	21.76	19.99	18.36	17.04	15.95	15.01	14.21	13.50	12.88	12.32
69	Tm	37.65	29.51	24.64	21.73	19.78	18.21	16.93	15.86	14.95	14.16	13.47	12.85	12.30

APPENDIX 5 continued

Z_2	element	energy of ^{1}H [keV]												
		200	400	600	800	1000	1200	1400	1600	1800	2000	2200	2400	2600
70	Yb	37.16	29.38	24.60	21.72	19.81	18.25	16.93	15.92	15.01	14.23	13.53	12.92	12.37
71	Lu	37.85	29.79	24.87	21.92	19.93	18.38	17.11	16.05	15.14	14.35	13.65	13.04	12.48
72	Hf	38.69	30.26	25.17	22.15	20.03	18.48	17.22	16.16	15.25	14.46	13.76	13.14	12.59
73	Ta	37.43	29.84	25.01	22.09	20.23	18.67	17.40	16.32	15.41	14.61	13.91	13.28	12.72
74	W	36.70	29.57	24.89	22.03	20.34	18.76	17.47	16.39	15.47	14.67	13.97	13.34	12.78
75	Re	40.02	31.21	25.90	22.77	20.76	19.11	17.78	16.67	15.72	14.90	14.18	13.54	12.97
76	Os	39.51	31.05	25.85	22.75	20.87	19.20	17.86	16.74	15.78	14.96	14.24	13.59	13.02
77	Ir	38.97	30.75	25.60	22.53	20.89	19.22	17.87	16.76	15.81	14.98	14.26	13.62	13.05
78	Pt	36.57	29.61	24.88	22.01	20.64	19.00	17.69	16.60	15.67	14.86	14.15	13.52	12.96
79	Au	39.65	31.66	26.29	23.11	20.83	19.20	17.88	16.78	15.85	15.04	14.33	13.69	13.12
80	Hg	36.40	30.08	25.35	22.42	21.12	19.44	18.10	16.99	16.04	15.22	14.49	13.85	13.28
81	Tl	39.44	32.04	26.64	23.36	21.70	19.94	18.54	17.38	16.40	15.55	14.81	14.15	13.56
82	Pb	43.40	34.28	27.97	24.25	21.66	19.91	18.52	17.37	16.39	15.55	14.82	14.16	13.57
83	Bi	48.05	36.97	29.39	25.04	21.90	20.12	18.71	17.54	16.55	15.70	14.96	14.29	13.70
84	Po	48.10	36.57	29.31	25.18	22.26	20.43	18.99	17.80	16.79	15.92	15.16	14.49	13.89
85	At	48.48	36.83	29.49	25.32	22.41	20.56	19.11	17.91	16.90	16.02	15.26	14.58	13.98
86	Rn	48.52	36.94	29.59	25.41	22.68	20.80	19.32	18.11	17.08	16.20	15.42	14.74	14.12
87	Fr	52.43	38.70	30.60	26.11	23.02	21.09	19.58	18.33	17.29	16.39	15.60	14.90	14.28
88	Ra	55.16	39.93	31.32	26.62	23.35	21.38	19.83	18.56	17.49	16.58	15.78	15.07	14.44
89	Ac	55.99	40.36	31.59	26.81	23.65	21.64	20.06	18.77	17.69	16.76	15.95	15.23	14.59
90	Th	56.42	40.53	31.68	26.86	23.63	21.62	20.05	18.77	17.69	16.77	15.96	15.24	14.61
91	Pa	55.25	40.22	31.53	26.77	23.80	21.78	20.19	18.90	17.82	16.88	16.07	15.35	14.71
92	U	54.88	40.18	31.54	26.80	23.97	21.92	20.32	19.02	17.93	16.99	16.17	15.45	14.80

APPENDIX 6 ^{4}He stopping cross section ε [10^{-15}eVcm^{2}] (Ref.[31])

Z_2	element	energy of ^{4}He [keV]												
		200	400	600	800	1000	1200	1400	1600	1800	2000	2200	2400	2600
1	H	8.07	9.48	9.58	9.15	8.53	7.90	7.31	6.80	6.35	5.97	5.64	5.35	5.09
2	He	9.39	11.05	11.80	12.05	11.99	11.73	11.36	10.91	10.44	9.97	9.51	9.08	8.67
3	Li	17.95	22.35	23.46	23.01	21.89	20.57	19.25	18.02	16.91	15.93	15.05	14.27	13.58
4	Be	25.17	26.92	26.48	25.39	24.14	22.88	21.69	20.58	19.57	18.64	17.80	17.02	16.32
5	B	30.90	36.66	37.71	36.64	34.69	32.48	30.30	28.28	26.45	24.82	23.37	22.09	20.94
6	C	31.28	37.03	38.38	37.68	36.07	34.13	32.17	30.30	28.59	27.04	25.65	24.40	23.28
7	N	29.96	38.36	41.66	42.10	40.99	39.13	36.99	34.85	32.81	30.94	29.25	27.73	26.37
8	O	27.70	36.86	40.97	42.09	41.50	40.03	38.18	36.24	34.36	32.59	30.98	29.51	28.18
9	F	24.86	33.74	38.36	40.31	40.59	39.86	38.59	37.06	35.46	33.89	32.39	30.99	29.69
10	Ne	21.29	29.46	34.03	36.29	37.05	36.88	36.17	35.18	34.06	32.91	31.78	30.69	29.66
11	Na	44.42	48.62	49.15	48.30	46.88	45.23	43.51	41.82	40.19	38.63	37.17	35.80	34.51
12	Mg	47.12	54.66	55.23	53.78	51.73	49.56	47.41	45.37	43.45	41.66	39.99	38.44	37.00
13	Al	50.88	57.34	58.11	56.87	54.87	52.61	50.33	48.13	46.06	44.13	42.34	40.67	39.13
14	Si	59.12	72.35	71.61	67.52	63.21	59.32	55.91	52.90	50.25	47.87	45.74	43.81	42.05
15	P	51.81	67.42	69.74	67.34	63.82	60.31	57.09	54.21	51.62	49.30	47.20	45.29	43.55
16	S	48.37	66.61	71.76	70.68	67.42	63.72	60.19	57.00	54.14	51.60	49.33	47.28	45.42
17	Cl	44.66	64.37	71.86	72.33	69.69	66.06	62.37	58.95	55.89	53.17	50.76	48.60	46.67
18	Ar	41.03	61.12	70.23	72.22	70.52	67.35	63.82	60.41	57.30	54.51	52.03	49.82	47.84
19	K	62.29	75.40	80.01	80.19	78.01	74.72	71.02	67.30	63.77	60.50	57.53	54.84	52.42
20	Ca	69.01	84.84	88.70	87.06	83.23	78.80	74.45	70.44	66.82	63.58	60.67	58.06	55.71
21	Sc	68.06	85.30	90.08	88.97	85.42	81.12	76.81	72.77	69.09	65.77	62.78	60.08	57.62
22	Ti	66.07	83.54	88.44	87.50	84.19	80.17	76.16	72.39	68.95	65.83	63.00	60.43	58.08
23	V	65.25	82.94	90.38	92.10	90.59	87.39	83.44	79.30	75.26	71.45	67.93	64.73	61.81
24	Cr	62.68	78.77	85.47	87.01	85.66	82.82	79.32	75.64	72.05	68.67	65.54	62.68	60.06
25	Mn	59.95	76.24	83.26	85.12	84.07	81.52	78.30	74.90	71.56	68.40	65.46	62.77	60.30
26	Fe	58.82	76.19	84.22	86.88	86.39	84.17	81.11	77.75	74.37	71.12	68.08	65.26	62.68
27	Co	57.19	74.14	81.22	82.97	81.96	79.67	76.88	73.99	71.16	68.48	65.98	63.65	61.48
28	Ni	51.01	66.14	73.78	77.09	77.75	76.84	75.07	72.86	70.49	68.09	65.76	63.54	61.44
29	Cu	48.21	64.05	71.40	74.12	74.39	73.42	71.87	70.05	68.14	66.22	64.34	62.53	60.78
30	Zn	48.21	64.09	71.58	74.53	75.08	74.39	73.07	71.44	69.67	67.86	66.05	64.29	62.57
31	Ga	48.21	64.13	71.74	74.88	75.65	75.17	74.03	72.54	70.88	69.15	67.40	65.67	63.97
32	Ge	62.90	75.65	79.92	81.06	80.71	79.57	77.99	76.17	74.24	72.28	70.32	68.40	66.53
33	As	70.67	84.60	86.96	86.45	84.88	82.83	80.56	78.21	75.87	73.58	71.37	69.26	67.24
34	Se	66.41	84.86	90.40	89.86	86.99	83.50	80.05	76.86	73.97	71.36	68.99	66.82	64.84

Z_2	element	energy of ^{4}He [keV]												
		200	400	600	800	1000	1200	1400	1600	1800	2000	2200	2400	2600
35	Br	65.38	84.77	89.87	90.08	88.61	86.51	84.15	81.72	79.29	76.92	74.63	72.44	70.34
36	Kr	61.35	85.98	93.65	93.89	91.61	88.64	85.60	82.66	79.89	77.30	74.86	72.58	70.44
37	Rb	85.17	101.86	106.80	106.19	103.10	99.08	94.88	90.84	87.08	83.62	80.45	77.55	74.90
38	Sr	86.96	110.03	117.50	117.04	112.98	107.66	102.19	97.02	92.32	88.10	84.32	80.92	77.86
39	Y	91.12	111.38	118.63	118.99	115.73	110.81	105.34	99.92	94.82	90.16	85.95	82.17	78.77
40	Zr	93.42	115.91	124.78	126.12	123.25	118.26	112.43	106.47	100.78	95.50	90.71	86.39	82.51
41	Nb	86.46	109.22	119.25	121.99	120.44	116.55	111.56	106.22	100.96	95.98	91.39	87.20	83.40
42	Mo	82.56	103.75	113.11	115.77	114.51	111.11	106.70	101.96	97.27	92.82	88.70	84.92	81.48
43	Tc	94.75	118.36	122.94	121.51	118.07	113.94	109.64	105.39	101.30	97.43	93.78	90.36	87.16
44	Ru	82.04	107.91	116.73	117.35	114.52	110.46	106.11	101.84	97.79	94.01	90.48	87.21	84.17
45	Rh	83.76	104.94	112.50	114.16	112.91	110.22	106.84	103.19	99.50	95.88	92.39	89.08	85.94
46	Pd	75.56	99.21	110.13	113.46	112.30	108.75	104.13	99.21	94.39	89.87	85.72	81.96	78.55
47	Ag	72.97	96.04	107.19	111.35	111.38	109.17	105.85	102.12	98.32	94.64	91.17	87.94	84.94
48	Cd	82.48	106.23	112.28	112.09	109.81	106.84	103.65	100.44	97.30	94.27	91.37	88.60	85.98
49	In	82.68	104.37	113.22	116.13	115.96	114.13	111.43	108.30	105.00	101.68	98.41	95.25	92.23
50	Sn	84.78	108.59	116.75	118.11	116.60	113.88	110.66	107.27	103.87	100.55	97.35	94.29	91.37
51	Sb	102.33	119.45	122.25	121.20	118.56	115.21	111.57	107.87	104.24	100.74	97.39	94.22	91.22
52	Te	98.09	122.44	127.02	125.88	122.88	119.20	115.29	111.38	107.56	103.89	100.40	97.09	93.97
53	I	93.03	124.40	131.25	128.17	122.41	116.49	111.09	106.29	102.03	98.22	94.78	91.65	88.78
54	Xe	86.33	121.20	131.44	130.69	126.34	121.20	116.17	111.50	107.20	103.27	99.66	96.32	93.23
55	Cs	110.66	134.10	141.84	141.68	137.66	132.03	125.98	120.08	114.58	109.56	105.00	100.87	97.13
56	Ba	119.02	145.66	151.70	148.88	142.86	136.14	129.65	123.68	118.27	113.39	108.97	104.95	101.27
57	La	118.81	144.68	154.60	155.85	152.26	146.22	139.19	132.01	125.13	118.76	112.95	107.71	103.00
58	Ce	111.37	138.42	146.65	145.74	140.87	134.71	128.45	122.57	117.20	112.35	107.98	104.02	100.42
59	Pr	107.69	134.42	143.31	143.24	139.07	133.40	127.46	121.79	116.56	111.81	107.50	103.60	100.04
60	Nd	104.37	130.77	140.20	140.88	137.36	132.15	126.52	121.04	115.95	111.29	107.05	103.20	99.69
61	Pm	101.10	127.08	136.96	138.34	135.46	130.73	125.44	120.19	115.24	110.69	106.52	102.73	99.26
62	Sm	98.00	123.61	133.91	135.94	133.65	129.35	124.35	119.28	114.46	109.99	105.89	102.14	98.72
63	Eu	95.06	116.48	126.40	129.93	129.57	126.93	123.08	118.67	114.13	109.70	105.48	101.55	97.92
64	Gd	95.61	121.82	134.01	138.02	137.13	133.52	128.61	123.24	117.90	112.83	108.13	103.84	99.93
65	Tb	90.23	114.81	125.97	129.46	128.54	125.29	121.00	116.39	111.85	107.56	103.59	99.95	96.62
66	Dy	88.98	112.56	123.84	128.02	127.90	125.31	121.46	117.10	112.65	108.34	104.29	100.54	97.09
67	Ho	85.39	108.87	120.01	124.04	123.84	121.31	117.63	113.53	109.40	105.43	101.71	98.28	95.11
68	Er	86.85	108.20	118.97	123.68	124.55	123.02	120.07	116.39	112.40	108.37	104.46	100.75	97.28
69	Tm	84.94	108.00	118.09	121.01	120.07	117.23	113.63	109.84	106.16	102.68	99.45	96.46	93.70

APPENDIX 6 continued

Z_2	element	energy of ^{4}He [keV]												
		200	400	600	800	1000	1200	1400	1600	1800	2000	2200	2400	2600
70	Yb	83.04	105.70	115.72	118.74	117.95	115.28	111.85	108.23	104.70	101.36	98.26	95.39	92.74
71	Lu	83.36	107.88	117.91	120.25	118.83	115.84	112.33	108.80	105.41	102.25	99.30	96.56	94.02
72	Hf	82.22	107.69	118.48	121.27	120.04	117.09	113.54	109.93	106.46	103.21	100.19	97.39	94.79
73	Ta	80.51	104.48	116.33	120.95	121.18	118.95	115.52	111.63	107.69	103.91	100.38	97.13	94.14
74	W	79.79	105.25	117.72	122.32	122.23	119.63	115.90	111.80	107.74	103.90	100.36	97.12	94.16
75	Re	79.63	108.37	121.50	125.44	124.55	121.48	117.63	113.67	109.86	106.32	103.05	100.03	97.25
76	Os	79.28	109.00	122.94	127.36	126.63	123.51	119.51	115.36	111.37	107.65	104.23	101.09	98.20
77	Ir	79.10	109.44	123.95	128.76	128.23	125.16	121.11	116.85	112.73	108.88	105.31	102.04	99.03
78	Pt	73.76	100.10	113.61	119.07	119.65	117.58	114.29	110.56	106.81	103.23	99.90	96.83	94.02
79	Au	71.16	100.70	116.66	123.62	125.00	123.28	120.08	116.32	112.46	108.74	105.24	102.00	99.01
30	Hg	75.25	106.96	122.56	128.10	128.15	125.59	122.02	118.17	114.39	110.81	107.46	104.34	101.44
81	Tl	85.70	116.79	131.47	136.28	135.62	132.34	128.04	123.51	119.11	114.99	111.19	107.69	104.46
82	Pb	93.69	123.23	135.63	139.00	137.96	134.99	131.28	127.38	123.52	119.81	116.28	112.94	109.78
83	Bi	104.76	133.65	144.77	146.57	143.86	139.31	134.23	129.21	124.48	120.11	116.10	112.41	109.01
84	Po	91.10	131.88	148.32	151.04	147.80	142.57	136.99	131.67	126.76	122.27	118.18	114.42	110.96
85	At	89.76	132.88	151.11	154.55	151.31	145.77	139.80	134.10	128.86	124.11	119.80	115.87	112.26
86	Rn	87.06	132.39	153.44	158.21	154.86	148.56	141.76	135.38	129.68	124.68	120.29	116.40	112.94
87	Fr	115.51	149.08	164.13	168.33	166.24	160.87	154.10	147.01	140.14	133.75	127.91	122.63	117.87
88	Ra	120.47	159.28	175.52	178.66	174.80	167.80	159.80	151.88	144.49	137.78	131.75	126.35	121.51
89	Ac	119.74	162.25	180.07	183.32	178.97	171.40	162.92	154.66	147.02	140.15	134.00	128.50	123.58
90	Th	118.39	163.48	182.88	186.78	182.52	174.78	166.06	157.53	149.65	142.55	136.21	130.55	125.47
91	Pa	113.81	156.65	175.79	180.47	177.28	170.54	162.64	154.75	147.37	140.66	134.63	129.21	124.34
92	U	111.00	153.54	173.18	178.59	176.05	169.77	162.17	154.46	147.20	140.56	134.57	129.20	124.36

APPENDIX 7 Electronic straggling function w_e (Eq.(1.109)) for H and He ions

Z_2	element	energy of H and He ions [MeV]						
		H^+ 0.1	0.2	0.3	0.4	0.5	0.75	1.0
		He^+ 0.4	0.8	1.2	1.6	2.0	3.0	4.0
2	He	.742	.966	1.06	1.09	1.10	1.10	1.08
3	Li	.651	.795	.887	.948	.989	1.04	1.06
4	Be	.706	.782	.839	.883	.919	.979	1.01
5	B	.737	.807	.842	.872	.898	.945	.977
6	C	.667	.824	.858	.881	.899	.934	.960
7	N	.576	.782	.862	.889	.905	.934	.955
8	O	.491	.707	.823	.881	.907	.937	.955
9	F	.419	.629	.757	.838	.887	.936	.956
10	Ne	.361	.557	.687	.777	.841	.924	.953
11	Na	.357	.516	.633	.719	.784	.888	.937
12	Mg	.378	.506	.604	.680	.741	.849	.912
13	Al	.404	.509	.589	.656	.710	.813	.881
14	Si	.428	.520	.587	.644	.692	.786	.852
15	P	.433	.534	.592	.641	.682	.766	.829
16	S	.418	.548	.601	.643	.679	.752	.811
17	Cl	.394	.552	.610	.648	.680	.745	.798
18	Ar	.367	.538	.617	.655	.683	.742	.789
19	K	.355	.519	.611	.656	.684	.738	.781
20	Ca	.357	.505	.599	.656	.685	.736	.775
21	Sc	.345	.487	.585	.649	.687	.738	.774
22	Ti	.330	.468	.566	.635	.682	.739	.774
23	V	.315	.448	.545	.617	.670	.739	.774
24	Cr	.285	.421	.520	.595	.653	.738	.776
25	Mn	.285	.410	.503	.576	.634	.729	.773
26	Fe	.271	.392	.483	.555	.614	.716	.769
27	Co	.257	.375	.463	.534	.592	.698	.762
28	Ni	.245	.358	.444	.514	.571	.680	.750
29	Cu	.219	.332	.418	.490	.548	.660	.735
30	Zn	.222	.327	.407	.475	.530	.640	.717
31	Ga	.229	.326	.400	.464	.515	.622	.699
32	Ge	.237	.329	.398	.456	.505	.606	.682
33	As	.243	.334	.398	.452	.498	.593	.667
34	Se	.245	.340	.400	.450	.493	.583	.654
35	Br	.242	.345	.404	.450	.491	.575	.642
36	Kr	.237	.348	.408	.452	.490	.570	.633
37	Rb	.235	.347	.410	.453	.489	.564	.625
38	Sr	.239	.346	.411	.454	.488	.559	.618
39	Y	.240	.346	.414	.458	.491	.558	.613
40	Zr	.239	.344	.415	.461	.493	.557	.610
41	Nb	.229	.339	.414	.464	.497	.558	.608
42	Mo	.223	.334	.410	.464	.500	.559	.607
43	Tc	.224	.331	.406	.461	.500	.560	.605
44	Ru	.209	.320	.399	.457	.500	.563	.606
45	Rh	.202	.312	.391	.451	.497	.564	.606
46	Pd	.189	.300	.382	.444	.493	.566	.608
47	Ag	.188	.294	.374	.436	.486	.565	.608
48	Cd	.190	.291	.367	.429	.478	.563	.608
49	In	.193	.289	.362	.422	.470	.559	.607
50	Sn	.199	.290	.359	.416	.464	.554	.606
51	Sb	.204	.292	.357	.412	.458	.548	.604
52	Te	.207	.295	.357	.410	.454	.542	.602
53	I	.208	.298	.358	.408	.451	.537	.598

Z_2	element	energy of H and He ions [MeV]						
		H^+ 0.1	0.2	0.3	0.4	0.5	0.75	1.0
		He^+ 0.4	0.8	1.2	1.6	2.0	3.0	4.0
54	Xe	.207	.301	.360	.408	.449	.532	.593
55	Cs	.207	.302	.360	.406	.446	.527	.588
56	Ba	.210	.304	.362	.406	.445	.523	.583
57	La	.211	.306	.364	.407	.444	.520	.579
58	Ce	.199	.292	.352	.396	.434	.511	.571
59	Pr	.193	.286	.346	.391	.428	.504	.564
60	Nd	.188	.280	.340	.385	.422	.498	.558
61	Pm	.183	.274	.334	.379	.416	.491	.551
62	Sm	.178	.267	.328	.373	.409	.484	.544
63	Eu	.173	.261	.321	.366	.403	.477	.537
64	Gd	.175	.261	.321	.365	.401	.473	.532
65	Tb	.165	.249	.309	.353	.390	.464	.523
66	Dy	.160	.244	.302	.347	.384	.457	.515
67	Ho	.156	.238	.296	.341	.377	.450	.508
68	Er	.152	.232	.290	.334	.371	.443	.501
69	Tm	.149	.227	.284	.328	.364	.435	.494
70	Yb	.145	.222	.278	.322	.358	.429	.487
71	Lu	.147	.222	.278	.321	.356	.425	.482
72	Hf	.149	.223	.278	.320	.355	.423	.478
73	Ta	.149	.223	.278	.320	.354	.421	.474
74	W	.148	.222	.278	.320	.354	.420	.472
75	Re	.147	.222	.277	.320	.355	.419	.469
76	Os	.145	.220	.276	.320	.355	.418	.468
77	Ir	.143	.218	.275	.319	.354	.418	.466
78	Pt	.136	.213	.272	.318	.354	.419	.466
79	Au	.133	.210	.269	.316	.353	.419	.465
80	Hg	.134	.210	.268	.314	.352	.419	.465
81	Tl	.138	.210	.266	.312	.350	.418	.464
82	Pb	.141	.212	.266	.311	.348	.418	.464
83	Bi	.145	.214	.266	.310	.347	.416	.463
84	Po	.148	.216	.267	.310	.346	.415	.463
85	At	.151	.219	.269	.310	.345	.414	.462
86	Rn	.152	.222	.270	.310	.345	.413	.462
87	Fr	.152	.224	.271	.310	.344	.411	.460
88	Ra	.155	.226	.273	.311	.344	.410	.460
89	Ac	.157	.229	.276	.313	.345	.410	.459
90	Th	.159	.232	.278	.314	.345	.409	.458
91	Pa	.152	.226	.274	.311	.343	.408	.458
92	U	.149	.223	.272	.310	.342	.407	.457
93	Np	.146	.220	.270	.308	.340	.405	.456
94	Pu	.139	.212	.264	.303	.336	.403	.455
95	Am	.136	.209	.261	.301	.334	.401	.453
96	Cm	.138	.210	.261	.300	.333	.399	.451
97	Bk	.135	.206	.258	.297	.330	.396	.449

APPENDIX 8 Factors c_φ and c_z for 1 MeV H ions and He ions incident on Si^z (Eqs.(1.136) and (1.137))

Z_2	element	H ions c_φ	H ions c_z [$10^{-17}cm^2$]	He ions c_φ	He ions c_z [$10^{-17}cm^2$]
1	H	11503	3.448	5056	2.665
2	He	5056	2.665	2282	2.172
3	Li	3089	2.239	1415	1.880
4	Be	2167	1.959	1003	1.679
5	B	1642	1.757	765.9	1.528
6	C	1307	1.603	613.1	1.410
7	N	1076	1.480	507.3	1.314
8	O	909.4	1.379	430.1	1.234
9	F	783.2	1.295	371.6	1.166
10	Ne	684.9	1.222	325.9	1.107
11	Na	606.5	1.160	289.3	1.055
12	Mg	542.6	1.105	259.4	1.010
13	Al	489.7	1.056	234.5	0.969
14	Si	445.3	1.013	213.6	0.932
15	P	407.5	0.974	195.8	0.899
16	S	375.0	0.938	180.4	0.869
17	Cl	346.9	0.906	167.1	0.841
18	Ar	322.2	0.876	155.4	0.815
19	K	300.5	0.849	145.1	0.792
20	Ca	281.2	0.824	135.9	0.770
21	Sc	264.0	0.801	127.7	0.750
22	Ti	248.5	0.779	120.3	0.731
23	V	234.6	0.759	113.7	0.713
24	Cr	222.0	0.740	107.7	0.696
25	Mn	210.6	0.722	102.2	0.680
26	Fe	200.1	0.705	97.18	0.665
27	Co	190.5	0.690	92.59	0.651
28	Ni	181.7	0.675	88.36	0.638
29	Cu	173.6	0.661	84.47	0.625
30	Zn	166.1	0.647	80.86	0.613
31	Ga	159.2	0.634	77.52	0.602
32	Ge	152.7	0.622	74.42	0.591
33	As	146.7	0.611	71.53	0.581
34	Se	141.1	0.600	68.82	0.571
35	Br	135.9	0.589	66.30	0.561
36	Kr	131.0	0.579	63.93	0.552
37	Rb	126.4	0.570	61.71	0.543
38	Sr	122.0	0.560	59.62	0.535
39	Y	118.0	0.552	57.65	0.527
40	Zr	114.1	0.543	55.79	0.519
41	Nb	110.5	0.535	54.04	0.512
42	Mo	107.1	0.527	52.38	0.504
43	Tc	103.8	0.519	50.81	0.497
44	Ru	100.8	0.512	49.32	0.491
45	Rh	97.84	0.505	47.90	0.484
46	Pd	95.07	0.498	46.56	0.478
47	Ag	92.43	0.492	45.28	0.472
48	Cd	89.91	0.485	44.06	0.466
49	In	87.52	0.479	42.89	0.460
50	Sn	85.23	0.473	41.78	0.455
51	Sb	83.05	0.467	40.72	0.450
52	Te	80.96	0.462	39.71	0.444
53	I	78.96	0.456	38.74	0.439
54	Xe	77.05	0.451	37.81	0.434

Z_2	element	H ions c_φ	H ions c_z $[10^{-17}cm^2]$	He ions c_φ	He ions c_z $[10^{-17}cm^2]$
55	Cs	75.22	0.446	36.92	0.430
56	Ba	73.46	0.441	36.06	0.425
57	La	71.78	0.436	35.24	0.421
58	Ce	70.16	0.431	34.45	0.416
59	Pr	68.60	0.427	33.69	0.412
60	Nd	67.10	0.422	32.96	0.408
61	Pm	65.66	0.418	32.26	0.404
62	Sm	64.27	0.414	31.59	0.400
63	Eu	62.94	0.410	30.93	0.396
64	Gd	61.65	0.406	30.31	0.392
65	Tb	60.41	0.402	29.70	0.388
66	Dy	59.21	0.398	29.11	0.385
67	Ho	58.05	0.394	28.55	0.381
68	Er	56.93	0.390	28.00	0.378
69	Tm	55.85	0.387	27.47	0.375
70	Yb	54.80	0.383	26.96	0.371
71	Lu	53.79	0.380	26.47	0.368
72	Hf	52.81	0.377	25.99	0.365
73	Ta	51.86	0.373	25.53	0.362
74	W	50.94	0.370	25.08	0.359
75	Re	50.05	0.367	24.64	0.356
76	Os	49.18	0.364	24.22	0.353
77	Ir	48.34	0.361	23.81	0.350
78	Pt	47.53	0.358	23.41	0.348
79	Au	46.74	0.355	23.02	0.345
80	Hg	45.97	0.352	22.65	0.342
81	Tl	45.23	0.350	22.28	0.340
82	Pb	44.50	0.347	21.93	0.337
83	Bi	43.80	0.344	21.58	0.334
84	Po	43.11	0.342	21.25	0.332
85	At	42.44	0.339	20.92	0.330
86	Rn	41.80	0.337	20.60	0.327
87	Fr	41.16	0.334	20.30	0.325
88	Ra	40.55	0.332	19.99	0.323
89	Ac	39.95	0.329	19.70	0.320
90	Th	39.37	0.327	19.41	0.318
91	Pa	38.80	0.325	19.14	0.316
92	U	38.24	0.323	18.86	0.314

APPENDIX 9 Dechanneling function $\tilde{P}(\tilde{\psi}_c, \tau)$ according to Eqs.(1.257, 1.258, 1.259) calculated using the multiple scattering angular distribution of Meyer [38] (for $\tilde{\psi}_c$ and τ see Eqs.(1.136,1.137) and Appendix 8)

reduced thick-ness τ	reduced critical angel $\tilde{\psi}_c$									
	0.2	0.4	0.6	0.8	1.0	1.2	1.4	1.6	1.8	2.0
0.0	0.0000	0.0000	0.0000	0.0000	0.0000	0.0000	0.0000	0.0000	0.0000	0.0000
0.2	0.3291	0.1461	0.0784	0.0460	0.0306	0.0288	0.0178	0.0140	0.0110	0.0084
0.4	0.5806	0.2969	0.1691	0.1032	0.0694	0.0515	0.0400	0.0321	0.0259	0.0208
0.6	0.7205	0.4170	0.2503	0.1584	0.1077	0.0789	0.0609	0.0490	0.0400	0.0325
0.8	0.8056	0.5230	0.3332	0.2190	0.1515	0.1113	0.0858	0.0689	0.0561	0.0460
1.0	0.8574	0.6077	0.4099	0.2795	0.1969	0.1457	0.1132	0.0906	0.0739	0.0610
1.2	0.8904	0.6726	0.4770	0.3352	0.2406	0.1791	0.1390	0.1110	0.0902	0.0742
1.4	0.9125	0.7237	0.5358	0.3889	0.2855	0.2150	0.1673	0.1337	0.1087	0.0897
1.6	0.9281	0.7633	0.5854	0.4372	0.3262	0.2483	0.1938	0.1548	0.1255	0.1035
1.8	0.9397	0.7948	0.6287	0.4817	0.3670	0.2827	0.2219	0.1776	0.1444	0.1190
2.0	0.9484	0.8206	0.6657	0.5220	0.4049	0.3158	0.2494	0.1995	0.1623	0.1336
2.5	0.9627	0.8647	0.7362	0.6048	0.4907	0.3939	0.3171	0.2577	0.2117	0.1758
3.0	0.9715	0.8942	0.7868	0.6697	0.5588	0.4606	0.3786	0.3124	0.2592	0.2166
3.5	0.9771	0.9137	0.8225	0.7185	0.6143	0.5182	0.4341	0.3629	0.3042	0.2558
4.0	0.9312	0.9280	0.8497	0.7571	0.6612	0.5688	0.4848	0.4115	0.3495	0.2970
4.5	0.9341	0.9389	0.8709	0.7885	0.6996	0.6119	0.5296	0.4558	0.3913	0.3353
5.0	0.9863	0.9472	0.8872	0.8133	0.7313	0.6482	0.5685	0.4954	0.4308	0.3737
6.0	0.9894	0.9589	0.9114	0.8506	0.7812	0.7079	0.6348	0.5640	0.4985	0.4385
7.0	0.9914	0.9665	0.9271	0.8761	0.8167	0.7523	0.6855	0.6192	0.5554	0.4956
8.0	0.9928	0.9719	0.9389	0.8956	0.8442	0.7869	0.7259	0.6644	0.6039	0.5462
10.0	0.9947	0.9792	0.9543	0.9210	0.8804	0.8345	0.7847	0.7323	0.6790	0.6259
12.0	0.9959	0.9839	0.9644	0.9381	0.9058	0.8680	0.8264	0.7816	0.7350	0.6876
14.0	0.9966	0.9367	0.9707	0.9439	0.9219	0.8902	0.8548	0.8162	0.7750	0.7322
16.0	0.9971	0.9888	0.9752	0.9566	0.9334	0.9059	0.8748	0.8408	0.8045	0.7662
18.0	0.9975	0.9903	0.9785	0.9623	0.9422	0.9182	0.8909	0.8607	0.8280	0.7936
20.0	0.9979	0.9916	0.9813	0.9672	0.9494	0.9281	0.9038	0.87658	0.8474	0.8162

APPENDIX 9 continued

reduced thick-ness τ	reduced critical angel $\tilde{\psi}_c$									
	2.2	2.4	2.6	2.8	3.0	3.2	3.4	3.6	3.8	4.0
0.0	0.0000	0.0000	0.0000	0.0000	0.0000	0.0000	0.0000	0.0000	0.0000	0.0000
0.2	0.0061	0.0040	0.0022	0.0006	0.0000	0.0000	0.0000	0.0000	0.0000	0.0000
0.4	0.0163	0.0126	0.0092	0.0063	0.0038	0.0016	0.0000	0.0000	0.0000	0.0000
0.6	0.0261	0.0207	0.0161	0.0121	0.0087	0.0057	0.0032	0.0009	0.0000	0.0000
0.8	0.0378	0.0311	0.0256	0.0212	0.0176	0.0146	0.0121	0.0101	0.0085	0.0085
1.0	0.0505	0.0418	0.0345	0.0285	0.0235	0.0194	0.0159	0.0129	0.0105	0.0105
1.2	0.0616	0.0513	0.0428	0.0357	0.0299	0.0250	0.0209	0.0174	0.0145	0.0145
1.4	0.0748	0.0628	0.0532	0.0453	0.0385	0.0329	0.0282	0.0242	0.0207	0.0207
1.6	0.0862	0.0722	0.0608	0.0515	0.0438	0.0375	0.0321	0.0276	0.0238	0.0238
1.8	0.0991	0.0832	0.0702	0.0596	0.0509	0.0438	0.0378	0.0327	0.0284	0.0284
2.0	0.1110	0.0931	0.0794	0.0670	0.0577	0.0496	0.0426	0.0371	0.0323	0.0280
2.5	0.1473	0.1244	0.1069	0.0912	0.0793	0.0689	0.0599	0.0529	0.0466	0.0410
3.0	0.1825	0.1548	0.1331	0.1136	0.0983	0.0859	0.0747	0.0660	0.0582	0.0512
3.5	0.2163	0.1837	0.1582	0.1351	0.1176	0.1022	0.0889	0.0784	0.0689	0.0606
4.0	0.2531	0.2167	0.1876	0.1608	0.1404	0.1223	0.1065	0.0941	0.0830	0.0732
4.5	0.2878	0.2476	0.2149	0.1850	0.1616	0.1408	0.1224	0.1081	0.0952	0.0837
5.0	0.3234	0.2801	0.2446	0.2117	0.1857	0.1624	0.1417	0.1254	0.1108	0.0979
6.0	0.3848	0.3376	0.2972	0.2594	0.2290	0.2015	0.1768	0.1570	0.1390	0.1230
7.0	0.4408	0.3912	0.3479	0.3071	0.2730	0.2418	0.2134	0.1902	0.1691	0.1502
8.0	0.4917	0.4407	0.3938	0.3516	0.3138	0.2797	0.2494	0.2225	0.1987	0.1776
10.0	0.5741	0.5244	0.4773	0.4333	0.3925	0.3550	0.3207	0.2896	0.2615	0.2361
12.0	0.6401	0.5933	0.5479	0.5043	0.4627	0.4237	0.3875	0.3539	0.3231	0.2949
14.0	0.6835	0.6448	0.6012	0.5590	0.5182	0.4794	0.4425	0.4078	0.3753	0.3450
16.0	0.7271	0.6873	0.6473	0.6076	0.5686	0.5311	0.4948	0.4600	0.4270	0.3960
18.0	0.7577	0.7210	0.6838	0.6462	0.6091	0.5728	0.5373	0.5030	0.4700	0.4386
20.0	0.7834	0.7495	0.7149	0.6799	0.6451	0.6104	0.5764	0.5430	0.5104	0.4792

APPENDIX 10 Rutherford scattering cross section $d\sigma/d\Omega$ (Eq.3.29) for 1 MeV ^1H [$10^{-24}cm^2sr^{-1}$]

Z_2 element	scattering angle θ									
	90°	100°	110°	120°	130°	140°	150°	160°	170°	180°
2 He	0.0201	0.0144	0.0108	0.0086	0.0070	0.0060	0.0053	0.0049	0.0046	0.0045
3 Li	0.0462	0.0334	0.0254	0.0202	0.0168	0.0145	0.0129	0.0119	0.0114	0.0112
4 Be	0.0824	0.0597	0.0455	0.0363	0.0302	0.0261	0.0233	0.0215	0.0205	0.0202
5 B	0.1290	0.0935	0.0714	0.0570	0.0475	0.0410	0.0367	0.0339	0.0323	0.0318
6 C	0.1860	0.1348	0.1030	0.0823	0.0685	0.0592	0.0529	0.0489	0.0467	0.0460
7 N	0.2533	0.1837	0.1404	0.1122	0.0935	0.0808	0.0723	0.0669	0.0638	0.0628
8 O	0.3311	0.2402	0.1835	0.1468	0.1223	0.1057	0.0946	0.0875	0.0836	0.0823
9 F	0.4193	0.3042	0.2325	0.1860	0.1550	0.1340	0.1200	0.1110	0.1060	0.1044
10 Ne	0.5177	0.3757	0.2872	0.2297	0.1914	0.1656	0.1482	0.1371	0.1309	0.1289
11 Na	0.6266	0.4547	0.3477	0.2782	0.2318	0.2005	0.1795	0.1661	0.1586	0.1562
12 Mg	0.7458	0.5413	0.4138	0.3311	0.2759	0.2387	0.2137	0.1978	0.1888	0.1860
13 Al	0.8754	0.6354	0.4858	0.3887	0.3240	0.2803	0.2510	0.2322	0.2218	0.2184
14 Si	1.0153	0.7369	0.5635	0.4509	0.3758	0.3251	0.2911	0.2694	0.2572	0.2533
15 P	1.1657	0.8461	0.6470	0.5177	0.4316	0.3733	0.3343	0.3094	0.2954	0.2910
16 S	1.3264	0.9627	0.7362	0.5891	0.4911	0.4248	0.3804	0.3520	0.3362	0.3311
17 Cl	1.4975	1.0870	0.8312	0.6652	0.5545	0.4797	0.4296	0.3976	0.3797	0.3739
18 Ar	1.6790	1.2188	0.9320	0.7459	0.6218	0.5380	0.4818	0.4459	0.4258	0.4193
19 K	1.8707	1.3579	1.0384	0.8311	0.6928	0.5994	0.5368	0.4967	0.4744	0.4672
20 Ca	2.0728	1.5046	1.1506	0.9209	0.7677	0.6642	0.5948	0.5504	0.5257	0.5177
21 Sc	2.2854	1.6590	1.2687	1.0154	0.8465	0.7324	0.6559	0.6070	0.5797	0.5709
22 Ti	2.5083	1.8209	1.3925	1.1145	0.9291	0.8039	0.7200	0.6663	0.6363	0.6267
23 V	2.7416	1.9902	1.5220	1.2182	1.0155	0.8787	0.7870	0.7283	0.6955	0.6850
24 Cr	2.9852	2.1671	1.6573	1.3265	1.1058	0.9568	0.8569	0.7930	0.7574	0.7459
25 Mn	3.2393	2.3515	1.7983	1.4394	1.1999	1.0382	0.9299	0.8605	0.8218	0.8094
26 Fe	3.5036	2.5434	1.9451	1.5568	1.2979	1.1230	1.0058	0.9308	0.8889	0.8755
27 Co	3.7783	2.7428	2.0976	1.6790	1.3997	1.2111	1.0847	1.0038	0.9587	0.9442
28 Ni	4.0634	2.9498	2.2559	1.8056	1.5053	1.3024	1.1665	1.0796	1.0310	1.0154
29 Cu	4.3589	3.1643	2.4200	1.9370	1.6148	1.3972	1.2514	1.1581	1.1061	1.0893
30 Zn	4.6648	3.3864	2.5898	2.0729	1.7281	1.4953	1.3393	1.2394	1.1837	1.1658
31 Ga	4.9810	3.6160	2.7654	2.2135	1.8453	1.5967	1.4301	1.3235	1.2640	1.2449
32 Ge	5.3076	3.8530	2.9463	2.3536	1.9664	1.7014	1.5239	1.4103	1.3469	1.3265
33 As	5.6445	4.0977	3.1338	2.5084	2.0912	1.8094	1.6207	1.4999	1.4324	1.4107
34 Se	5.9918	4.3498	3.3267	2.6528	2.2199	1.9208	1.7204	1.5922	1.5206	1.4976

APPENDIX 10 continued

Z_2 element	scattering angle θ									
	90°	100°	110°	120°	130°	140°	150°	160°	170°	180°
35 Br	6.3495	4.6095	3.5253	2.8217	2.3524	2.0355	1.8231	1.6872	1.6114	1.5870
36 Kr	6.7176	4.8767	3.7297	2.9853	2.4888	2.1535	1.9288	1.7851	1.7048	1.6790
37 Rb	7.0960	5.1514	3.9398	3.1535	2.6290	2.2748	2.0375	1.8856	1.8009	1.7736
38 Sr	7.4847	5.4336	4.1556	3.3263	2.7731	2.3995	2.1492	1.9890	1.8996	1.8708
39 Y	7.8839	5.7234	4.3772	3.5037	2.9210	2.5274	2.2638	2.0951	2.0009	1.9706
40 Zr	8.2934	6.0207	4.6046	3.6857	3.0727	2.6587	2.3814	2.2039	2.1048	2.0730
41 Nb	8.7132	6.3255	4.8377	3.8723	3.2283	2.7933	2.5020	2.3155	2.2114	2.1779
42 Mo	9.1435	6.6379	5.0766	4.0635	3.3877	2.9313	2.6255	2.4298	2.3206	2.2855
43 Tc	9.5841	6.9577	5.3213	4.2593	3.5510	3.0726	2.7521	2.5470	2.4325	2.3956
44 Ru	10.0351	7.2851	5.5717	4.4598	3.7181	3.2172	2.8816	2.6668	2.5470	2.5084
45 Rh	10.4964	7.6201	5.8278	4.6648	3.8890	3.3651	3.0141	2.7894	2.6641	2.6237
46 Pd	10.9682	7.9628	6.0901	4.8745	4.0638	3.5164	3.1496	2.9149	2.7838	2.7434
47 Ag	11.4502	8.3125	6.3574	5.0887	4.2425	3.6709	3.2880	3.0430	2.9062	2.8622
48 Cd	11.9427	8.6700	6.6309	5.3076	4.4250	3.8288	3.4294	3.1739	3.0312	2.9853
49 In	12.4455	9.0351	6.9101	5.5311	4.6113	3.9900	3.5739	3.3075	3.1588	3.1110
50 Sn	12.9587	9.4076	7.1950	5.7592	4.8015	4.1546	3.7212	3.4439	3.2891	3.2393
51 Sb	13.4823	9.7877	7.4857	5.9919	4.9955	4.3224	3.8716	3.5831	3.4220	3.3702
52 Te	14.0162	10.1754	7.7822	6.2292	5.1933	4.4937	4.0250	3.7250	3.5576	3.5037
53 I	14.5605	10.5705	8.0844	6.4711	5.3950	4.6681	4.1813	3.8696	3.6957	3.6398
54 Xe	15.1152	10.9732	8.3924	6.7176	5.6005	4.8460	4.3406	4.0171	3.8365	3.7784
55 Cs	15.6802	11.3834	8.7061	6.9687	5.8099	5.0271	4.5028	4.1672	3.9800	3.9197
56 Ba	16.2556	11.8011	9.0256	7.2245	6.0231	5.2116	4.6681	4.3202	4.1260	4.0636
57 La	16.8413	12.2263	9.3508	7.4848	6.2401	5.3994	4.8363	4.4759	4.2747	4.2100
58 Ce	17.4374	12.6591	9.6818	7.7497	6.4610	5.5906	5.0075	4.6343	4.4260	4.3590
59 Pr	18.0439	13.0994	10.0185	8.0193	6.6857	5.7850	5.1816	4.7955	4.5799	4.5106
60 Nd	18.6608	13.5472	10.3610	8.2934	6.9143	5.9828	5.3588	4.9594	4.7365	4.6648
61 Pm	19.2880	14.0026	10.7093	8.5722	7.1467	6.1839	5.5389	5.1262	4.8958	4.8217
62 Sm	19.9256	14.4654	11.0633	8.8556	7.3830	6.3883	5.7221	5.2956	5.0576	4.9811
63 Eu	20.5736	14.9358	11.4231	9.1435	7.6231	6.5961	5.9081	5.4678	5.2221	5.1430
64 Gd	21.2319	15.4138	11.7886	9.4362	7.8670	6.8072	6.0972	5.6428	5.3892	5.3076
65 Tb	21.9006	15.8993	12.1599	9.7333	8.1148	7.0216	6.2893	5.8206	5.5590	5.4748
66 Dy	22.5797	16.3922	12.5370	10.0352	8.3664	7.2393	6.4843	6.0010	5.7314	5.6446
67 Ho	23.2691	16.8928	12.9198	10.3416	8.6219	7.4604	6.6823	6.1843	5.9063	5.8169
68 Er	23.9689	17.4008	13.3083	10.6526	8.8812	7.6847	6.8832	6.3703	6.0840	5.9919

APPENDIX 10 continued

Z_2	element	scattering angle θ									
		90°	100°	110°	120°	130°	140°	150°	160°	170°	180°
69	Tm	24.6790	17.9163	13.7026	10.9682	9.1443	7.9124	7.0872	6.5590	6.2642	6.1694
70	Yb	25.3996	18.4394	14.1027	11.2885	9.4113	8.1435	7.2941	6.7505	6.4471	6.3496
71	Lu	26.1305	18.9701	14.5085	11.6133	9.6822	8.3778	7.5040	6.9448	6.6326	6.5323
72	Hf	26.8717	19.5082	14.9201	11.9428	9.9568	8.6155	7.7169	7.1418	6.8208	6.7176
73	Ta	27.6234	20.0539	15.3375	12.2768	10.2353	8.8565	7.9328	7.3416	7.0116	6.9055
74	W	28.3854	20.6071	15.7606	12.6155	10.5177	9.1008	8.1516	7.5441	7.2051	7.0960
75	Re	29.1577	21.1678	16.1894	12.9588	10.8039	9.3484	8.3734	7.7494	7.4011	7.2891
76	Os	29.9405	21.7361	16.6240	13.3066	11.0939	9.5994	8.5982	7.9575	7.5998	7.4848
77	Ir	30.7336	22.3118	17.0644	13.6591	11.3878	9.8537	8.8260	8.1683	7.8011	7.6831
78	Pt	31.5371	22.8951	17.5105	14.0162	11.6855	10.1113	9.0567	8.3818	8.0051	7.8839
79	Au	32.3509	23.4860	17.9624	14.3779	11.9871	10.3722	9.2905	8.5981	8.2117	8.0874
80	Hg	33.1751	24.0843	18.4200	14.7443	12.2925	10.6365	9.5272	8.8172	8.4209	8.2934
81	Tl	34.0097	24.6902	18.8834	15.1152	12.6017	10.9041	9.7669	9.0390	8.6328	8.5021
82	Pb	34.8546	25.3036	19.3526	15.4907	12.9148	11.1750	10.0095	9.2636	8.8473	8.7133
83	Bi	35.7099	25.9245	19.8275	15.8708	13.2318	11.4492	10.2551	9.4909	9.0644	8.9272
84	Po	36.5756	26.5530	20.3081	16.2556	13.5525	11.7268	10.5037	9.7210	9.2841	9.1436
85	At	37.4516	27.1890	20.7945	16.6449	13.8771	12.0076	10.7553	9.9538	9.5065	9.3626
86	Rn	38.3381	27.8325	21.2867	17.0389	14.2056	12.2919	11.0099	10.1895	9.7316	9.5842
87	Fr	39.2348	28.4836	21.7846	17.4375	14.5379	12.5794	11.2674	10.4278	9.9592	9.8084
88	Ra	40.1420	29.1421	22.2883	17.8407	14.8740	12.8703	11.5280	10.6689	10.1895	10.0352
89	Ac	41.0595	29.8082	22.7978	18.2484	15.2140	13.1644	11.7915	10.9128	10.4224	10.2645
90	Th	41.9874	30.4819	23.3130	18.6608	15.5578	13.4619	12.0580	11.1594	10.6579	10.4965
91	Pa	42.9256	31.1630	23.8339	19.0778	15.9055	13.7628	12.3274	11.4088	10.8961	10.7311
92	U	43.8742	31.8517	24.3606	19.4994	16.2570	14.0669	12.5998	11.6609	11.1368	10.9682

APPENDIX 11 Rutherford scattering cross section $d\sigma/d\Omega$ (Eq.3.29) for 1 MeV ^4He [$10^{-24}cm^2sr^{-1}$]

Z_2 element	scattering angle θ									
	90°	100°	110°	120°	130°	140°	150°	160°	170°	180°
3 Li	0.1524	0.1016	0.0707	0.0514	0.0391	0.0312	0.0261	0.0230	0.0213	0.0208
4 Be	0.2972	0.2065	0.1504	0.1144	0.0907	0.0750	0.0647	0.0582	0.0546	0.0534
5 B	0.4815	0.3396	0.2515	0.1946	0.1569	0.1317	0.1149	0.1043	0.0984	0.0965
6 C	0.7038	0.4993	0.3722	0.2900	0.2355	0.1988	0.1744	0.1589	0.1502	0.1475
7 N	0.9736	0.6953	0.5221	0.4099	0.3352	0.2849	0.2514	0.2300	0.2180	0.2142
8 O	1.2848	0.9212	0.6949	0.5481	0.4503	0.3845	0.3404	0.3122	0.2966	0.2915
9 F	1.6418	1.1816	0.8950	0.7091	0.5852	0.5015	0.4456	0.4098	0.3898	0.3834
10 Ne	2.0322	1.4641	1.1102	0.8806	0.7276	0.6243	0.5552	0.5109	0.4863	0.4783
11 Na	2.4706	1.7831	1.3548	1.0770	0.8917	0.7666	0.6829	0.6293	0.5994	0.5898
12 Mg	2.9450	2.1269	1.6172	1.2865	1.0660	0.9171	0.8173	0.7536	0.7180	0.7065
13 Al	3.4654	2.5052	1.9070	1.5188	1.2599	1.0851	0.9681	0.8931	0.8513	0.8379
14 Si	4.0225	2.9089	2.2151	1.7649	1.4647	1.2619	1.1261	1.0392	0.9907	0.9751
15 P	4.6262	3.3478	2.5513	2.0345	1.6898	1.4570	1.3011	1.2013	1.1456	1.1277
16 S	5.2665	3.8120	2.9058	2.3177	1.9255	1.6607	1.4832	1.3697	1.3063	1.2860
17 Cl	5.9540	4.3120	3.2889	2.6250	2.1822	1.8832	1.6829	1.5547	1.4831	1.4601
18 Ar	6.6842	4.8434	3.6964	2.9520	2.4556	2.1203	1.8957	1.7520	1.6718	1.6459
19 K	7.4459	5.3948	4.1168	3.2875	2.7344	2.3608	2.1106	1.9504	1.8610	1.8323
20 Ca	8.2524	5.9797	4.5637	3.6448	3.0319	2.6179	2.3407	2.1632	2.0642	2.0323
21 Sc	9.1077	6.6020	5.0408	4.0277	3.3520	2.8956	2.5899	2.3942	2.2850	2.2499
22 Ti	10.0005	7.2505	5.5371	4.4252	3.6835	3.1826	2.8471	2.6324	2.5125	2.4739
23 V	10.9347	7.9291	6.0564	4.8411	4.0305	3.4830	3.1163	2.8815	2.7506	2.7084
24 Cr	11.9077	8.6351	6.5959	5.2727	4.3901	3.7939	3.3946	3.1390	2.9964	2.9505
25 Mn	12.9247	9.3737	7.1611	5.7252	4.7675	4.1207	3.6874	3.4100	3.2553	3.2055
26 Fe	13.9806	10.1398	7.7466	6.1936	5.1578	4.4581	3.9895	3.6895	3.5221	3.4682
27 Co	15.0807	10.9387	8.3579	6.6832	5.5661	4.8116	4.3062	3.9827	3.8022	3.7441
28 Ni	16.2182	11.7637	8.9882	7.1871	5.9858	5.1744	4.6308	4.2829	4.0888	4.0263
29 Cu	17.4033	12.6249	9.6476	7.7156	6.4269	5.5565	4.9734	4.6002	4.3919	4.3249
30 Zn	18.6262	13.5127	10.3265	8.2589	6.8798	5.9483	5.3243	4.9250	4.7021	4.6304
31 Ga	19.8932	14.4330	11.0309	8.8232	7.3506	6.3560	5.6897	5.2633	5.0253	4.9487
32 Ge	21.2001	15.3820	11.7568	9.4043	7.8352	6.7754	6.0654	5.6111	5.3575	5.2759
33 As	22.5478	16.3604	12.5052	10.0034	8.3346	7.2075	6.4525	5.9693	5.6996	5.6128
34 Se	23.9385	17.3704	13.2780	10.6222	8.8509	7.6544	6.8529	6.3400	6.0537	5.9616

Z_2 element	scattering angle θ									
	90°	100°	110°	120°	130°	140°	150°	160°	170°	180°
35 Br	25.3681	18.4080	14.0713	11.2570	9.3799	8.1120	7.2627	6.7191	6.4157	6.3182
36 Kr	26.8415	19.4780	14.8899	11.9125	9.9266	8.5852	7.6867	7.1116	6.7907	6.6874
37 Rb	28.3547	20.5764	15.7298	12.5848	10.4870	9.0701	8.1209	7.5135	7.1744	7.0653
38 Sr	29.9096	21.7052	16.5932	13.2758	11.0631	9.5686	8.5674	7.9267	7.5690	7.4540
39 Y	31.5055	22.8636	17.4789	13.9847	11.6540	10.0798	9.0252	8.3503	7.9736	7.8524
40 Zr	33.1436	24.0528	18.3885	14.7127	12.2610	10.6050	9.4957	8.7857	8.3894	8.2620
41 Nb	34.8227	25.2717	19.3206	15.4588	12.8829	11.1431	9.9776	9.2317	8.8154	8.6814
42 Mo	36.5441	26.5215	20.2766	16.2241	13.5211	11.6953	10.4723	9.6896	9.2527	9.1121
43 Tc	38.3065	27.8009	21.2551	17.0073	14.1740	12.2603	10.9783	10.1579	9.6999	9.5526
44 Ru	40.1109	29.1110	22.2572	17.8095	14.8429	12.8392	11.4969	10.6378	10.1583	10.0041
45 Rh	41.9560	30.4505	23.2816	18.6295	15.5265	13.4306	12.0266	11.1281	10.6265	10.4652
46 Pd	43.8445	30.8211	24.3301	19.4696	16.2263	14.0365	12.5603	11.6304	11.1061	10.9232
47 Ag	45.7714	33.2204	25.4002	20.3253	16.9404	14.6541	13.1225	12.1423	11.5952	11.4192
48 Cd	47.7425	34.6517	26.4951	21.2020	17.6715	15.2869	13.6894	12.6671	12.0965	11.9129
49 In	49.7538	36.1119	27.6119	22.0960	18.4168	15.9318	14.2671	13.2017	12.6070	12.4157
50 Sn	51.8073	37.6029	28.7525	23.0091	19.1782	16.5907	14.8574	13.7480	13.1289	12.9297
51 Sb	53.9018	39.1236	29.9156	23.9402	19.9545	17.2625	15.4591	14.3050	13.6608	13.4536
52 Te	56.0390	40.6756	31.1029	24.8909	20.7474	17.9488	16.0740	14.8742	14.2044	13.9891
53 I	58.2148	42.2548	32.3104	25.8571	21.5528	18.6455	16.6979	15.4515	14.7558	14.5320
54 Xe	60.4343	43.8663	33.5431	26.8441	22.3757	19.3577	17.3359	16.0420	15.3198	15.0875
55 Cs	62.6940	45.5068	34.7977	27.8482	23.2128	20.0820	17.9846	16.6424	15.8932	15.6521
56 Ba	64.9964	47.1785	36.0764	28.8719	24.0665	20.8207	18.6464	17.2549	16.4783	16.2284
57 La	67.3391	48.8791	37.3770	29.9129	24.9343	21.5716	19.3190	17.8773	17.0726	16.8138
58 Ce	69.7231	50.6097	38.7005	30.9722	25.8174	22.3356	20.0033	18.5106	17.6774	17.4094
59 Pr	72.1484	52.3702	40.0468	32.0498	26.7156	23.1128	20.6993	19.1547	18.2927	18.0152
60 Nd	74.6162	54.1619	41.4172	33.1468	27.6303	23.9043	21.4083	19.8109	18.9193	18.6325
61 Pm	77.1252	55.9835	42.8104	34.2620	28.5601	24.7089	22.1290	20.4779	19.5563	19.2598
62 Sm	79.6759	57.8353	44.2268	35.3958	29.5055	25.5269	22.8618	21.1561	20.2040	19.8978
63 Eu	82.2675	59.7167	45.6656	36.5475	30.4656	26.3576	23.6058	21.8447	20.8617	20.5454
64 Gd	84.9018	61.6294	47.1288	37.7189	31.4423	27.2030	24.3631	22.5456	21.5312	21.2048
65 Tb	87.5763	63.5710	48.6137	38.9074	32.4332	28.0603	25.1310	23.2562	22.2098	21.8732
66 Dy	90.2930	65.5433	50.1222	40.1150	33.4400	28.9316	25.9114	23.9786	22.8997	22.5526
67 Ho	93.0507	67.5453	51.6534	41.3406	34.4619	29.8157	26.7034	24.7115	23.5998	23.2421
68 Er	95.8498	69.5774	53.2076	42.5847	35.4991	30.7132	27.5073	25.4554	24.3103	23.9418
69 Tm	98.6902	71.6394	54.7846	43.8469	36.5514	31.6237	28.3228	26.2102	25.0310	24.6517

APPENDIX 11 continued

Z_2	element	scattering angle θ 90°	100°	110°	120°	130°	140°	150°	160°	170°	180°
70	Yb	101.5728	73.7323	56.3854	45.1284	37.6199	32.5483	29.1510	26.9768	25.7632	25.3728
71	Lu	104.4963	75.8546	58.0085	46.4276	38.7030	33.4855	29.9905	27.7537	26.5051	26.1035
72	Hf	107.4617	78.0075	59.6552	47.7457	39.8020	34.4365	30.8423	28.5420	27.2583	26.8451
73	Ta	110.4682	80.1902	61.3245	49.0819	40.9160	35.4005	31.7058	29.3412	28.0214	27.5967
74	W	113.5163	82.4031	63.0170	50.4367	42.0456	36.3779	32.5812	30.1514	28.7952	28.3588
75	Re	116.6057	84.6460	64.7324	51.8098	43.1903	37.3684	33.4685	30.9724	29.5792	29.1312
76	Os	119.7371	86.9194	66.4713	53.2018	44.3509	38.3727	34.3681	31.8051	30.3744	29.9144
77	Ir	122.9094	89.2224	68.2325	54.6116	45.5263	39.3897	35.2790	32.6481	31.1798	30.7073
78	Pt	126.1234	91.5557	70.0171	56.0401	46.7172	40.4203	36.2021	33.5024	31.9956	31.5109
79	Au	129.3785	93.9188	71.8245	57.4867	47.9233	41.4639	37.1368	34.3675	32.8218	32.3245
80	Hg	132.6757	96.3125	73.6553	58.9523	49.1452	42.5212	38.0839	35.2440	33.6590	33.1490
81	Tl	136.0143	98.7364	75.5092	60.4363	50.3825	43.5919	39.0430	36.1316	34.5066	33.9839
82	Pb	139.3941	101.1901	77.3858	61.9385	51.6349	44.6756	40.0136	37.0300	35.3644	34.8289
83	Bi	142.8151	103.6737	79.2853	63.4588	52.9025	45.7723	40.9960	37.9391	36.2328	35.6841
84	Po	146.2775	106.1871	81.2075	64.9974	54.1852	46.8822	41.9901	38.8591	37.1114	36.5494
85	At	149.7810	108.7304	83.1525	66.5542	55.4830	48.0051	42.9958	39.7898	38.0003	37.4248
86	Rn	153.3289	111.3068	85.1235	68.1323	56.7991	49.1442	44.0163	40.7345	38.9030	38.3135
87	Fr	156.9157	113.9106	37.1149	69.7262	58.1279	50.2939	45.0461	41.6875	39.8130	39.2099
88	Ra	160.5443	116.5450	89.1297	71.3390	59.4725	51.4574	46.0883	42.6521	40.7341	40.1171
89	Ac	164.2140	119.2090	91.1671	72.9698	60.8321	52.6338	47.1419	43.6271	41.6655	41.0343
90	Th	167.9261	121.9040	93.2284	74.6199	62.2079	53.8243	48.2084	44.6142	42.6080	41.9627
91	Pa	171.6783	124.6278	95.3115	76.2871	63.5977	55.0268	49.2854	45.6109	43.5598	42.9001
92	U	175.4737	127.3834	97.4193	77.9745	65.0047	56.2444	50.3761	46.6203	44.5238	43.8497

APPENDIX 12 Nuclear reactions used for the detection of light elements (particle emission) according to Ref.[100]

nucleus	reaction	Q-value Q [MeV]	incident energy E_1 [MeV]	emitted energy[1] E_1' [MeV]	cross section $(d\sigma/d\Omega)_{NR}$ [mb/sr]	Mylar thickness z_{sf} [μm]
^{2}H	$^{2}H(d,p)^{3}H$	4.032	1.0	2.3	5.2	14
^{2}H	$^{2}H(^{3}He,p)^{4}He$	18.352	0.7	13.0	61	6
^{3}He	$^{3}He(d,p)^{4}He$	18.352	0.45	13.6	64	8
^{6}Li	$^{6}Li(d,\alpha)^{4}He$	22.374	0.7	9.7	6	8
^{7}Li	$^{7}Li(p,\alpha)^{4}He$	17.347	1.5	7.7	1.5	35
^{9}Be	$^{9}Be(d,\alpha)^{7}Li$	7.153	0.6[2]	4.1	≈1	6
^{11}B	$^{11}B(p,\alpha)^{8}Be$	8.586	0.65	5.57(α_0)	0.12(α_0)	10
		5.65	0.65	3.70(α_1)	90 (α_1)	10
^{12}C	$^{12}C(d,p)^{13}C$	2.722	1.20	3.1	35	16
^{13}C	$^{13}C(d,p)^{14}C$	5.951	0.64	5.8	0.4	6
^{14}N	$^{14}N(d,\alpha)^{12}C$	13.574	1.5	9.9(α_0)	0.6(α_0)	23
		9.146	1.2	6.7(α_1)	1.3(α_1)	16
^{15}N	$^{15}N(p,\alpha)^{12}C$	4.964	0.8[3]	3.9	≈15	12
^{16}O	$^{16}O(d,p)^{17}O$	1.917	0.90	2.4(p_0)[4]	0.74(p_0)	12
		1.05	0.90	1.6(p_1)	4.5 (p_1)	12
^{18}O	$^{18}O(p,\alpha)^{15}N$	3.980	0.73[3]	3.4	15	11
^{19}F	$^{19}F(p,\alpha)^{16}O$	8.114	1.25	6.9	0.5	25
^{23}Na	$^{23}Na(p,\alpha)^{20}Ne$	2.379	0.592	2.238	4	6
^{31}P	$^{31}P(p,\alpha)^{28}Si$	1.917	1.514	2.734	16	[5]

1) For laboratory emission angle 150° with recoil nucleus in ground state (excited state)

2) 0.6 MeV is optimum for Be in light Z matrix and 1.6 MeV for high Z matrix

3) Maximum energy for Mylar to stop backscattered proton

4) Measured at $\theta = 164°$

5) Range of α < range of p

APPENDIX 13 Nuclear reactions used for the detection of light elements (γ-ray emission) according to Refs.[101 - 104]

nucleus	reaction	incident energy E_1 [MeV]	emitted energy E_γ [MeV]	cross section σ_{NR} [mb]	resonance width Γ [keV]	emission percentage per reaction
^7Li	$^7Li(p,\gamma)\,^8Be$	0.441	17.65	6	12	63
			14.75			37
		1.030	18.15	-	168	40
			15.25			60
^9Be	$^9Be(p,\gamma)^{10}B$	0.330	6.15	-	160	21
			5.15			55
			4.75			11
			2.15			3.5
			1.7			13
			1.43			4.5
			1.02			58
			0.72			84
			0.41			3.5
		0.991	7.5	-	89	≈100
		1.084	6.85	-	3.8	85
			4.45			4.5
			3.01			9.5
			2.4			15
			2.15			3
			1.43			4
			1.02			3
			0.72			96
			0.41			3
^{10}B	$^{10}B\ (p,\gamma)^{11}C$	1.146	9.7	0.0055	450	-
		1.180		0.0075	570	-
^{11}B	$^{11}B\ (p,\gamma)^{12}C$	0.163	16.11	0.157	7	3.5
			11.68			96.5
			4.43			96.3
		0.675	12.15	0.050	322	≈100
			4.43			≈100
		1.390	17.23	0.053	1270	66
			12.80			34
			4.43			34
^{12}C	$^{12}C\ (p,\gamma)^{13}N$	0.459	2.36	0.127	39.5	100
		1.698	3.51	0.035	70	95
			2.36			5
			1.14			5
^{13}C	$^{13}C\ (p,\gamma)^{14}N$	0.550	8.06	1.44	32.5	-
		1.160	8.62	0.56	6	-
		1.250	8.71	0.062	500	-
		1.470	5.83	0.074	20	-
		1.550	8.99	0.037	7	-
		1.748	9.17	340	0.075	-
^{14}N	$^{14}N(p,\gamma)^{15}O$	0.278	6.8	-	1.6	-
		0.700	8.0	-	100	-
		1.544	8.8	-	34	-
		1.742	9.0	-	5	-
		1.807	9.0	-	5	-

APPENDIX 13 continued

nucleus	reaction	incident energy E_1 [MeV]	emitted energy E_γ [MeV]	cross section σ_{NR} [mb]	resonance width Γ [keV]	emission percentage per reaction
^{15}N	$^{15}N(p,\gamma)^{16}O$	0.360	12.43	0.007	94	-
	$^{15}N(p,\alpha\gamma)^{12}C$	0.360	4.43	0.03	94	100
		0.429	4.43	300	0.9	100
	$^{15}N\ (p,\gamma)^{16}O$	0.429	6.46	0.001	0.9	-
		0.710	6.72	-	4.0	-
	$^{15}N(p,\gamma)^{12}C$	0.898	4.43	800	2.2	100
		1.040	4.43	15	130	100
	$^{15}N\ (p,\gamma)^{16}O$	1.040	13.09	1	130	-
		1.059	8.34	-	4	-
	$^{15}N(p,\alpha\gamma)^{12}C$	1.210	4.43	425	22.5	100
		1.640	4.43	340	68	100
		1.979	4.43	35	23	100
^{18}O	$^{18}O\ (p,\gamma)^{19}F$	0.630	8.5	-	2.6	-
		0.849	8.8	-	40	-
		1.167	6.3	-	0.6	≈100
			2.6			≈100
		1.399	9.3	-	15	-
		1.685	9.6	-	15	-
		1.769	9.6	-	4	-
		1.931	9.8	-	1.5	-
^{19}F	$^{19}F(p,\alpha\gamma)^{16}O$	0.224	7.12	0.2	1	-
			6.72			-
		0.340	6.13	102	2.4	96.5
		0.4836	6.13	32	0.9	79
		0.597	6.13	7	30	≈100
		0.672	6.13	57	6	81
		0.835	6.13	19	6.5	-
		0.872	6.13	661	4.5	68
		0.902	6.13	23	5.1	>90
		0.935	6.13	180	8.6	75
		1.090	6.13	13	0.7	-
		1.140	6.13	15	2.5	-
		1.189	6.13	19	110	-
		1.283	6.13	29	19	74
		1.348	6.13	89	5.6	55
		1.375	6.13	300	11	87
		1.591	6.13		35	-

APPENDIX 14 X-ray emission energies E_X [eV]

Z_2	element	$K_{\alpha 1}$	$K_{\beta 1}$	$L_{\alpha 1}$	$L_{\beta 1}$	$M_{\alpha 1}$
11	Na	1026	1068	–	–	–
12	Mg	1240	1299	–	–	–
13	Al	1474	1555	–	–	–
14	Si	1729	1834	–	–	–
15	P	2005	2136	–	–	–
16	S	2302	2463	–	–	–
17	Cl	2619	2813	–	–	–
18	Ar	2957	3188	–	–	–
19	K	3316	3586	–	–	–
20	Ca	3695	4009	–	–	–
21	Sc	4096	4455	–	–	–
22	Ti	4517	4926	–	–	–
23	V	4959	5421	–	–	–
24	Cr	5423	5940	–	–	–
25	Mn	5907	6484	635	637	–
26	Fe	6412	7052	703	707	–
27	Co	6939	7645	774	781	–
28	Ni	7487	8263	927	942	–
29	Cu	8057	8906	927	942	–
30	Zn	8648	9574	1010	1028	–
31	Ga	9261	10267	1096	1119	–
32	Ge	9895	10985	1186	1214	–
33	As	10552	11729	1279	1313	–
34	Se	11231	12499	1377	1416	–
35	Br	11932	13295	1478	1524	–
36	Kr	12656	14117	1582	1635	–
37	Rb	13402	14965	1691	1751	–
38	Sr	14171	15839	1803	1872	–
39	Y	14963	16740	1919	1996	–
40	Zr	15778	17668	2039	2125	–
41	Nb	16617	18623	2163	2259	–
42	Mo	17479	19606	2290	2397	–
43	Tc	18365	20616	2421	2539	–
44	Ru	19275	21654	2556	2686	–
45	Rh	20210	22720	2694	2838	–
46	Pd	21169	23814	2837	2994	–
47	Ag	22153	24937	2983	3155	–
48	Cd	23163	26088	3133	3321	–
49	In	24197	27269	3287	3492	–
50	Sn	25258	28479	3445	3668	–
51	Sb	26344	29719	3606	3849	–
52	Te	27457	30989	3772	4035	–
53	I	28596	32289	3941	4226	–
54	Xe	29763	33621	4114	4423	–
55	Cs	30957	34983	4291	4626	–
56	Ba	32178	36376	4472	4833	–
57	La	33427	37801	4656	5047	–
58	Ce	34704	30257	4845	5266	881
59	Pr	36011	40746	5037	5491	930
60	Nd	37346	42268	5233	5722	980
61	Pm	38711	43823	5433	5959	1030
62	Sm	40105	45412	5637	6203	1080
63	Eu	41530	47034	5845	6452	1132
64	Gd	42985	48690	6057	6709	1184
65	Tb	44472	50381	6273	6972	1238
66	Dy	45990	52108	6492	7242	1292
67	Ho	47540	53869	6716	7518	1348
68	Er	49122	55667	6944	7802	1405

APPENDIX 14 continued

Z_2	element	$K_{\alpha 1}$	$K_{\beta 1}$	$L_{\alpha 1}$	$L_{\beta 1}$	$M_{\alpha 1}$
69	Tm	50737	57500	7175	8094	1463
70	Yb	52385	59371	7411	8392	1523
71	Lu	54067	61279	7650	8698	1584
72	Hf	55783	63224	7893	9012	1646
73	Ta	57534	65207	8141	9334	1710
74	W	59320	67229	8392	9664	1775
75	Re	61142	69290	8648	10002	1842
76	Os	63000	71391	8907	10349	1910
77	Ir	64895	73531	9171	10705	1979
78	Pt	66827	75712	9439	11069	2050
79	Au	68797	77934	9710	11442	2122
80	Hg	70805	80198	9986	11825	2195
81	Tl	72852	82503	10266	12217	2269
82	Pb	74939	84851	10550	12618	2345
83	Bi	77065	87242	10838	13030	2421
84	Po	79232	89455	11110	13452	2498
85	At	81441	92156	11426	13884	2576
86	Rn	–	–	11727	14326	2655
87	Fr	–	–	12031	14779	2734
88	Ra	–	–	12340	15244	2814
89	Ac	–	–	12653	15719	2893
90	Th	–	–	12970	16206	2973
91	Pa	–	–	13291	16705	3053
92	U	–	–	13617	17215	3132
93	Np	–	–	13946	17738	–
94	Pu	–	–	14280	18273	–
95	Am	–	–	14618	18822	–
96	Cm	–	–	14961	19383	–
97	Bk	–	–	15308	19957	–
98	Cf	–	–	15659	20545	–
99	Es	–	–	16014	21146	–
100	Fm	–	–	16374	21762	–

References

[1] J.W. Mayer and E. Rimini (Eds.), Catania Working Data, Catania, 1974.

[2] W.K. Chu, J.W. Mayer, and M.A. Nicolet, Backscattering Spectrometry, Academic Press, New York, San Francisco, London, 1978.

[3] D.V. Morgan (Ed.), Channeling. Theory, Observation and Applications, John Wiley & Sons, London, New York, Sydney, Toronto, 1973.

[4] K. Gärtner and K. Hehl, Phys. Stat. Sol. (b) 94 (1979) 231.

[5] L.H. Thomas, Proc. Cambridge Phys. Soc. 23 (1926) 542.
E. Fermi, Z. Phys. 48 (1928) 73.

[6] J. Lindhard, M. Scharff, and H.E. Schiøtt, K. Dan. Vidensk. Selsk. Mat.-Fys. Medd. 39 (1963) no. 14.

[7] V. Bush and S.H. Caldwell, Phys. Rev. 38 (1931) 1898.

[8] W. Lenz, Z. Phys. 77 (1932) 713.
H. Jensen, Z. Phys. 77 (1932) 722.

[9] K. Gärtner and K. Hehl, Wiss. Zeitschr. Friedrich-Schiller-Univ. Jena, Math.-Nat. R. 22 (1973) 257.

[10] J. L'Ecuyer, J.A. Davies, and N. Matsunami, in J. Gyulai, T. Lohner, and E. Pasztor (Eds.), Proc. Int. Conf. Ion Beam Modification of Materials, Budapest, 1978, KFKI, Budapest, 1979, Vol. 3, p. 1723.

[11] J. Lindhard, V. Nielsen, and M. Scharff, K. Dan. Vidensk. Selsk. Mat.-Fys. Medd. 36 (1968) no. 10.

[12] E. Clementi and C. Roetti, At. Data & Nucl. Data Tables 14 (1974) 177.

[13] P. Loftager and J.E. Andersen, Aarhus, personal communication.

[14] E. Merzbacher and H.W. Lewis, Handbuch der Physik 34, Springer-Verlag, Berlin, 1958, p. 166.

[15] J.D. Garcia, R.J. Fortner, and T.M. Kavanagh, Rev. Mod. Phys. 45 (1973) 111.

[16] J.C. Slater, Phys. Rev. 36 (1930) 57.

[17] N. Bohr, K. Dan. Vidensk. Selsk. Mat.-Fys. Medd. 18 (1948) no. 8.

[18] H. Bethe, Ann. Phys. 5 (1930) 325.

[19] F. Bloch, Ann. Phys. 16 (1933) 285.

[20] E. Bonderup, Lecture Notes on Penetration of Charged Particles Through Matter, University of Aarhus, Denmark, 1978.
[21] O.B. Firsov, Zh. Eksp. & Teor. Fiz. 36 (1959) 1517.
O.B. Firsov, Sov. Phys. - JETP, 9 (1959) 1076.
[22] J. Lindhard and M. Scharff, Phys. Rev. 124 (1961) 128.
[23] P. Hvelplund and B. Fastrup, Phys. Rev. 165 (1968) 408.
[24] D. Ward, H.R. Andrews, I.V. Mitchell, W.N. Lennard, R.B. Walker, and N. Rud, Can. J. Phys. 57 (1979) 645.
[25] I.M. Cheshire, G. Dearnaley, and J.M. Poate, Phys. Lett. A27 (1968) 318.
[26] D.J. Land, J.G. Brennan, D.G. Simons, and M.D. Brown, Phys. Rev. A16 (1977) 492.
[27] S.A. Cruz, C. Vargas, and D.K. Brice, Radiat. Eff. Lett. 43 (1979) 79.
[28] D.K. Brice, Phys. Rev. A6 (1972) 1971.
[29] J.P. Biersack and D. Fink, in S. Datz, B.R. Appleton, and C.D. Moak (Eds.), Proc. Int. Conf. Atomic Collisions in Solids, Gatlinburg, 1973, Plenum Press, New York, 1975, p. 737.
[30] J.F. Ziegler, The Stopping and Ranges of Ions in Matter, Pergamon Press, New York, Toronto, Oxford, Sydney, Frankfurt, Paris, 1977, Vol. 3.
[31] J.F. Ziegler, The Stopping and Ranges of Ions in Matter, Pergamon Press, New York, Toronto, Oxford, Sydney, Frankfurt, Paris, 1977, Vol. 4.
[32] J. Lindhard and M. Scharff, K. Dan. Vidensk. Selsk., Mat.-Fys. Medd., 27 (1953) no. 15.
[33] W.K. Chu and J.W. Mayer, in J.W. Mayer and E. Rimini, Catania Working Data, Energy Loss and Energy Straggling, Catania, 1974.
[34] E. Bonderup and P. Hvelplund, Phys. Rev. A4 (1971) 562.
[35] L. Landau, J. Phys. USSR 8 (1944) 201.
[36] P. Sigmund and K.B. Winterbon, Nucl. Instr. & Meth. B12 (1985) 1.
[37] W. Bothe, Z. Phys. 5 (1921) 63.
[38] L. Meyer, Phys. Stat. Sol. (b) 44 (1971) 253.
[39] P. Sigmund and K.B. Winterbon, Nucl. Instr. & Meth. 119 (1974) 541.
[40] N. Andersen, P. Sigmund, J. Heinemeier, P. Hvelplund, H. Knudsen, and G. Sidenius, Nucl. Instr. & Meth. 132 (1976) 703.
[41] J.H. Barret, Phys. Rev. B3 (1971) 1527.
[42] H.D. Carstanien, Phys. Stat. Sol. (a) 59 (1980) 11.
[43] J. Lindhard, K. Dan. Vidensk. Selsk., Mat.-Fys. Medd., 34 (1965) no. 14.
[44] U. Finger, Ph. D. Thesis, Jena, 1975.
[45] L.M. Howe and S. Schmied, Can. J. Phys. 49 (1971) 2321.
[46] M. Arendt, Diploma Thesis, Jena, 1976.

[47] J.U. Andersen, A.S. Jensen, K. Jorgensen, E. Laegsgaard, K.O. Nielsen, J.S. Forster, I.V. Mitchell, D. Ward, E.M. Gibson, and J.J. Cuomo, K. Dan. Vidensk. Selsk., Mat.-Fys. Medd., 40 (1980) no. 7.

[48] F. Fuschini, A. Uguzzoni, K. Gärtner, and K. Hehl, Nucl. Instr. & Meth. B12 (1985) 334.

[49] G. Sommer, Ph. D. Thesis, Jena, 1975.

[50] E. Bonderup, H. Esbensen, J.U. Andersen, and H.E. Schiøtt, Radiat. Eff. 12 (1972) 261.

[51] V.V. Beloshitsky, M.A. Kumakhov, and V.A. Muralev, Radiat. Eff. 13 (1972) 9.

[52] E. Bonderup, Aarhus, personal communication.

[53] H.E. Schiøtt, E. Bonderup, J.U. Andersen, and H. Esbensen, in S. Datz, B.R. Appleton and C.D. Moak (Eds.), Proc. Int. Conf. Atomic Collisions in Solids, Gatlinburg, 1973, Plenum Press, New York, 1975, Vol. 2, p. 843.

[54] M. Matsunami and L.M. Howe, Radiat. Eff. 51 (1980) 111.

[55] U. Finger, K. Gärtner, and K. Hehl, in Ju.V. Bulgakov and A.F. Tulinov (Eds.), Proc. Int. Conf. Atomic Collisions in Solids, Moscow, 1977, Moscow State University Publishing House, 1980, Vol. 2, p. 262.

[56] K. Gärtner, K. Hehl, and G. Schlotzhauer, Nucl. Instr. & Meth. 216 (1983) 275.

[57] J. Lindhard, K. Dan. Vidensk. Selsk., Mat.-Fys. Medd., 28 (1954) no. 8.

[58] A. Seeger, in S. Flügge (Ed.), Encyclopedia of Physics, Vol. VII/1, Theorie der Gitterfehlstellen, Springer-Verlag, Berlin, Göttingen, Heidelberg, 1955, p. 383.

[59] K. Gärtner, K. Hehl, and G. Schlotzhauer, Nucl. Instr. & Meth. B4 (1984) 55.

[60] Y. Quéré, Ann. Phys. 5 (1970) 105.
Y. Quéré, Phys. Stat. Sol. 30 (1968) 713.

[61] H. Kudo, J. Phys. Soc. Jpn. 40 (1976) 1645.

[62] D. Wielunska, L. Wielunski, and A.E. Turos, Phys. Stat. Sol. (a) 67 (1981) 413.

[63] K. Gärtner, K. Hehl, and G. Schlotzhauer, Nucl. Instr. & Meth. B4 (1984) 63.

[64] L.D. Landau and E.M. Liftshits, Lehrbuch der Theoretischen Physik, Vol. 1, Akademie Verlag, Berlin, 1970.

[65] K. Kimura, T. Oshiyoma, and M. Mannami, Jpn. J. Appl. Phys. 21 (1982) 1222.

[66] K. Gärtner and K. Hehl, Nucl. Instr. & Meth. B12 (1985) 205.

[67] G. Götz and B Gruska, Nucl. Instr. & Meth. 194 (1982) 199 and personal communication.

[68] E. Glaser, G. Götz, N. Sobelev, and W. Wesch, Phys. Stat. Sol. (a) 69 (1982) 603 and personal communication.
[69] E. Bøgh, Can. J. Phys., 46 (1968) 653.
[70] J.E. Westmoreland, J.W. Mayer, F.H. Eisen, and B. Welch, Radiat. Eff. 6 (1970) 161.
[71] S.T. Picraux, J. Appl, Phys. 44 (1973) 587.
[72] D. van Vliet, Phys. Stat. Sol. (a) 2 (1970) 521.
[73] S.U. Campisano, G. Foti, E. Rimini, and S.T.Picraux, Nucl. Instr. & Meth. 149 (1978) 371.
[74] J. Lindhard, Phys. Lett. 12 (1964) 126.
[75] J.H. Barrett, Phys. Rev. B3 (1971) 1527.
[76] L. Vályi, Atom and Ion Sources, Akademiai Kiadó, Budapest, 1977.
[77] H. Ewald and H. Hintenberger, Methoden und Anwendungen der Massenspektroskopie, Verlag Chemie, GMBH, Weinheim/Bergstrasse, 1953.
[78] Ch. Edelmann, Vakuumphysik und -technik, Akademische Verlagsgesellschaft Geest & Portig KG, Leipzig, 1978.
[79] G. Bertolini and A. Coche, Semiconductor Detectors, North-Holland Publishing Company, Amsterdam, 1968.
[80] Physical Sciences Division Catalog, ORTEC Incorporated, 1976.
[81] R.F. Lever, in O. Meyer, G. Linker, and F. Käppeler (Eds.), Proc. Int. Conf. Ion Beam Surface Layer Analysis, Karlsruhe, 1975, Plenum Press, New York, London, 1976, Vol. 1, p. 111.
[82] J.S. Williams, Nucl. Instr. & Meth. 126 (1975) 205.
[83] J.S. Williams, in O. Meyer, G. Linker, and F. Käppeler (Eds.), Proc. Int. Conf. Ion Beam Surface Layer Analysis, Karlsruhe, 1975, Plenum Press, New York, London, 1976, Vol. 1, p. 223.
[84] J.S. Williams, Nucl. Instr. & Meth. 149 (1978) 207.
[85] J.S. Williams and W. Möller, Nucl. Instr. & Meth. 157 (1978) 213.
[86] W. Möller, H. Hufschmidt, and D. Kamke, Nucl. Instr. & Method. 140 (1977) 157.
[87] A.D. Marwick and P. Sigmund, Nucl. Instr. & Meth. 126 (1975) 317.
[88] A. Feuerstein, H. Graham, S. Kalbitzer, and H. Oetzmann, in O. Meyer, G. Linker, and F. Käppeler (Eds.), Proc. Int. Conf. Ion Beam Surface Layer Analysis, Karlsruhe, 1975, Plenum Press, New York, London, 1975, Vol. 1, p. 471.
[89] E. Rutherford, Phil. Mag. 21 (1911) 669.
[90] J.B. Marion and F.C. Young, Nuclear Reaction Analysis, Graphs and Tables, Wiley, New York, 1968.
[91] J. L'Ecuyer, C. Brassard, C. Cardinal, J. Chabbal, L. Deschenes, J.P. Labrie, B. Terrault, J.G. Martel, and R.St.-Jaques, J. Appl. Phys. 47 (1976) 881.
[92] J. L'Ecuyer, C. Brassard, and C. Cardinal, Nucl.Instr. & Meth. 149 (1978) 271.
[93] A. Turos and O. Meyer, Nucl. Instr. & Meth. B4 (1984) 92.

[94] S. Nagata, S. Yamaguchi, Y. Fujino, Y. Hori, N. Sugiyama, and K. Kamada, Nucl. Instr. & Meth. B6 (1985) 533.
[95] G. Amsel, J.P. Nadai, E. D'Artemare, D. David, G. Gigard, and J. Moulin, Nucl. Instr. & Meth. 92 (1971) 481.
[96] J.R. Bird, B.L. Campbell, and R.B. Price, Atomic Energy Review 12 (1974) 275.
[97] E.A. Wolitzky, in J.F. Ziegler (Ed.), New Uses of Low Energy Accelerators, Plenum Press, New York, 1975.
[98] S.T. Picraux, G. Amsel, and L.C. Feldmann, in J.W. Mayer and E. Rimini (Eds.), Catania Working Data, Catania, 1974.
[99] J. Bøttiger, S.T. Picraux, and N. Rud, in O. Meyer, G. Linker, and F. Käppeler (Eds.), Proc. Int. Conf. Ion Beam Surface Layer Analysis, Karlsruhe, 1975, Plenum Press, New York, London, 1976, Vol. 2, p. 811.
[100] L.C. Feldmann and S.T. Picraux, in J.W. Mayer and E. Rimini (Eds.), Ion Beam Handbook for Material Analysis, Academic Press, New York, 1977, Ch. 4.
[101] F. Ajzenberg-Selove and T. Lauritsen, Nucl. Phys. 11 (1959) 1.
[102] J.W. Butler, NRL - 5282, 1959.
[103] J.B. Marion, Rev. Mod. Phys. 38 (1966) 660.
[104] G.M. Padawer, Nucl. Appl. Technol. 9 (1970) 856.
[105] S.A.E. Johannson and T.B. Johannson, Nucl. Instr. & Meth. 137 (1976) 473.
[106] R. Theisen and D. Vollath, in Tables of X-Ray Mass Attenuation Coefficients, Verlag Stahleisen M.B.H., Düsseldorf, 1967.
[107] I.V. Mitchell, T. Cahill, F.V. Frazoli, T.J. Gray, and J.F. Ziegler, in J.W. Mayer and E. Rimini (Eds.), Catania Working Data, Catania, 1974.
[108] L.C. Feldmann and P.J. Silvermann, in O. Meyer, G. Linker, and F. Käppeler (Eds.), Proc. Int. Conf. Ion Beam Surface Layer Analysis, Karlsruhe, 1975, Plenum Press, New York, London, 1976, Vol. 2, p. 735.
[109] K. Adler, A. Bohr, T. Huns, B. Mottelson, and A. Winter, Rev. Mod. Phys. 28 (1956) 432.
[110] Ch. Bauer, K. Hohmuth, P. Gippner, R. Mann, A. Nebelung, and W. Rudolph, Z. Phys. A284 (1978) 275.
[111] Technical Data of Silicon (Li) X-Ray Detectors, ORTEC Incorporated, 1979.
[112] Annual Report 1983/84, Dept. of Electronic and Electrical Engineering, University of Surrey, Guildford.
[113] F. Folkmann, in O. Meyer, G. Linker, and F. Käppeler (Eds.), Proc. Int. Conf. Ion Beam Surface Layer Analysis, Karlsruhe, 1975, Plenum Press, New York, London, 1976, Vol. 2, p. 695.

[114] W. Rudolph, in O. Brummer, J. Heydenreich, K.H. Krebs, and H.G. Schneider, Festkörperanalyse mit Elektronen, Ionen und Röntgenstrahlen, Deutscher Verlag der Wissenschaften, Berlin, 1980.
[115] W.K. Chu, J.W. Mayer, M.-A. Nicolet, T.M. Buck, G. Amsel, and F. Eisen, Thin Solid Films 17 (1973) 1.
[116] G. Götz, K. Hehl, J. Mittenbacher, Kh. Müller, and W. Trippensee, Exp. Tech. Phys. 18 (1970) 233.
[117] C.S. Barrett, R.M. Mueller, and W. White, J. Appl. Phys. 39 (1968) 4694.
[118] B.R. Appleton and G. Foti, in J.W. Mayer and E. Rimini (Eds.), Ion Beam Handbook for Material Analysis, Academic Press, New York, 1977.
[119] W. Wesch, E. Wilk, and K. Hehl, Phys. Stat. Sol. (a) 70 (1982) 243.
[120] H. Fischer, G. Götz, and H. Karge, Phys. Stat. Sol. (a) 76 (1983) 249.
[121] S. Jetschke, H. Karge, and K. Hehl, Phys. Stat. Sol. (a) 77 (1983) 207
[122] G. Götz and H. Karge, Nucl. Instr. & Meth. 209/210 (1983) 1079.
[123] E. Glaser, G. Götz, and K. Hehl in Ju.V. Bulgakov and A.F. Tulinov (Eds.), Int. Conf. Atomic Collisions in Solids, Moscow, 1977, Moscow State University Publishing House, 1980, Vol. 2, p. 258.
[124] E. Glaser, Ph. D. Thesis, Jena, 1978.
[125] B. Gruska, Ph. D. Thesis, Jena, 1981.
[126] E. Rimini, S.U. Campisano, G. Foti, P. Baeri, and S.T. Picraux, in O. Meyer, G. Linker, and F. Käppeler (Eds.), Proc. Int. Conf. Ion Beam Surface Layer Analysis, Karlsruhe, 1975, Plenum Press, New York, London, 1976, Vol. 2, p. 597.
[127] G. Foti, S.T. Picraux, S.U. Campisano, E. Rimini, and R.A. Kant in F. Chernow, D.K. Brice, and J.A. Borders (Eds.), Proc. Int. Conf. Ion Implantation in Semiconductors and Other Materials, Boulder, 1976, Plenum Press, New York, 1977, p. 246.
[128] B. Gruska and G. Götz, Phys. Stat. Sol. (a) 67 (1981) 129.
[129] H. Kudo, Phys. Rev. B18 (1978) 995.
[130] D.V. Morgan and D. van Vliet, in D.W. Palmer, M.W. Thompson and P.D. Townsend (Eds.), Proc. Int. Conf. Atomic Collisions in Solids Brighton, 1969, North-Holland, Amsterdam 1970, p. 476.
[131] D. Sigurd, R.W. Bower, W.F. van der Weg, and J.W. Mayer, Thin Solid Films 19 (1973) 319.
[132] H. Ishiwara and S. Furukawa, J. Appl. Phys. 47 (1976) 1686.
[133] H. Ishiwara and S. Furukawa, J. Vac. Sci. & Technol. 12 (1975) 1374.
[134] G. Foti, L. Csepregi, E.F. Kennedy, J.W. Mayer, P.P. Pronko, and M.D. Rechtin, Nucl. Instr. & Meth. 149 (1978) 381.

[135] W. Wesch, G. Götz, P. Ressel, D. Stock, and A. Unkroth, in K. Hennig (Ed.), Proc. Int. Conf. Energy Pulse Modification of Semiconductors and Related Materials, Dresden, 1984, ZfK-555, Rossendorf, 1984, p. 527.

[136] F.H. Eisen and E Uggerhoj, Radiat. Eff. 12 (1972) 233.

[137] J. Haskell, E. Rimini, and J.W. Mayer, J. Appl. Phys. 43 (1972) 3425.

[138] S.T. Picraux, W.L. Brown and W.M. Gibson, Phys. Rev. B6 (1972) 1382.

[139] J.U. Andersen, O. Andersen, J.A. Davies and E. Uggerhoj, Radiat. Eff. 7 (1971) 25.

[140] S.T. Picraux and F.L. Vook, Phys. Rev. Lett. 33 (1974) 1216.

[141] F.L. Vook and S.T. Picraux, in M. Kaminsky (Ed.), Radiation Effects on Solid Surfaces (Advances in Chemistry Series No. 158), American. Chemical Society, 1976.

[142] J.U. Andersen, E. Laegsgaard, and L.C. Feldman, Radiat. Eff. 12 (1972) 219.

[143] R.B. Alexander, P.T. Callaghan, and J.M. Poate, Phys. Rev. B9 (1974) 3022.

[144] R.B. Alexander, G. Dearnaley, D.V. Morgan, and J.M. Poate, Phys. Lett. A32 (1970) 365.

[145] G. Götz and G. Sommer, Radiat. Eff. 41 (1979) 195.

[146] W. Frank, Inst. Phys. Conf. Ser. No. 23 (1975) 23.

[147] W. Wesch, K. Gärtner, E. Wendler, and G. Gotz, Nucl. Instr. & Meth. B15 (1986) 431.

[148] M.J. Pedersen, J.U. Andersen, D.J. Elliott, and E. Laegsgaard, in S. Datz, B.R. Appleton, and C.D. Moak (Eds.), Proc. Int. Conf. Atomic Collision in Solids, Gatlinburg, 1973, Plenum Press, New York, 1975, Vol. 2, p. 863.

[149] L.M. Howe, M.L. Swanson, and A.F. Quenneville, Nucl. Instr. & Meth. 132 (1976) 241.

[150] G. Götz and G. Sommer, Phys. Stat. Sol. (a) 32 (1975) K 151.

[151] E. Wendler, W. Wesch, and K. Unger, in K. Hennig (Ed.), Proc. Int. Conf. Energy Pulse Modification of Semiconductors and Related Materials, Dresden, 1984, ZfK-555, Rossendorf, 1984, p. 535.

[152] W. Wesch, E. Wendler, G. Götz, K. Unger, H. Röppischer, and Chr. Resagk, Phys. Stat. Sol. (b) 130 (1985) 539.

[153] Periodic Table of the Elements, Sargent-Welch Scientific Company, 1968

List of Symbols

A	accessible area
A_o	area of the unit cell of the transverse space
$A^{(j)}$	Symbol characterizing element j
a_{TF}	Thomas-Fermi screening length
α	energy loss ratio
b	collision diameter
$\vec{b}$	Burgers vector
β	characteristic parameter for dechanneling due to dislocations
C_1	relative part of crystalline grains
C_2	relative part of the grain boundaries
C_1'	volume fraction of perfectly oriented regions
C_2'	volume fraction of twinned regions
$c_{i,A^{(j)}}$	relative concentration of element $A^{(j)}$ in layer i
χ	normalized Rutherford backscattering yield
χ_{min}	normalized Rutherford backscattering minimum yield
χ_a	aligned fraction of the beam
χ_r	random fraction of the beam
D	diffusion function, ion dose
d	interatomic spacing for a low index direction
d_a	foil thickness
d_{am}	thickness of an amorphous layer
$\vec{\delta}_r$	displacement vector of a string
d_{lam}	thickness of a twin lamella
d_p	distance of low index planes
ds	differential path length
dE/ds	mean energy loss per unit path length (stopping power)
$d\sigma$	differential cross section
$d\sigma_e$	differential cross section for electronic scattering
$d\sigma_n$	differential cross section for nuclear scattering
$d\sigma/d\Omega$	differential cross section per unit solid angle
Δ	shift term, standard deviation of the angular distribution of crystallites
ΔE	energy loss in a layer of thickness z

$\langle \Delta E \rangle$	mean energy loss in a layer of thickness Δz
ΔE_2	energy width
ΔE_{sf}	energy loss in the stopper foil
$\Delta E_{\perp i}$	width of the transverse energy interval i
$\Delta \Omega$	solid angle given by the detector geometry
ΔR_p	range straggling
Δz	layer thickness
Δz_i	thickness of the i-th layer
δE	change of the ion energy
$\delta E_\perp$	change of the transverse energy of the ions
δE_2	energy resolution
δE_d	energy resolution of the detector
δE_g	geometrical energy resolution
δE_{ms}	energy resolution caused by multiple scattering
δE_s	energy straggling
δE_{sr}	energy spread due to surface roughness
δM_2	mass resolution
δZ_2	atomic number resolution
δz	depth resolution
$\delta \psi^2$	angular spread
E	ion energy
E_o	incidence energy of the ions
E_1	ion energy before the backscattering event
E_2	energy of the emerging ions
$E_{B,i}$	binding energy of an electron in an shell i
E_{cm}	center-of-mass energy
E_ε	characteristic energy for nuclear scattering processes
E_η	characteristic energy for electronic scattering processes
$\bar{E}_{in}$	mean ion energy on the inward path
E_n	quantum mechanic energy of an atom in state n
$\bar{E}_{out}$	mean ion energy on the outward path
$E_\perp$	transverse energy
$E_{\perp c}$	critical transverse energy
$E_{\perp i}$	transverse energy partition
E_X	characteristic X-ray energy
e	elementary charge
ε	stopping cross section
ε_e	electronic stopping cross section
ε_n	nuclear stopping cross section
$[\varepsilon]$	stopping cross section factor
$[\varepsilon_o]$	stopping cross section factor in surface approximation
$[\bar{\varepsilon}]$	stopping cross section factor in mean energy approximation
ε_o	permittivity of vacuum
ε_r	reduced energy for nuclear scattering
ε_{ch}	channel width

η	X-ray detection efficiency
η_r	reduced energy for electronic scattering
$F(\vec{r})$	ion flux distribution in an axial channel
$F(y)$	ion flux distribution in a planar channel
$g(E_\perp)$	transverse energy distribution
g_i	relative number of ions in the transverse energy interval i
g_n	nuclear stopping function
γ	dechanneling probability
$\hbar$	Planck's constant
$h(\varphi)$	angular distribution of crystallites
I	mean excitation energy
K	kinematic factor for the ion
k	kinematic factor for the recoiled target atom
L_{dpl}	projected edge length of the displaced planes per unit volume
λ	dechanneling diameter of dislocation
λ_o	dechanneling diameter constant
$\vec{M}$	total magnetic moment
M_1	projectile mass
M_2	mass of a target atom
m	electron mass
$\vec{\mu}$	mean magnetic moment per atom
μ	X-ray absorption coefficient
N	atomic density
N_{disl}	integral density of dislocations
N_{dst}	integral density of displaced strings
N_{pd}	integral density of point defects
n	number of incident ions
n_{ch}	channel number
n_e	electron density
n_j	relative number of atoms of kind j
n_{disl}	total projected length of dislocation lines per unit volume
n_{dst}	number of displaced strings per unit volume
n_{pd}	relative concentration of displaced atoms
ν	Poisson's ratio
Ω^2	energy straggling
ω_i	fluorescence yield from an inner shell i
$P(\Delta E,\Delta z)$	energy loss distribution
$P(\varphi,\Delta z)$	angular distribution
$P(E_\perp',E_\perp)dE_\perp'$	probability per depth for transverse energy transition
$P(r)$	distribution of thermally vibrating lattice atoms
$\tilde{P}$	dechanneling function
$\vec{p}$	transverse momentum
Φ_e	electron flux
φ	deflection angle, screening function
$\tilde{\varphi}$	reduced deflection angle

φ_{LJ}	Lenz-Jensen screening function
φ_{TF}	Thomas-Fermi screening function
φ_1	angle of incidence with respect to the surface normal
φ_2	angle of emergence with respect to the surface normal
$\varphi_{1/2}$	half width of the angular distribution
$\tilde{\varphi}_{1/2}$	reduced half width of the angular distribution
ψ	actual angle of ion motion with respect to the atomic string
ψ_1	critical angle introduced by Lindhard
$\psi_{1/2}$	half width of the channeling dip
ψ_c	critical angle for channeling
ψ_{in}	angle of incidence with respect to a low index direction
ψ_{out}	angle of emergence with respect to a low index direction
ψ_s	standard deviation of the channeling dip
Π_i, Π_{in}	relative probabilities for hitting a thermally vibrating lattice atom
Q_{ij}	dechanneling matrix
R_p	projected range
$\vec{r}$	vector of the distance from the string
$\vec{r}_a$	vector of the displacement of an atom from the string
r_{min}	distance of minimum approach to a string
r_o	radius of the unit area of the transverse space
ρ	mass density, mean density of electrons of a string
$\rho_\perp^2$	mean square thermal vibration amplitude
S_{dpl}	dechanneling cross section per unit edge length of a displaced plane
s	impact parameter
σ^d	differential scattering cross section per unit solid angle
σ_{dst}	dechanneling cross section of one displaced string
$\sigma_{v,i}$	cross section for vacancy production in an inner shell i
$\sigma_{X,i}$	X-ray cross section contribution from an inner shell i
T	transferred energy, temperature
T_a	annealing temperature, transmission coefficient
T_i	implantation temperature
t	time
τ	reduced thickness
Θ	scattering angle of the ion
Θ_{rec}	scattering angle of the recoiled target atom
ϑ	scattering angle in the center-of-mass system
U	string potential
V	interatomic interaction potential
$\vec{v}$	velocity of the projectile
$\vec{v}_{cm}$	velocity of the center-of-mass system
W	straggling cross section
W_e	electronic straggling cross section
W_n	nuclear straggling cross section

w_e	electronic straggling function
w_n	nuclear straggling function
Y	plane potential, Rutherford backscattering yield
Y_{al}	backscattering yield for aligned incidence
Y_r	backscattering yield for random incidence
Y_i	backscattering yield per channel
Y_{tot}	integrated backscattering yield
y_{min}	distance of minimum approach to a plane
Z_1	atomic number of the projectile
Z_2	atomic number of a target atom
z	depth
z_a	accessible depth

Subject Index

FSC
www.fsc.org
MIX
Papier | Fördert
gute Waldnutzung
FSC® C083411

Kolibri 360 GmbH
Ferdinand-Jühlke-Straße 7
99095 Erfurt, Deutschland
produktsicherheit@kolibri360.de